Primate Responses to Environmental Change

Primate Responses to Environmental Change

Edited by

HILARY O. BOX
Lecturer
Department of Psychology
University of Reading

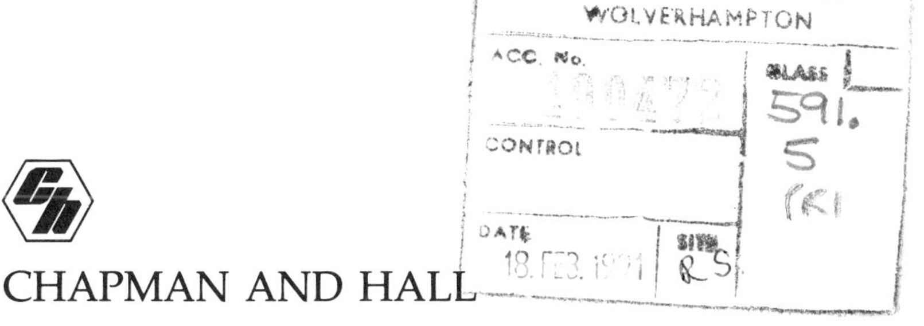

CHAPMAN AND HALL
LONDON • NEW YORK • TOKYO • MELBOURNE • MADRAS

UK	Chapman and Hall, 2–6 Boundary Row, London SE1 8HN
USA	Chapman and Hall, 29 West 35th Street, New York NY10001
JAPAN	Chapman and Hall Japan, Thomson Publishing Japan, Hirakawacho Nemoto Building, 7F, 1-7-11 Hirakawa-cho, Chiyoda-ku, Tokyo 102
AUSTRALIA	Chapman and Hall Australia, Thomas Nelson Australia, 480 La Trobe Street, PO Box 4725, Melbourne 3000
INDIA	Chapman and Hall India, R. Seshadri, 32 Second Main Road, CIT East, Madras 600 035

First edition 1991

Typeset in 10 on 12pt Palatino by Mayhew Typesetting, Bristol
Printed in Great Britain at the University Press, Cambridge

ISBN 0 412 29940 2 (HB)

British Library Cataloguing in Publication Data

Primate responses to environmental change.
1. Primates. Behaviour
I. Box, Hilary O. (Hilary Oldfield)
599.80451

ISBN 0–412–29940–2

Library of Congress Cataloging-in-Publication Data

Primate responses to environmental change / edited by Hilary O. Box – 1st ed.
p. cm.
Includes bibliographical references and index.
ISBN 0–412–29940–2
1. Primates – Behavior – Climatic factors. I. Box, Hilary O. (Hilary Oldfield), 1935–
QL737.P9P73 1990 90–37504
599.8′045–dc20 CIP

For Jo

Contents

Contributors

DAVID H. ABBOTT
Institute of Zoology, The Zoological Society of London, Regents Park, London NW1 4RY, UK.

LEAH E. ADAMS-CURTIS
Department of Psychology, Washington State University, Pullman, Washington 99164–4830, USA.

SIMON K. BEARDER
Anthropology Unit, Oxford Polytechnic, Oxford OX3 0BP, UK.

HILARY O. BOX
Department of Psychology, The University of Reading, Whiteknights, Reading RG6 2AL, UK.

DAVID J. CHIVERS
Sub-Department of Veterinary Anatomy, University of Cambridge, Tennis Court Road, Cambridge CB2 1QS, UK.

ANNE B. CLARK
Department of Biological Sciences, State University of New York at Binghamton, Binghamton, New York, 13901 USA.

JAMES G. ELSE
Yerkes Regional Primate Research Center, Emory University, Atlanta, Georgia 30322 USA.

JOHN E. FA
Institute of Mediterranean Ecology, The Gibraltar Museum, PO Box 303, Gibraltar.

MICHEL FERNANDEZ
Centre Internationale de Recherches Medicales de Franceville, BP 769, Franceville, Gabon, West Africa, and Department of Psychology, University of Stirling, Stirling FK9 4LA, UK.

DOROTHY M. FRAGASZY
Department of Psychology, Washington State University, Pullman, Washington 99164–4830, USA.

JEFFREY A. FRENCH
Department of Psychology, University of Nebraska at Omaha, Omaha, Nebraska 68182–0274, USA.

LYNNE M. GEORGE
Institute of Zoology, The Zoological Society of London, Regents Park, London NW1 4RY, UK.

ALISON C. HANNAH
Department of Psychology, University of Stirling, Stirling FK9 4LA, UK.

BETTY J. INGLETT
Department of Psychology, University of Nebraska at Omaha, Omaha, Nebraska 68182–0274, USA.

ANDREW D. JOHNS
Makerere University Biological Field Station, PO Box 10066, Kampala, Uganda.

PHYLLIS C. LEE
Department of Biological Anthropology, Downing Street, Cambridge CB2 3DZ, UK.

JEREMY E. MALLINSON
Jersey Wildlife Preservation Trust, Les Augres Manor, Jersey, Channel Islands, UK.

WILLIAM C. McGREW
Department of Psychology, University of Stirling, Stirling FK9 4LA, UK.

SALLY P. MENDOZA
Department of Psychology and California Primate Research Center, University of California, Davis, California 95616, USA.

TREVOR B. POOLE
Universities Federation for Animal Welfare, 8 Hamilton Close, South Mimms, Potters Bar, Hertfordshire, EN6 3QD, UK.

MARGARET E. REDSHAW
Department of Child Health, University of Bristol, Bristol BS2 8BJ, UK.

LEAH SCOTT
Ministry of Defence, Porton Down, Salisbury, Wiltshire SP4 0JQ, UK.

CAROLINE E.G. TUTIN
Centre Internationale de Recherches Medicales de Franceville, BP 769, Franceville, Gabon, West Africa, and Department of Psychology, University of Stirling, Stirling FK9 4LA, UK.

DAVID M. WARBURTON
Department of Psychology, University of Reading, Whiteknights, Reading RG6 2AL, UK.

Introduction

This book concerns the various ways that primates respond to environmental change. By studying these patterns of responsiveness we not only gain useful knowledge about the structural, physiological and behavioural propensities of different species, but also acquire important information relating to issues of contemporary concern, such as conservation and the management of animals in the wild as well as in various forms of captivity.

For example, there is growing concern among biologists and conservationists about the influence of habitat destruction, such as logging, on the fitness and survival of wild primates. There is also increased awareness of the need to improve the care of primates in zoos and laboratories, including the enrichment of captive environments. Further, because an increasing number of primate species are becoming endangered, knowledge of their responsiveness to new environments is an essential requirement for effective breeding programmes in captivity, and for the translocation and rehabilitation of species in the wild.

In theory, studies of many closely related species are required in order to consider relevant evolutionary processes, as well as to develop functional hypotheses about the adaptive significance of various biological propensities and their interrelationships in the short and longer terms.

In the case of primates, we have a sound body of information about the physiological responsiveness, social organization, cognitive capacities and life strategies of different species. However, we need a much wider comparative data base and, so far, there have been few long-term comparative studies. Moreover, it is only relatively recently that researchers have turned to studying the interrelationships of primate characteristics and how these influence the ways in which different species, and individuals of different age and gender within species, respond to changes in their environment.

The idea for this volume developed from that of my last single-author book entitled *Primate Behaviour and Social Ecology*, which was also much concerned with aspects of responsiveness to environmental change. Since that publication in 1984 there have been notable developments in this research area and I believe that the time is now right for a volume devoted solely to the topic. The aim has been to provide information in novel contexts on a wide variety of theoretical and practical issues, and to stimulate further interest in an area with enormous research potential. The result is an unusual contribution in biology, and, at the time of writing, there is no similar text. Students in a variety of disciplines, including zoology, psychology, anthropology, conservation biology and wildlife management, should find it valuable.

I have organized the book into three parts, for each of which I give an introductory overview of the contents. The first part presents a series of discussion papers. Briefly, these generally involve characteristics of primates that underpin their responsiveness to change. These include morphological and sensory characteristics, behavioural flexibility in the short and long term, cognitive abilities, interrelationships between physiological and behavioural indices of responsiveness, reproductive status and social change, and, last but not least, the delineation and potential functions of individual variation in responsiveness.

The second part is devoted to environmental change in nature. Many topics could have been chosen. I made the selection from studies that address critical and frequently occurring examples of environmental change, and the contributors provide new perspectives and, most often, new information. The topics included are deforestation, provisioning by supplemental food, the problem of primates as pests, rehabilitation of apes into a free-ranging environment, the responsiveness of different species of ape to attempts by observers to habituate them to their presence, and, finally, problems of primate conservation.

Similar perspectives determined the choice of topics in the third part. An overriding aim in this case was to present principles and techniques that will, hopefully, stimulate different kinds of research in captivity. For example, the means whereby natural patterns of behaviour may be encouraged are discussed in various contexts. Several empirical studies also demonstrate the comparative interface between reproductive status and social behaviour in species of marmosets and tamarins. 'Stress' responses are considered with reference to the nature of the physiological mechanisms involved and the environmental stressors that influence them. Long-term and interdisciplinary comparative studies of two species of New World monkeys present an exemplary case of the interrelationships between behavioural organization and physiological responsiveness. The final concluding remarks on my part are an attempt to highlight some advances in studies of environmental change generally, as well as to indicate future developments.

With two relevant exceptions (Chapters 1 and 5), all the material in the book refers to simian primates (monkeys and apes) exclusively. This bias reflects my own interests, but there is also a good case at this relatively early stage for not confusing attempts to draw out many of the comparative perspectives by including studies on prosimian species. Primate responsiveness within the whole order has to recognize that prosimians 'share distinct patterns of anatomy, physiology and behaviour which make them quite unlike simians . . . that prosimians are specialised along very different lines from simians' (S.K. Bearder, personal communication). There is also much interest in simian species in the wild as important indicators of the state of whole ecosystems. Moreover, in captivity, they elicit special concern for their environmental needs, given their natural social, cognitive and manipulative skills. It is also the case, of course, that higher primates are appropriate models for many biomedical interests, as well as for investigations of comparative principles of responsiveness to environmental change.

Finally, as an editor, I have inevitably found that editing a book really *is* all the things that one is told it is going to be! In some cases there have indeed been many hours of apparently thankless work. Nevertheless, there has also been a sustained interest and enthusiasm for the whole project on my part, as well as on that of the contributors. It is my pleasure to thank each and every one of them with warmth and respect. Undertaking any such project presupposes not only a willingness to work at it, but the availability of appropriate secretarial assistance. 'Appropriate' in this context ideally implies a very high level of speed and typographical accuracy, together with a sufficient interest in the whole enterprise to make it a pleasure to work with (rather than despite) the person, and on whose help in numerous ways one may completely rely. My thanks as ever go to Joan M. Morris, whose work on all these counts has been superb.

It is also a pleasure to acknowledge the skill, tact, and friendly advice of Sarah Bunney, who copy-edited the manuscript. I thank her most warmly on behalf of all of us.

Hilary O. Box

Part One
General Perspectives

A white-handed gibbon (*Hylobates lar*) feeding on the terminal branches of a tree in Malaysia. (Photo: David Chivers/Anthrophoto.)

In this first part of the book is a series of discussion papers on the characteristics of primates that influence their responsiveness to environmental change. David Chivers (Chapter 1) lays a sound foundation by considering the importance of feeding strategies, as the means whereby species can adjust to short-term and long-term changes that occur naturally, as well as those brought about by human destruction of habitats. To this end, he describes primate characteristics in terms of their special senses, locomotion and posture, masticulatory apparatus, digestive systems, and the ways in which adaptations to different types of diet influence potential responsiveness to environmental perturbations.

Phyllis Lee (Chapter 2) considers sources of environmental change in nature and the various ways in which these may be studied. In particular, she provides some examples to establish principles of adaptive behavioural responsiveness. As an example of a short-term change, she considers changes in behaviour among the chimpanzees of Gombe Stream National Park, Tanzania, with reference to the supply of provisioned food. Innovation and traditions of behaviour are also described. Ecological changes in the longer term are discussed with reference to life history variables and, most unusually, she further considers attributes of behavioural flexibility that have contributed to the survival and extinction of different primate genera.

The editor (Chapter 3) considers aspects of cognitive processes among the primates that are implied, but not empirically well understood, in the context of responsiveness, with special reference to the complexities of social change. She then considers how responsiveness is expressed differently among species, according to their different lifestyles and physiological propensities.

In line with an increasing interest in the complex parameters that influence primate reproduction, David Abbott (Chapter 4) provides a comparative review to emphasize the social control of fertility. Recent research, such as his own, into the influences of the social status of females, gives clear insights into the ways in which fertility in different species may be influenced to varying degrees by social environment.

Finally, it is an unfortunate omission in our knowledge and thinking about responsiveness to change that so little attention is given to issues which concern the nature of, and especially the functions of, individual variation. Anne Clark (Chapter 5) redresses this balance admirably, by drawing together information from various disciplines. She proposes that the differences among behavioural characteristics between one individual and its conspecifics are adaptive rather than the exact value of the characteristics in a continuum of variation.

1 Species differences in tolerance to environmental change

DAVID J. CHIVERS

1.1 INTRODUCTION

The central issue in considering tolerance to environmental change – the secret of survival for both individuals and species – lies in the ability to locate, consume and process adequate food; only then can individuals breed successfully to perpetuate the species. Food location involves the use of special senses and postcranial musculoskeletal system in locomotor aspects of positional behaviour. Food consumption or intake involves postural behaviour and manual activity and/or oral movements. Food processing involves mechanical breakdown of food in the mouth and chemical breakdown in the gastrointestinal tract, as well as transport through the whole digestive system. These three crucial facets of behaviour will be examined in turn for primates living in different environments, with special emphasis on their capabilities to cope with changes, whether natural or induced by humans.

Primates are unusual among mammals in their lack of anatomical specialization and in their behavioural flexibility. Their success lies in avoiding specializations for faunivory (eating animal matter) or folivory (eating leaves or grasses), and in escaping the reproductive constraints of the oestrous cycle. Although the behavioural, especially social, advantages of the menstrual cycle of haplorhine primates (tarsiers, monkeys, apes and humans) are of prime importance, the focus in this chapter is on diet and feeding behaviour in the widest sense – on the ecological aspects of behaviour.

Most orders of mammals are specialized for eating either animal matter (vertebrate and/or invertebrate) or foliage. Faunivores include the Insectivora (insectivores), Carnivora (carnivores) and Cetacea (whales and dolphins). Folivores include the Rodentia (rodents), Proboscidea (elephants), Sirenia (manatees and dugongs), Perissodactyla (odd-toed ungulates) and Artiodactyla (even-toed ungulates). Only the Chiroptera (bats) approach the Primates in dietary diversity in consuming insects, fish, amphibians, reptiles, birds, mammals, fruit, pollen and nectar (e.g. Walker, 1964; Romer, 1966; Wilson, 1975).

Specializations for eating animal matter involve, for example, elaboration of the special senses for detection, sprinting ability (on land or in water) to catch the prey, powerful jaws to hold it, puncture-crushing and/or shearing ability of the teeth, and elaboration of the small intestine (rather than stomach or colon) to digest these very nutritious but rare (relative to plants) and elusive (because of their mobility) foods.

In contrast, specializations for leaf-eating centre on large body size to accommodate adequate guts for the bacterial fermentation of foliage (either in an expanded sac of the forestomach or in an expanded tube of the caecum and colon). Small body size can be retained if the first faeces are reingested, that is, if the small guts have two chances to process the very common but difficult-to-digest foods. The musculoskeletal system and special senses of folivores reflect the ability to travel long distances in search of food, and to sustain running away from sprinting predators once detected. The dentition reflects the hard work done through chewing to cut up the plant food so as to maximize the surface area available for bacterial action, through the large size, complex structure and oral rearrangement of teeth, with concomitant increase in size of the jaws and skull.

In their radiation from an insectivorous, nocturnal ancestor, primates have gone on to exploit the reproductive parts of plants (frugivory), and then, to varying degrees with further increases in body size, the abundant vegetative parts of plants (folivory). This radiation has been achieved by both strepsirhine primates (lemurs and lorises) and haplorhine primates, and without the more extreme adaptations shown by other mammals. In most cases, especially among haplorhine primates, dietary diversity is based on arboreal abilities (in posture and locomotion), moderate body size and a diet based on fruit; they have, thereby, ascended into most parts of the 'trophic tree' (Ripley, 1979, 1984), evolving a morphology intermediate in most respects to the specialist faunivores and folivores (e.g. Chivers *et al.*, 1984). Subsequently, after this prolonged arboreal apprenticeship, evolutionary success seems to have involved a return to a more terrestrial way of life – humans, then apes and finally Old World monkeys (e.g. Wood *et al.*, 1986).

Thus, questions centred on food are important for the emphasis placed on differences among species in structural, physiological and behavioural characteristics, which underpin the responsiveness to environmental change. The special attributes of primates, and their variety, for food location, ingestion and processing will now be examined in relation to environmental differences and change; these are special senses, locomotion and posture, mastication and digestion, and their anatomical basis, with the focus on nose, eyes, ears, limbs (and trunk), jaws, teeth and guts.

1.2 SPECIAL SENSES

The basis for the increased socialization and behavioural flexibility that characterizes the complex social strategies of primates is provided by a greater reliance on vision (stereoscopic) and vocal communication, which, with greater intelligence and increased manipulative skills, has led to an expanded communicative repertoire of complex and graded signals (e.g. Smuts *et al.*, 1987). While primates have retained the basic conservative reproductive strategies of mammals in the widespread occurrence of polygny in some form or other, male dominance systems and matrilineal influences between generations, with a moderate increase in body size and the acquisition of diurnal habit, they have developed their visual and auditory systems at the expense of the olfactory system (Wilson, 1975).

The special senses and brain, however, also play a basic part in the ecological aspects of behaviour. The more primitive sense of smell has a central role in the location of prey and plant foods by nocturnal strepsirhine primates, as well as in their social interactions, such as urine and scent-marking (Schilling, 1979). Hearing also plays an important role in the location of animal prey by strepsirhines. Vision is poorly developed, although the tapetum lucidum (the reflective layer behind the retina) maximizes the stimulation of rods in the eye, and, hence, their visual ability. Nevertheless, the use of vision means that strepsirhines are more active early or late in the night, rather than at times of complete darkness. Such crepuscular activity has led to some species becoming diurnal in Madagascar, in the absence of competition from haplorhine primates (Doyle and Martin, 1979).

The lack of a tapetum lucidum in haplorhine primates indicates that those species that are nocturnal – tarsiers (*Tarsius*) and the night monkey (*Aotus*) – are secondarily so: the former has exploited the insectivorous niche in South-East Asia (MacKinnon and MacKinnon, 1980a), and the latter has escaped the intense diurnal competition from the great diversity of Neotropical frugivorous primates by day (there not having been the opportunity to escape into different, nonforest habitats as happened in the Old World). For *Aotus*, smell is probably important in locating the fruit on which it mostly feeds (Wright, 1985).

Cartmill (1974) tried to overturn the established idea that the unique stereoscopic vision of primates resulted from moving around in the three-dimensional arboreal environment (Martin, 1986), by suggesting that it evolved through the improved manual trapping of mobile invertebrate prey. The answer probably lies in a combination of the two, but the more important point is that it has been crucial in the long periods through which primates have lived in trees.

Vision has been elaborated among diurnal haplorhine primates for locating plant foods, as well as for catching the animal component of their diet. They may locate fruiting trees visually on their travels around their home ranges or territories, or indirectly from seeing the intense activity of birds around such trees from a distance. Auditory cues may also play an important part in the dense vegetation in which many species live; calls, and hearing, are especially important in the spacing of monogamous social groups, which are often frugivorous and territorial (e.g. Smuts *et al.*, 1987). Smell is apparently less important in haplorhine primates, although it may still play a role in food selection; it is certainly important in reproductive interactions in enhancing female attractiveness, along with visual cues, around ovulation (Keverne, 1987).

The special senses probably combine with the greater cerebral development that characterizes primates (Martin, 1986), especially haplorhines (and hominoids), in this case with respect to 'maps' of food distribution through the home range, accumulated through previous experience.

Reference should be made here to the special sense of taste, which will link in with later discussion of food selection. Ten years ago, Hladik and Chivers (1978) drew attention to the possible role of taste in regulating food intake, both in terms of selecting primary (nutrient) compounds and avoiding secondary compounds (digestion-inhibitors and toxins). Little has been done since to investigate such topics, which have perhaps increased in relevance. Primates play a part in the coevolution of plants and animals (Janzen, 1970). Forest trees present attractive food packages to promote the dispersal of their seeds by the animals who consume them, but at the same time these trees have to protect their foliage and seeds against those animals that would abuse the relationship (Waterman, 1984), hence the proliferation of secondary compounds in plants and the distinction, even among primates, between seed dispersers and seed predators – respectively, Atelinae and Pithecinae in the New World (Ayres, 1986) and Hominoidea (and Cercopithecinae) and Colobinae in the Old World (Davies *et al.*, 1984).

While these general points can be made about the use of the special senses in the location of food, and about smell (and taste) in its consumption, they are so basic to the animal that little can be deduced about their roles in changed environments. All that can be said for primates is that the elaboration of vision and hearing, in particular, in relation to marked cerebral expansion, confers on them special behavioural abilities and flexibilities to cope with such changes, in ways unparalleled among animals.

1.3 LOCOMOTION AND POSTURE

The ability to adapt to changes in the environment is also reflected in the diverse repertoires of positional behaviour that characterize most primates. The range of locomotor and postural modes exploited by primates will now be considered, along with anatomical and ecological correlates, indicating the potential for adaptation to the differing kinds of environmental change.

Locomotion relates to travel around the home range between food sources, whereas posture is involved in the actual procuring of food within these sources. The latter has previously been obscured by the greater glamour of the former, but they are inextricably linked, behaviourally and anatomically, under the heading of positional behaviour. While locomotion has acute survival value in terms of movement about the habitat for food and avoidance of predators, posture is of chronic importance, given the amount of time spent feeding (and resting). For example, among five species of Malayan forest primate (Raemaekers and Chivers, 1980), locomotion occupied only 12% of the day on average, whereas postures (other than resting occupying about 35% of the daytime) involves more than 56% of the day, most while feeding.

Locomotion in mammals is basically quadrupedal, with the hindlimb being the main propulsive limb. Quadrupedal movement, whether in the trees or on the ground, and whether walking, running or jumping, is very common among primates. Leaping and terrestrial activity increase with body size. Vertical clinging and leaping has become a common mode of locomotion among strepsirhine primates (Napier and Walker, 1967), especially the larger ones (e.g. Indriidae), so much so that it has been regarded as the primitive mode for primates. The discovery of two different patterns (Stern and Oxnard, 1973), however, as reflected by behaviour and anatomy, implies separate origins from different quadrupedal forms. For example, the pelvic and hindlimb anatomy of indriids indicates the development of a pattern seen in lemurs (*Lemur*), whereas those of tarsiers and bushbabies

(Galaginae) relate to mouse lemurs (*Microcebus*). Such movements require considerable elongation of the hindlimb, especially in the foot region, and elaboration of muscles for powerful propulsion (rapid extension of all joints). While quadrupedalism relates mainly to horizontal supports, vertical clinging and leaping relate to vertical ones.

Suspensory behaviour has become more important among the larger primates, especially in relation to procuring food from the flexible terminal branches of forest trees. Thus, we find bimanual progression (brachiation) in the larger New World monkeys (spider monkeys Atelinae), in the larger leaf monkeys (Colobinae) of the Old World, especially the largest (the proboscis monkey *Nasalis*, to a lesser extent), and, most dramatic of all, in the apes, especially the smallest (i.e. the gibbons, *Hylobates*). Such upright locomotion (and posture) is pre-adaptive for the bipedalism seen generally but briefly in these primates, and most elaborately in humans. The brachiation of gibbons, quadrumanualism of orang-utans (*Pongo*), knuckle-walking of chimpanzees (*Pan*) and gorillas (*Gorilla*) and the bipedalism of humans (*Homo*) (Andrews and Groves, 1976) have their origins in a diffuse and complex component of locomotion that has been overlooked until recently – namely, climbing (Fleagle *et al.*, 1981).

Climbing, along with hanging while feeding or resting, appear to be the key pre-adaptive activities for upright body posture, as reflected in the various positional activities of the larger living primates today. Because of the time involved in feeding, and the greater stability afforded in feeding in the terminal branches where the most nutritious foods (e.g. fruit) are to be found, hanging merits greater attention, along with climbing, than it has so far achieved. Andrews and Groves (1976) point out that it is similarity in posture rather than locomotion that unites the hominoid primates both behaviourally and anatomically.

Thus, the transition from arboreal to terrestrial niches involves a transition from suspensory behaviour to bipedalism (or knuckle-walking in the case of the outsized great apes). For quadrupeds, there is transition from palmigrade to digitigrade quadrupedalism, to increase stride length and running efficiency. Within the Neotropical forests the great range in body size is reflected by a comparable diversity in positional behaviour, shown by the small, scurrying marmosets and tamarins (Callitrichidae), the walking and running squirrel monkeys (Cebinae), the climbing sakis (Pithecinae), the suspensory howler monkeys (*Alouatta*) and the brachiating spider monkeys (Atelinae), where the prehensile tail acts as a fifth limb to help spread the weight across flexible supports. Stern (1971) found this to be rich ground for seeking anatomical correlations with positional behaviour.

Postures are less diverse but equally, if not more, important. Sitting in various ways is the most widespread, and common to all primates; standing, quadrupedally or, less often, bipedally (and tripedally) is also important (e.g. Rose, 1974). Hanging, briefly during travel or more prolonged during feeding, by forelimb or hindlimbs (and tail in larger New World monkeys) or various combinations of one to four limbs, is of special importance, as indicated above in the context of suspensory (below-branch) behaviour.

What are the anatomical bases of these locomotor and postural repertoires? Because of the apparently continuous spectrum between wholly suspensory (below-branch) and wholly supportive (above-branch) activities, interest centred first on the range of anatomy of the forelimb (e.g. Ashton and Oxnard, 1963; Ashton *et al.*, 1965). At the quadrupedal extreme of the spectrum (Figure 1.1a), shoulder movement is more restricted to the lower quadrant by the shape of the bones, relationship to the body wall (bilaterally compressed) and the arrangement of muscles, whereas in those species with greatest suspensory activities the range of shoulder movement encompasses all four quadrants, with associated changes in bones and muscles and the trunk compressed dorsoventrally (so that the scapula shifts dorsally and the shoulder joint projects laterally). Quadrupedal activity reflects adaptations for two-dimensional movement (flexion and extension in the sagittal plane) under compression, whereas suspension involves three-dimensional movement, mainly in the upper quadrant, under tension (e.g. Oxnard, 1987).

Such contrasts in bones and muscles continue right down the forelimb, to encompass more powerful elbow extension, modifications of the wrist and the grasping abilities of the hand (e.g. Tuttle, 1969). Real progress has been made through electromyographic studies (Stern *et al.*, 1976, 1980), which identify the specific roles of muscles in different modes of locomotion and/or posture.

The hindlimb morphology, by contrast, radiates from a central position relating to the basic propulsive role in quadrupedalism, to contrasting bone and muscle assemblages that reflect more or less compressive forces or more or less tensile ones. Thus, the contrasting roles of the hindlimb in vertical clinging and leaping (tarsier and indri types), climbing (loris and monkey types), suspension (atelines and apes), leaping (colobines), terrestrial quadrupedalism (some cercopithecines, e.g. *Papio, Macaca, Erythrocebus*) and bipedalism can be displayed (Oxnard, 1979, 1987) (Figure 1.1b). The bipedal attributes of humans and other primates (macaque, gibbon and chimpanzee) have

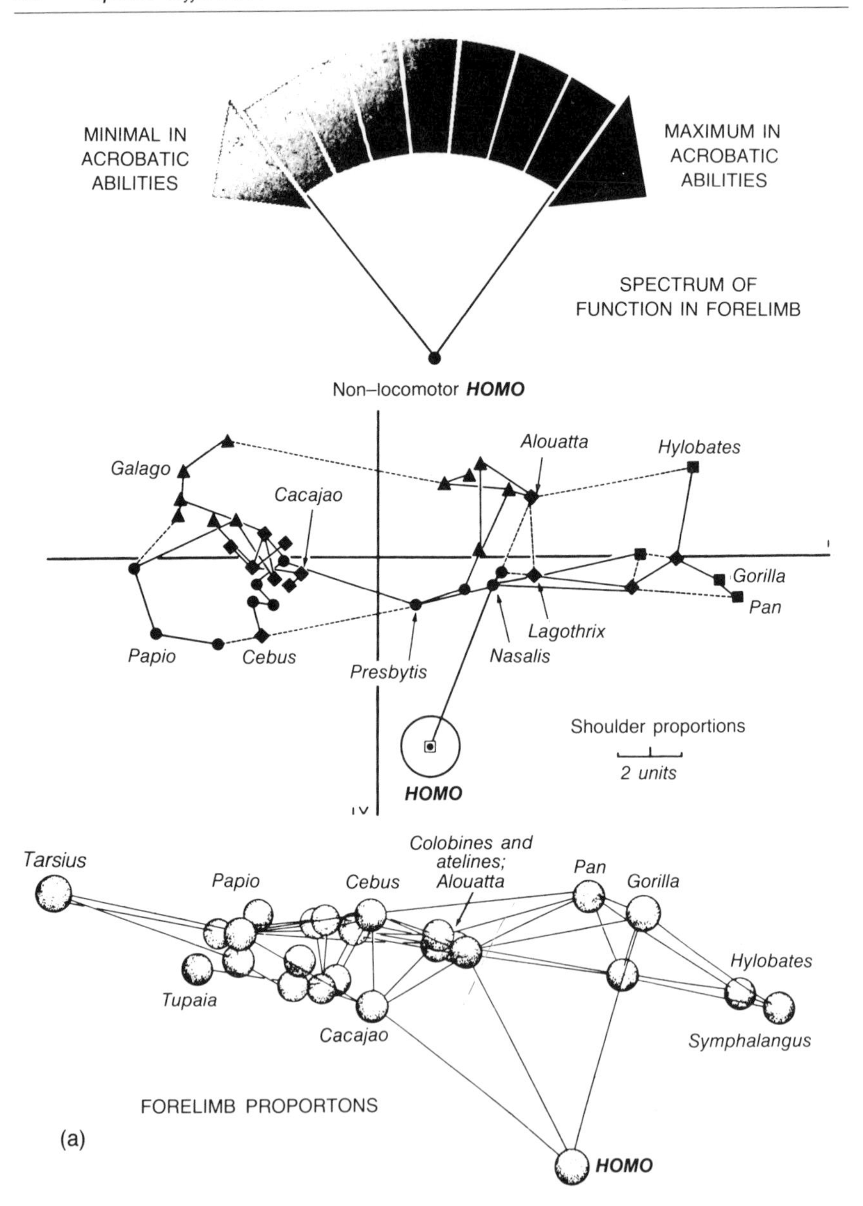

Figure 1.1 Locomotor functions in relation to anatomy in primates. (a) Upper limb, reflecting results of multivariate analyses of shoulder dimensions and forelimb proportions (from Oxnard, 1975). (b) (opposite) Lower limb, reflecting results of multivariate analyses of pelvic and hindlimb proportions (from Oxnard, 1975). (c) (overleaf) All four limbs, based on a model of generalized distances among primate genera resulting from multivariate analysis of combined proportions of upper and lower limbs. (From Oxnard, 1987.)

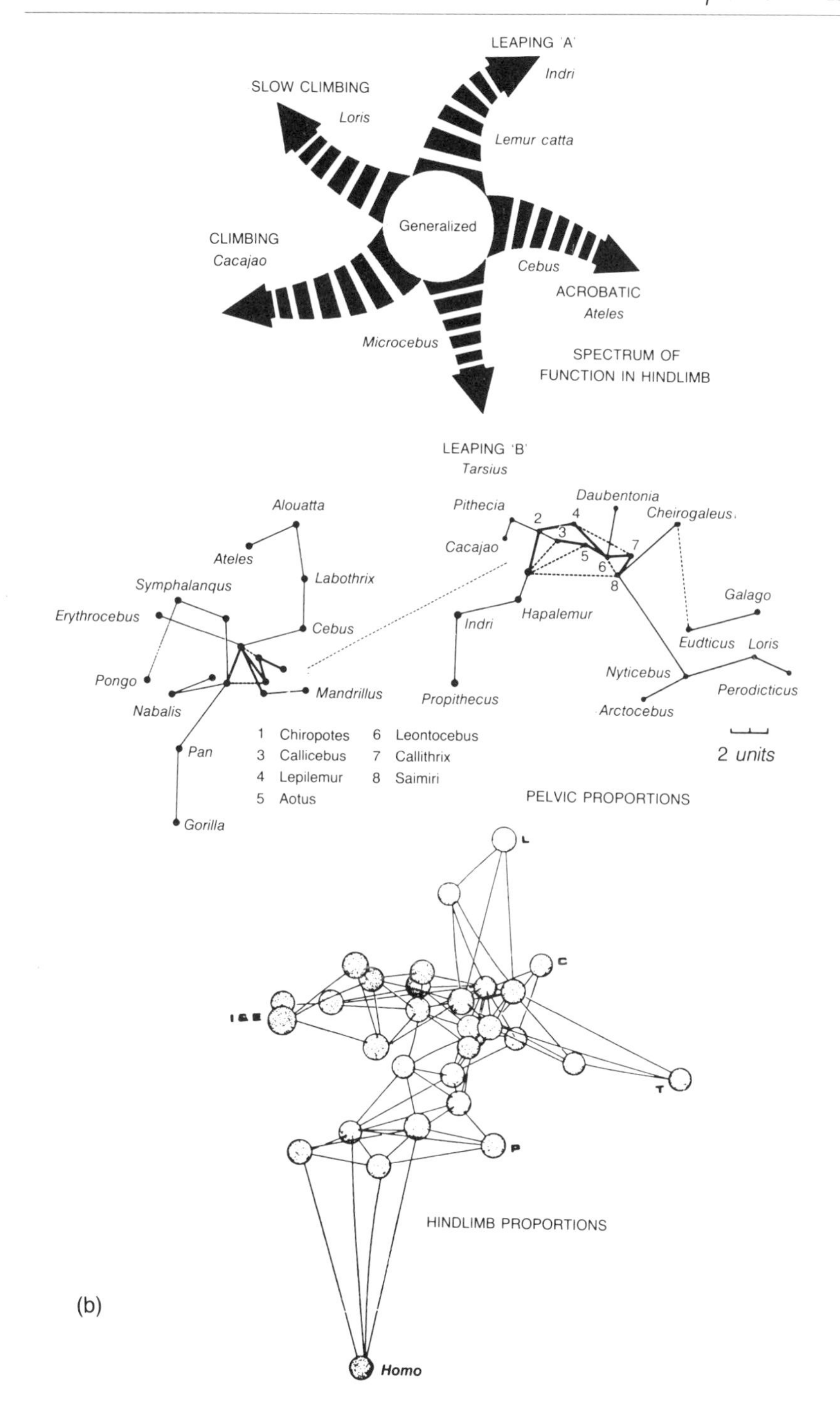
LEAPING 'A'
Indri
Lemur catta
SLOW CLIMBING
Loris
Generalized
CLIMBING
Cacajao
Cebus
ACROBATIC
Ateles
Microcebus
SPECTRUM OF
FUNCTION IN HINDLIMB
LEAPING 'B'
Tarsius
Alouatta
Ateles
Labothrix
Symphalanqus
Erythrocebus
Cebus
Pongo
Nabalis
Mandrillus
Pan
Gorilla
Pithecia
Cacajao
Daubentonia
Cheirogaleus
Hapalemur
Indri
Propithecus
Galago
Eudticus
Loris
Nyticebus
Perodicticus
Arctocebus
1 Chiropotes
3 Callicebus
4 Lepilemur
5 Aotus
6 Leontocebus
7 Callithrix
8 Saimiri
2 units
PELVIC PROPORTIONS
L
C
I & E
T
P
HINDLIMB PROPORTIONS
Homo

(b)

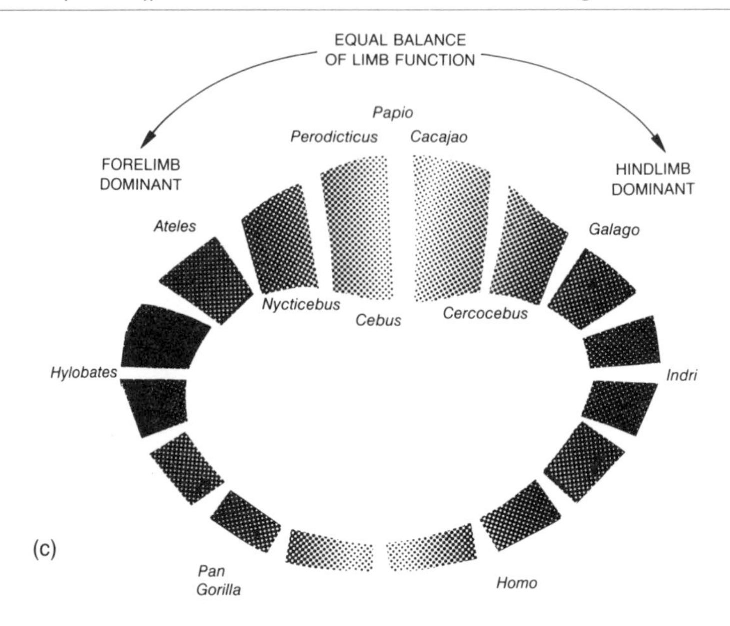

been investigated electromyographically by Ishida *et al.* (1975), who found that those nonhuman primates best adapted to upright posture through their positional repertoire (e.g. *Hylobates* and *Ateles*) are the most efficient at bipedal movement. Contrasts in foot anatomy, including the position of the load line and relative lengths of the load and power arms, have been investigated by, for example, Campbell (1966) and Lessertisseur and Jouffroy (1975).

Thus, for forelimbs and hindlimbs there is a spectrum from the least to the most acrobatic and from the least to the most arboreal, from the firmest to the most flexible supports. Old World monkeys are complex in their recent mix of locomotor activities, such as leaping, running and climbing; other primate families show more distinctive diversity according to body size and the part of the forest habitat most exploited. Vertebral columns are longer and thus more flexible in quadrupeds; they are shorter and more rigid in those indulging most in suspensory and other upright activities. Oxnard (1987) blended the results of the classic morphological and multivariate morphometric approaches for the whole body, and produced a 'signet-ring' arrangement of species based on the relative predominance of forelimb or hindlimb activity (Figure 1.1c). He concluded that the functional aspects of regions summate to evolutionary aspects of the whole body.

In relation to environmental changes, or changes in behaviour, quadrupedalism transfers most easily from trees to ground, vertical clinging and leaping less so; climbing aids the alternation between trees and ground in more open habitats, and leads to bipedalism on the ground. Such are the diversity and flexibility of positional behaviours that primates have responded successfully to both long-term and short-term environmental changes, as well as to competition producing shifts in habitat, in relation to changes in diet (see below).

Palaeontological studies show an increase in woodland savannah habitats during the Miocene epoch (24–5 million years ago), and a decrease in forests. While the apes remained in the forests, increasingly adapting their suspensory behaviour for frugivory, postcranial remains of monkeys indicate increased terrestrial activity, and there is evidence of adaptations for folivory (Andrews and Aiello, 1984), with a developing tolerance for secondary compounds (Andrews, 1981). The latter arises in part from the greater seasonality in more open habitats, making frugivory less viable. Longer forelimbs, and other anatomical features, in Old World monkeys relate to the greater stability required in terrestrial movement. Having gained a digestive advantage over apes, many Old World monkeys (cercopithecines) returned to the frugivorous niche (often in forests), while others (colobines) became increasingly specialized for folivory in forests (see below).

Short-term environmental changes relate mainly to disturbance. When the forest canopy is disrupted, flexibility in locomotion is advantageous in crossing gaps between trees (e.g. leaping, climbing, ground-walking and running). Long-term competition in Neotropical forests would seem to have selected for diverse positional behaviours in New World monkeys (see above, and Fleagle and Mittermeier, 1980). Comparable diversification may now be occurring in South-East Asian (and African) forests. While siamang (*Hylobates syndactylus*) and lar gibbons (*H. lar*) are confined to high forests, along with the dusky langur (*Presbytis obscura*) and the pig-tail macaque (*Macaca nemestrina*, which travels mainly on the ground, as reflected in anatomy), the long-tailed macaque (*M. fascicularis*) and the banded langur (*P. melalophos*) live more at the forest edge, in smaller trees and on more flexible supports, coming more often to the ground (Chivers, 1980). Resulting divergence in postcranial anatomy has been well documented for the langurs (Fleagle, 1977, 1978).

1.4 FEEDING: GAPE, TEETH AND JAWS

Various features of the masticatory apparatus are affected by activities other than feeding (e.g. calling, cleaning and fighting). This applies also to gape (the ability to open the mouth widely), which also involves modifications to the temporomandibular joint, jaw muscles and forward condylar translation (Smith, 1984). The only correlations with diet are that folivores have larger gapes, and frugivores have a more extensive origin for the temporalis muscle. The powerful mandible for crushing nuts and seeds does not affect gape.

Hiiemae (1984) found that face length, molar height, canine size and the presence or absence of a dental comb did not affect the operation of a common functional design to the oropharyngeal system of primates (or indeed in mammals), the jaw apparatus having been established early in mammalian evolution.

The transport of food through the oral cavity involves assessment and bolus preparation (Figure 1.2); it depends on the action of the tongue against hard and soft palates, as well as muscles of the jaws and mouth floor. Tongue movements depend in part on the hyoid apparatus, studies of which have been neglected. Masticatory cycles vary in pattern, muscular activity and force, according to diet, but they fall into two groups – puncture-crushing and chewing cycles. In both cases, there are closing power (with buccal and lingual phases) and closing strokes, affected by the bulk or resistance of the food. It is the power stroke that is crucial in maximizing dental occlusion, the efficiency of meeting of the cusps, crests and basins of the upper and lower teeth (especially the premolars and molars, the cheek teeth).

Teeth are renowned in vertebrate studies for their abundance in the fossil record, and their value as taxonomic markers. However, within taxonomic groups they vary to a surprising extent in terms of details related to different diets, as adaptive responses to changes in diet and/or niche. It is useful to distinguish between the anterior dentition (incisors and canines), which are concerned with food procurement and preparation (e.g. Maier, 1984) as well as with nonfeeding activities (hence dietary correlations may be reduced), and the posterior dentition (premolars and molars), which is concerned solely with food processing – mastication of various types (e.g. Kay, 1975; Kay and Covert, 1984). Such oral processing is crucial to digestive efficiency.

Dental design is best understood from a mechanical viewpoint (Lucas and Luke, 1984). While tooth size is proportional to basic metabolic rate (metabolic body size), tooth shape depends on the mechanical characteristics of the food. Foods are best categorized as:

1. Hard, brittle – elastic, stiff, of high strength, low toughness and low strain at failure (e.g. seeds, unripe fruit, root storage organs, nuts and some invertebrate parts).
2. Juicy – variable deformability of compartment wall, plastic flow of contents, low strength, variable toughness and variable strain at failure (e.g. ripe juicy fruits and some invertebrates).

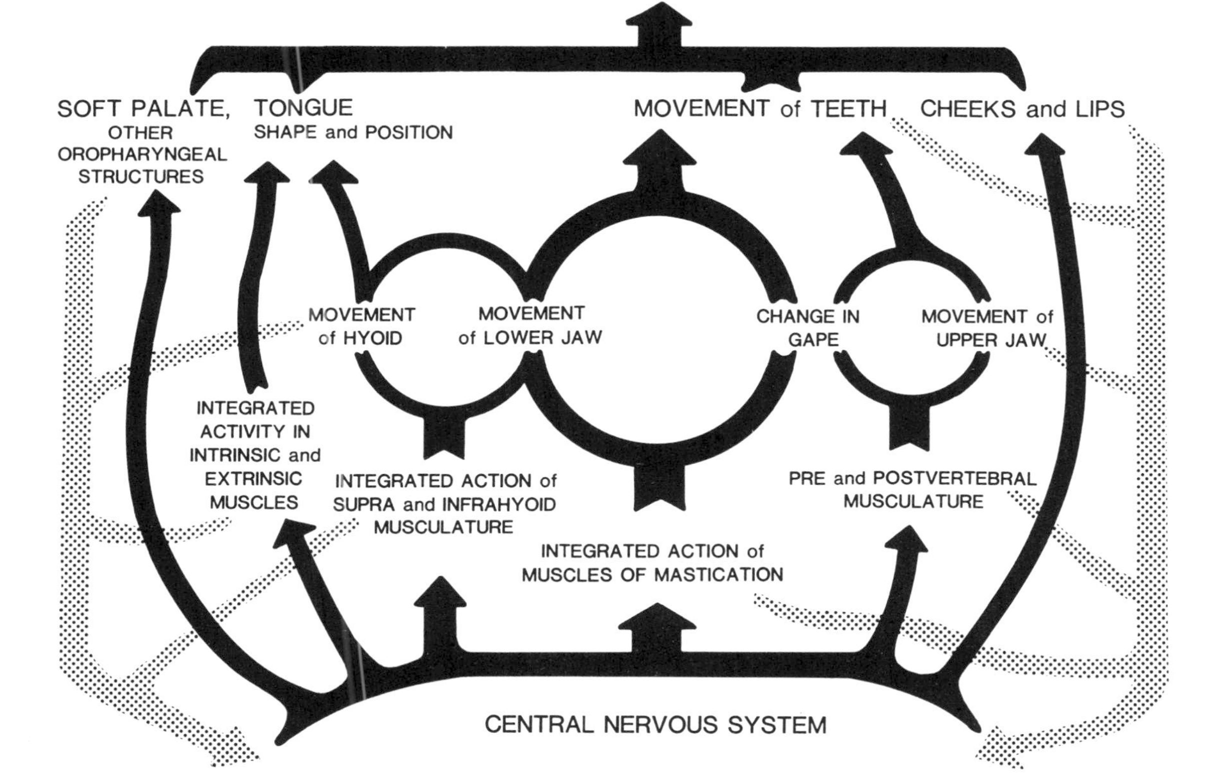

Figure 1.2 The jaw apparatus, showing effector system (solid) and sources of sensory input (dotted), minus tongue and palate inputs. (From Hiiemae, 1984.)

3. Tough and/or soft – viscoelastic, pliant, moderate strength, high toughness, often high strain at failure (e.g. soft animal tissues, invertebrates, leaves and grasses).

The key features of tooth design are 'opposing blades' which are sharp and crack, with a localized high occlusal pressure, and 'pestle and mortar', which is blunt and shatters, with a dispersed occlusal pressure. Hard, brittle foods require multiple fracture and a reasonable zone of actions – blunt cusps, thick enamel and crushing basins; the same is true for juicy foods. With tough and/or soft foods, cracks are not easily spread by crushing, so one needs a system of multiple blades to cut against each other – high cusps, long shearing crests and thin enamel (that can be worn away to create dentine windows and increase the extent of cutting edges).

Thus, faunivores have short, pointed incisors and high-cusped cheek teeth, with well-developed shearing blades – for trapping and cutting prey, and then cutting up the exoskeleton to release the contents and, among primates, to maximize the surface area of chitin for digestion. Folivores also have small anterior teeth for cutting leaves, and high-crowned cheek teeth with an impressive series of cutting edges to partition several times over – to maximize the surface area for the bacterial fermentation of cellulose. By contrast, frugivores have low, broad, procumbent incisors for opening fruit, and low-crowned cheek teeth with rounded cusps and thick enamel to crush and grind the fruit pulp. They are particularly well developed, along with the jaw and their muscles, for breaking up seeds. The pithecine monkeys (*Pithecia*, *Cacajao* and *Chiropotes* spp.) of the New World show all these features; the large incisors are designed to reach seeds that are mechanically protected. This contrasts with the small incisors of colobine monkeys (e.g. *Colobus* and *Presbytis* spp.), which, when they eat seeds, seem to be going for those that are inaccessible to other primates because they are chemically protected, a protection that the sacculated stomach and its contents can penetrate.

Incisor design reflects both food quality and external features. For example, dental combs may be fine and elongated for scraping exudates (e.g. the fork-crowned lemur, *Phaner*, and bushbabies, *Galago*) or short and stout for gouging bark to release such exudates (e.g. sifakas, *Propithecus*, and marmosets, callitrichids). Similarly, baboons (*Papio*) have large incisors because they pull grass rhizomes from the ground with their teeth, which are subjected to abrasive wear, whereas the gelada baboon (*Theropithecus*) has small incisors, because it pulls the same food out of the ground with its hands (Kay and Covert, 1984).

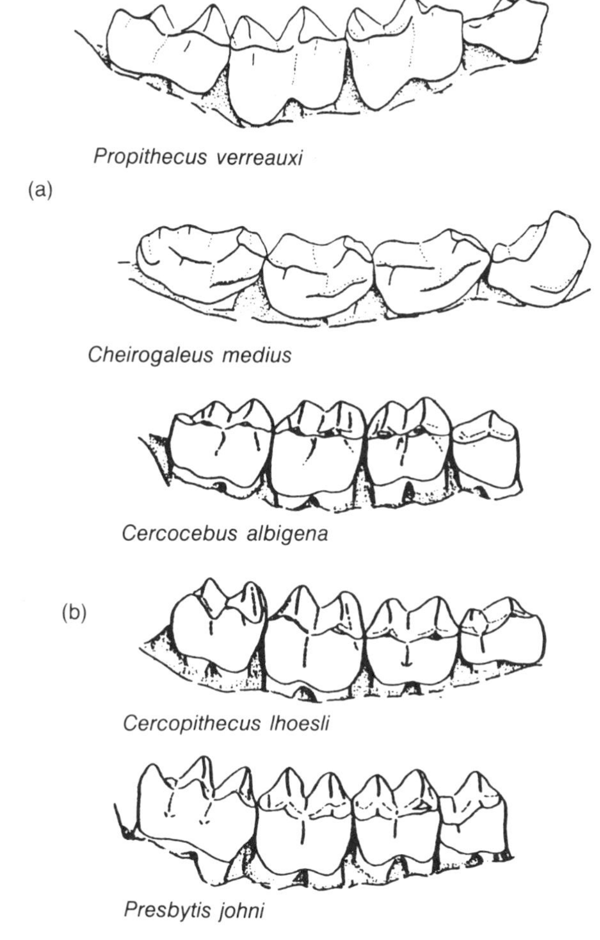

Figure 1.3 Lower cheek teeth of (a) two strepsirhine primates (*Propithecus* and *Cheirogaleus*) and (b) three cercopithecids (*Cercocebus*, *Cercopithecus* and *Presbytis*), viewed from laterally. ((a) From Kay and Hylander, 1978; (b) from Kay, 1975.)

Diet can be inferred from molar structure, because the cheek teeth are concerned solely with food processing. One can then evaluate the role of incisors (and canines) within this dietary context. Although the molar morphology is very similar in faunivores and folivores (Figure 1.3), there is an absolute size difference between them (because of such differences in body size). Furthermore, while crushing and grinding surfaces increase disproportionately with body size, shearing blades decrease relatively in length (Kay, 1975). For a given body size,

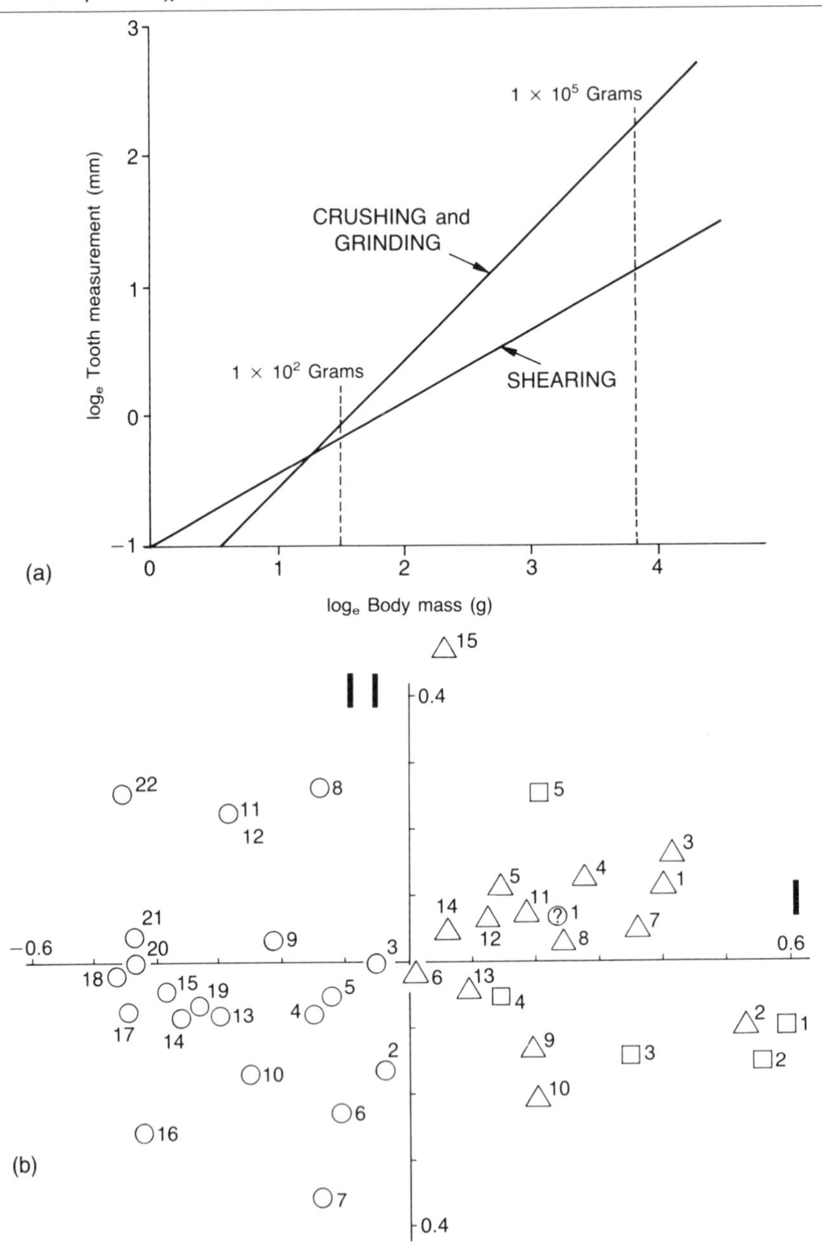

Figure 1.4 (a) Areas of crushing and grinding surfaces, and lengths of shearing blades (cristid obliqua) plotted against body size (from Kay, 1975). (b) First two coordinates of a principal coordinates analysis of molar dimensions on non-cercopithecoid primates (all standardized for body size, showing insectivores (1–5), folivores (6–15) and frugivores (16–22). (From Kay, 1975.)

however, folivores and faunivores have larger grinding surfaces and longer shearing blades than do frugivores (Figure 1.4).

Within genera such dimensions will vary with diet, reflecting long-term adaptations to differing environments or niches. For example, the siamang (*Hylobates syndactylus*), which eats mostly leaves, has higher cusps and longer shearing blades than most other gibbon species, whereas the Kloss gibbon (*H. klossi*), which eats virtually no leaves, has the lowest and most rounded cusps (Kay, 1975). The same pattern can be detected in species of *Presbytis* (leaf-eating dusky langurs compared with seed-eating banded or red langurs). Frugivorous cercopithecine monkeys that eat more animal matter have better developed cusps (e.g. *Macaca* spp.).

Inevitably, with such durable and complex structures as teeth, it is variety of function associated with a particular morphology rather than structural change that confers on a primate the ability to cope with a dietary shift in the short term. Such shifts are possible only if the dietary staple is fruit; faunivory and folivory require specializations that tend to prohibit such dietary shifts, a common feature among most of the topics discussed here.

1.5 FEEDING: GASTROINTESTINAL TRACTS

No body system is as malleable as the digestive system. The cells of the gastrointestinal mucosa turn over more rapidly than any other tissue of the body, so the proportions of different cell types can change markedly within a relatively short time. The whole tract can, through the generous arrangement of longitudinal and circular muscle, widen or narrow, elongate or shorten markedly (but not synchronously) during the normal digestive process. As with the frequency of different cell types, the overall proportions of each compartment can change markedly and rapidly in response to a change in diet, as will be discussed below.

Gastrointestinal adaptations also contrast with dental ones in that faunivores and folivores are at opposite ends of the structural spectrum, with frugivores intermediate, as first reviewed by Hill (1958) and investigated in more detail by Hladik and Chivers (1978), Martin *et al.* (1985) and MacLarnon *et al.* (1986). A long, small intestine is required for the absorption of the products of the simple digestive processes for animal matter, with a simple stomach and reduction of the large intestine (and often loss of the caecum in nonprimate mammals). By contrast, a diet of leaves requires a greatly expanded stomach and/or caecum and right colon (first part of large intestine) for the bacterial fermentation of cellulose. Fruits require an intermediate morphology:

relative expansion of stomach and caecum and colon, and relative reduction of the small intestine.

Among strepsirhine primates, for example, the faunivore *Arctocebus* (angwantibo), the frugivore *Perodicticus* (potto), and the folivores *Galago* (*Euoticus*) and *Lepilemur* (bushbabies and sportive lemur) span this dietary and structural spectrum (Figure 1.5). None shows the extreme adaptation characteristic of the nonprimate specialists – for example, carnivores and ungulates (Chivers and Hladik, 1980). Among primates only the colobine monkeys converge on ruminants with sacculation (and compartmentalization) of the stomach; only folivorous strepsirhines show dramatic enlargement of the caecum and colon, although folivorous ceboids (howler monkeys), hominoids (siamang, chimpanzees and gorilla) and colobine monkeys show expansion of the colon. Most primates are frugivores; hence their guts are of intermediate proportions, veering towards faunivorous features among the smaller species, and folivorous ones among the larger species.

In considering the dimensions of each gut compartment, the surface area is taken to reflect absorptive ability, the volume fermenting capacity and the weight, the muscular activity involved in promoting digestion.

Surface area of the small intestine, the main absorbing chamber, scales closest to metabolic body size (to the power 0.75 of body weight; Kleiber, 1961), but larger faunivores (nonprimate) have less absorptive area than expected and larger folivores have more, especially the forestomach fermenters (Figure 1.6a). Better resolution is achieved by adding in half the areas of stomach and large intestine, as a more accurate indicator of absorptive ability (Figure 1.6b).

The volumes of the potential fermenting chambers – stomach and caecum/colon – scale to actual body size closely for faunivores (with no fermentation); larger frugivores have more voluminous chambers than expected, and large caeco-colic fermenters even more so, but larger forestomach fermenters have smaller chambers than expected (Figure 1.7). The latter can be explained by the absorptive inefficiency of larger stomachs, of the relative decrease in surface area with increasing volume of an expanded 'sac', a problem not encountered when expanding a 'tube' (Chivers and Hladik, 1980).

Further allometric analyses were based on the surface areas of each gut compartment (Martin *et al.*, 1985; MacLarnon *et al.*, 1986). Compartmental quotients were calculated according to deviations from the expected slope of 0.75 for each species (Figure 1.8) and subjected to multivariate analysis. Dendrograms were found to be too labile with the subtraction or addition of data and, although caeco-colic and forestomach fermenters were consistently clearly separated, the rest

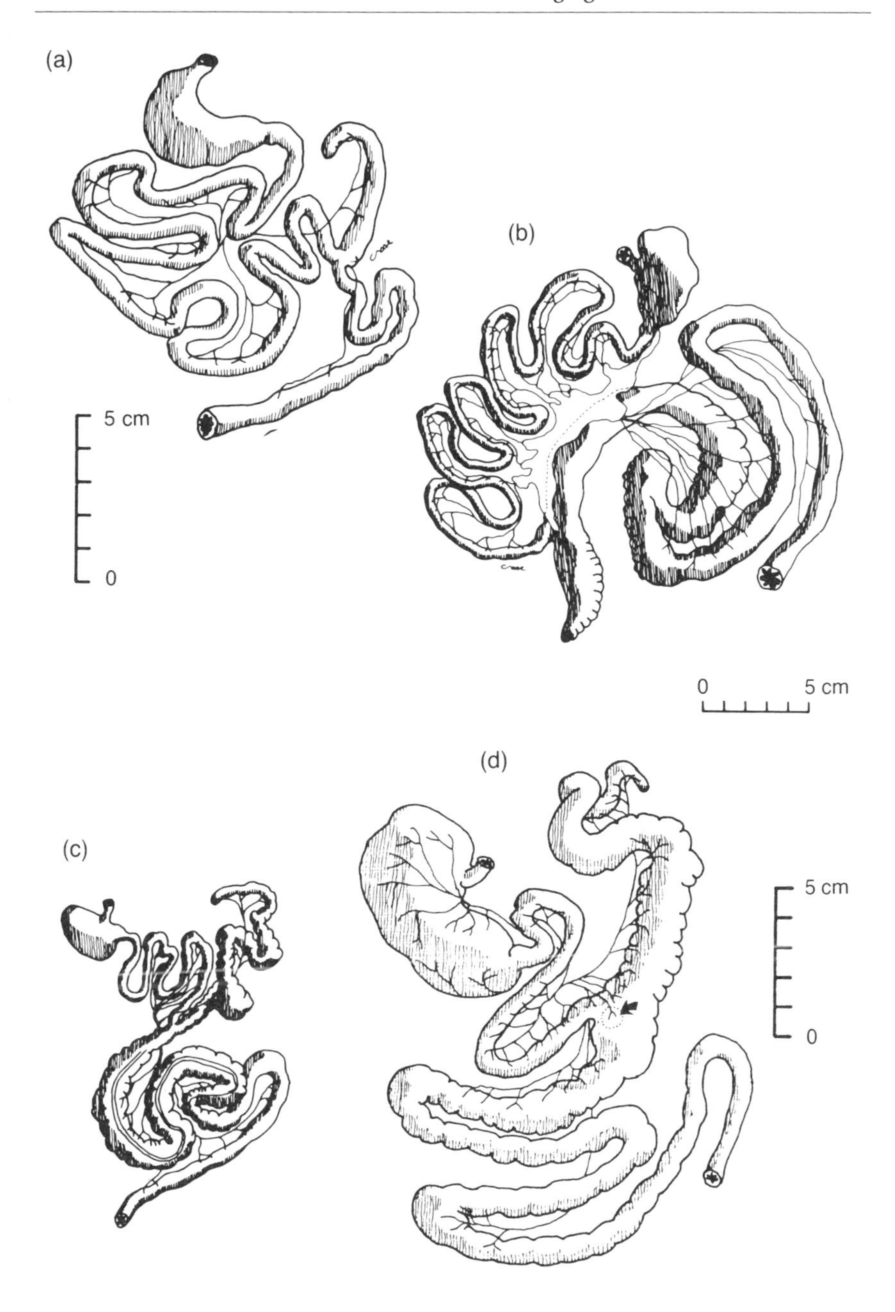

Figure 1.5 Gastrointestinal tracts of (a) faunivore, *Arctocebus*; (b) frugivore, *Perodicticus*; (c) gummivore, *Galago* (*Euoticus*); and (d) folivore, *Lepilemur*.

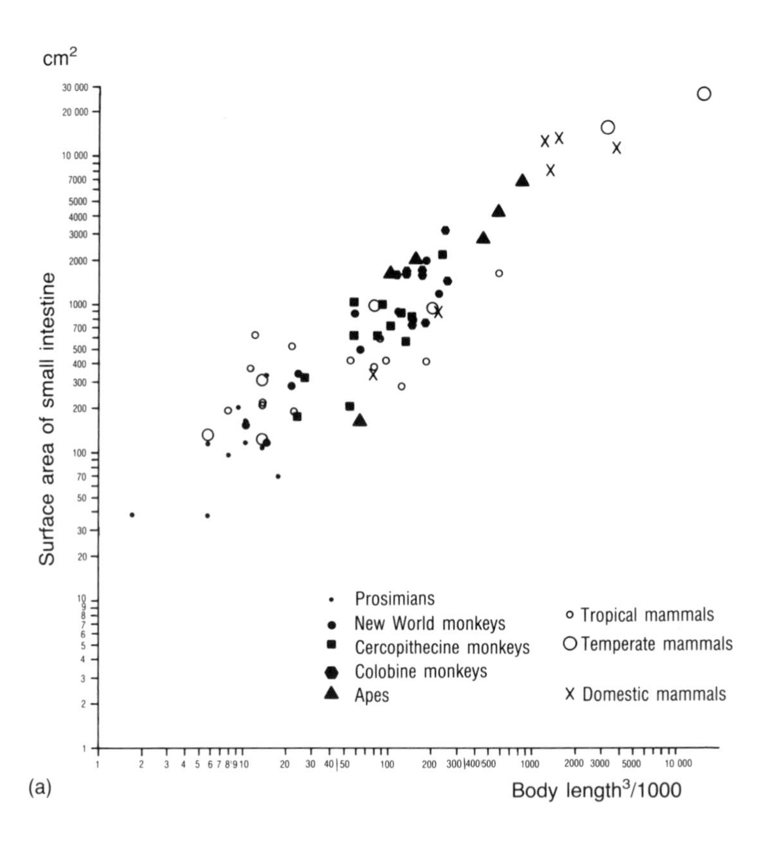

Figure 1.6 (a) Relationship between area of small intestine and body size in various mammals (from Chivers and Hladik, 1980). (b) Regression lines for the relationship between potential area of gut for absorption and body size in folivores, frugivores and faunivores. (From Chivers and Hladik, 1980.)

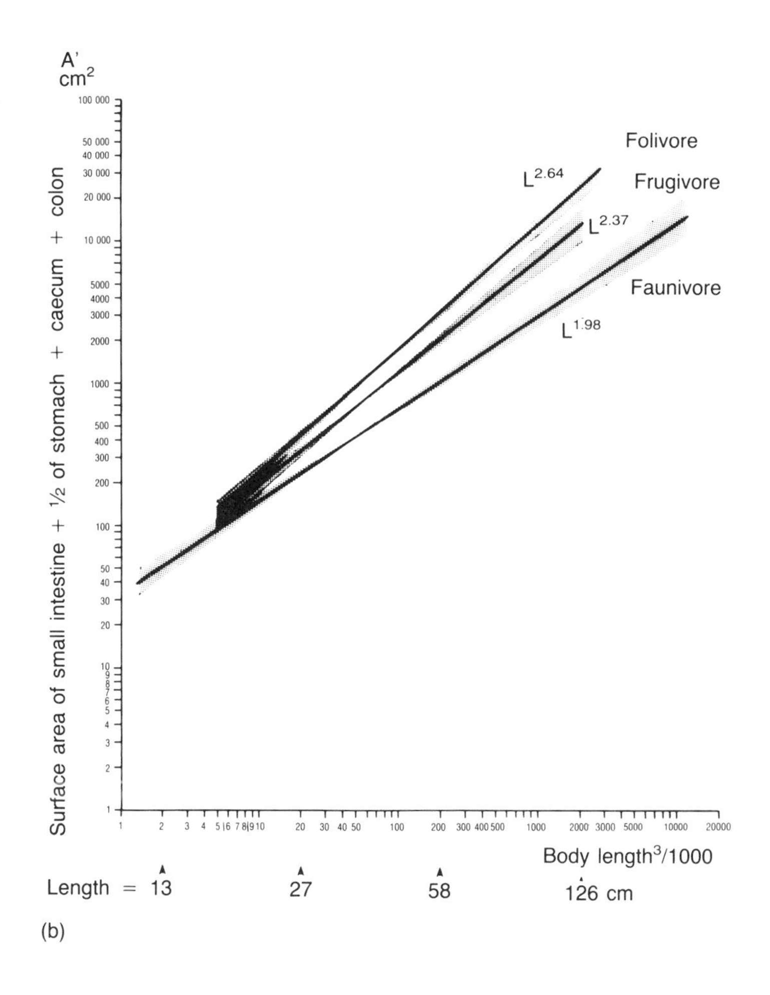
A' cm²
Surface area of small intestine + ½ of stomach + caecum + colon
Folivore
Frugivore
Faunivore
L2.64
L2.37
L1.98
Body length³/1000
Length = 13
27
58
126 cm

(b)

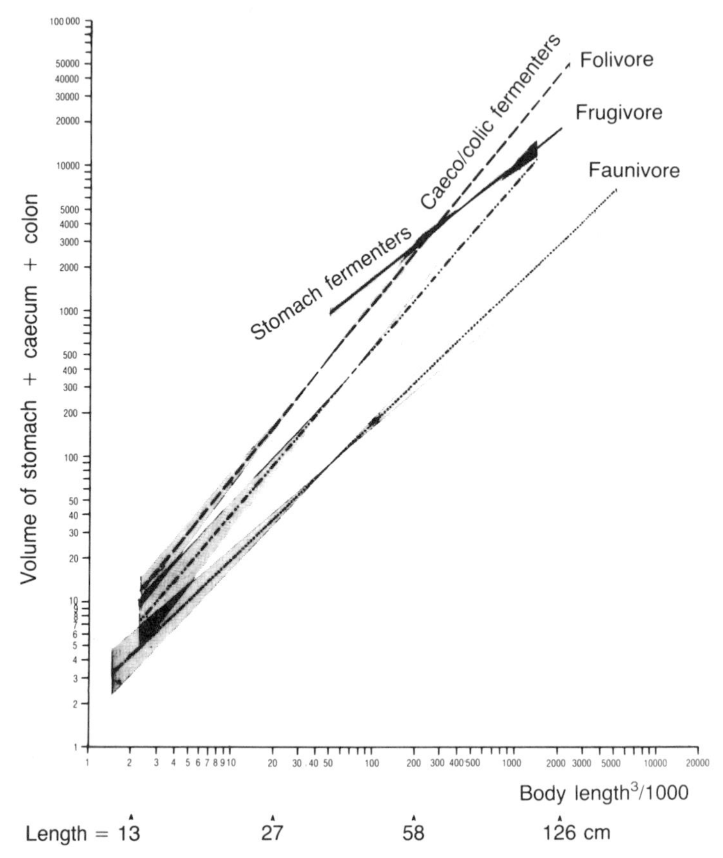

Figure 1.7 Regression lines for the relationship between volume of potential fermenting chambers and body size in folivores (stomach and caeco-colic fermenters), frugivores and faunivores. (From Chivers and Hladik, 1980.)

(frugivores and faunivores) were not. By contrast, multidimensional scaling was found to be sufficiently robust and informative.

Mammalian species separated into a central cluster of frugivores, with outlying groups of caeco-colic and forestomach fermenters (both with some compartments larger than expected) and of faunivores (with some compartments smaller than expected) (Figure 1.9). Having eliminated the effects of body size, the degrees of similarity between species are shown to reflect dietary adaptation rather than phylogeny.

While the horse is the most extreme of caeco-colic fermenters, and ruminants show the most extreme forestomach elaboration, primates fall in both groups; for example, the strepsirhines *Euoticus, Lepilemur* and *Perodicticus* are caeco-colic fermenters whereas the colobine monkeys *Colobus, Presbytis* and *Nasalis* are forestomach fermenters.

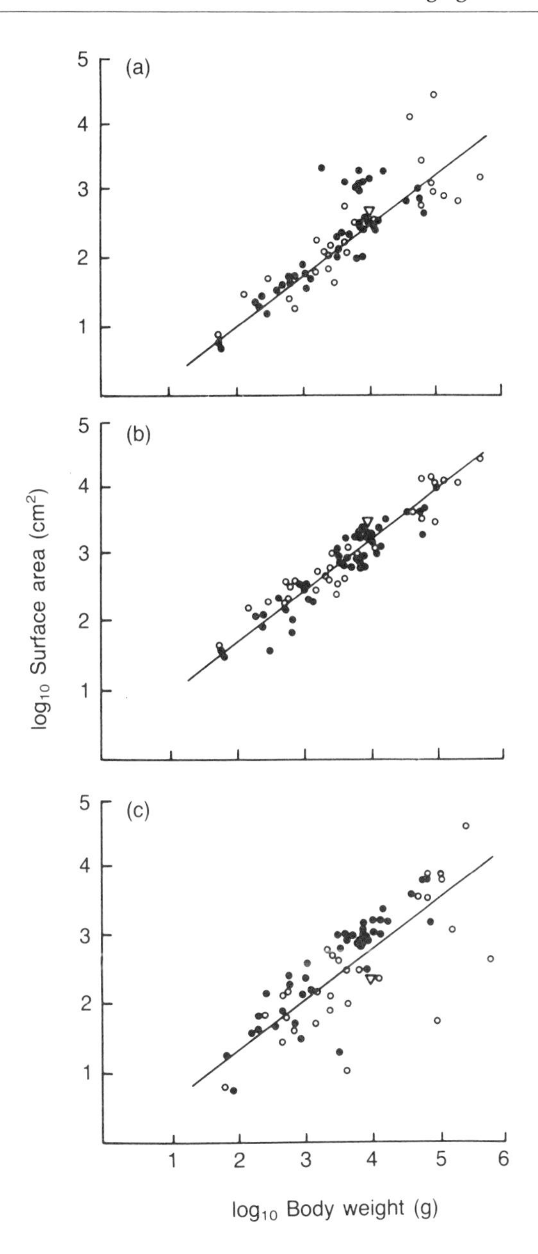

Figure 1.8 Logarithmic plots of surface areas of (a) stomach, (b) small intestine and (c) colon for 80 mammalian species in relation to the best-fit line of slope 0.75, from which compartment quotients are calculated (Martin *et al.*, 1985): primates (●), other mammals (○; ▽ badger, *Meles meles*). (From Stark *et al.*, 1987.)

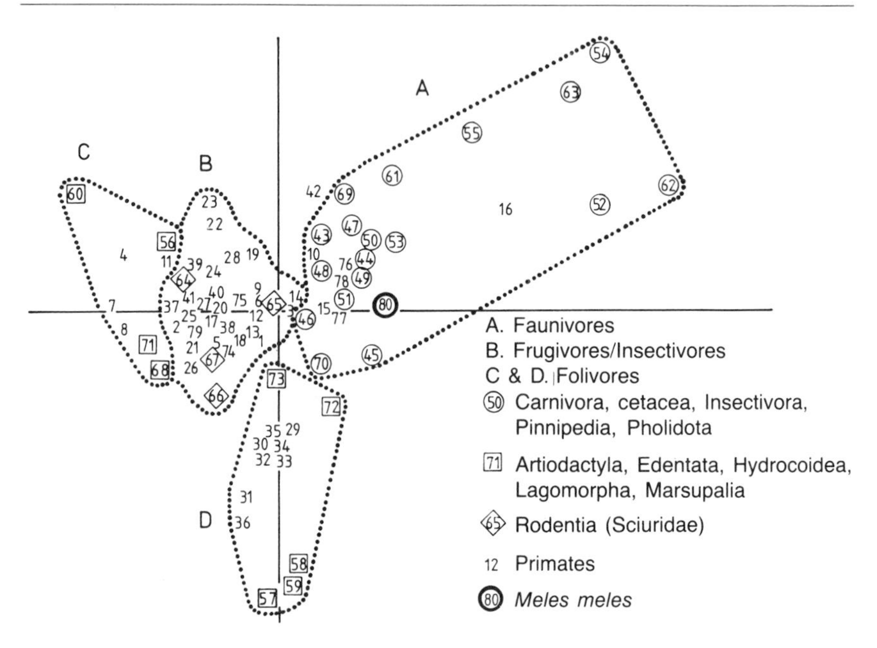

Key (*, caecum-less species; C, captive specimens; all other specimens were wild caught): 1, *Arctocebus calabarensis*; 2, *Avahi laniger*; 3, *Cheirogaleus major*; 4, *Euoticus elegantulus*; 5, *Galago alleni*; 6, *Galago demidovii*; 7, *Lepilemur mustelinus* (C); 8, *Lepilemur leucopus*; 9, *Loris tardigradus*; 10, *Microcebus murinus*; 11, *Perodicticus potto*; 12, *Saguinus geoffroyi*; 13, *Aotus trivirgatus*; 14, *Ateles belzebuth* (C); 15, *Saimiri oerstedii*; 16, *Cebus capucinus*; 17, *Alouatta palliata*; 18, *Lagothrix lagotricha* (C); 19, *Miopithecus talapoin*; 20, *Cercopithecus cephus*; 21, *Cercopithecus neglectus*; 22, *Cercopithecus nicitans*; 23, *Cercocebus albigena*; 24, *Macaca sylvanus*; 25, *Macaca sinica*; 26, *Macaca fascicularis*; 27, *Papio sphinx*; 28, *Erythrocebus patas* (C); 29, *Colobus polykomos*; 30, *Presbytis entellus*; 31, *Presbytis cristata*; 32, *Presbytis obscura*; 33, *Presbytis melalophos*; 34, *Presbytis rubicunda* (C); 35, *Nasalis larvatus* (C); 36, *Pygathrix nemaeus* (C); 37, *Hylobates pileatus* (C); 38, *Hylobates syndactylus* (C); 39, *Pongo pygmaeus* (C); 40, *Pan troglodytes*; 41, *Gorilla gorilla*; 42, *Homo sapiens* (C); 43, *Felis domestica* (C); 44, *Canis familiaris* (C); 45, *Mustela nivalis**; 46, *Vulpes vulpes*; 47, *Atilax paludinosus*; 48, *Nandinia binotota**; 49, *Poiana richardsoni*; 50, *Genetta servalina*; 51, *Mustela* sp.*; 52, *Ailurus fulgens** (C); 53, *Nasua narica** (C); 54, *Genetta* sp. (C); 55, *Panthera tigris* (C); 56, *Sus scrofa* (C); 57, *Capra hircus* (C); 58, *Ovis aries* (C); 59, *Cervus elaphus*; 60, *Equus caballus* (C): 61, *Halichoerus grypus* (C): 62, *Phocaena phocaena**; 63, *Tursiops truncatus**; 64, *Sciurus vulgaris*; 65, *Epixerus ebii*; 66, *Heliosciurus rufobrachium*; 67, *Sciurus carolinensis*; 68, *Oryctolagus cuniculus*; 69, *Potamogale velox**; 70, *Manis tricuspis**; 71, *Dendrohyrax dorsalis*; 72, *Bradypus tridactylus**; 73, *Macropus rufus* (C); 74, *Cacajao calvus*; 75, *Cacajao melanocephalus*; 76, *Cebus apella*; 77, *Saimiri sciureus*; 78, *Saimiri vanzolinii*; 79, *Alouatta seniculus*; 80, *Meles meles**.

Figure 1.9 Multidimensional scaling plot using log indices for three gut compartment surface areas (stomach, small intestine, caecum plus colon) for 80 mammalian species: clusters A, faunivores; B, frugivores; C, caeco-colic fermenting folivores; D, forestomach fermenting folivores. (From MacLarnon *et al.*, 1986; Stark *et al.*, 1987.)

Among faunivores, the vertebrate consumers show more marked adaptations (and are more peripheral) than insectivorous species, which include *Microcebus*, *Saimiri* and *Cebus* among primates, with *Homo* nearby. Otherwise, primates form the bulk of the frugivore cluster, with contrasting positions towards the periphery according to the proportions of foliage or animal matter added to the diet. Thus, *Ateles* (curiously), *Cheirogaleus*, *Loris*, *Galago demidovii*, *Arctocebus*, *Saguinus* and *Aotus* are positioned close to the faunivores, whereas *Avahi*, *Hylobates pileatus* (curiously), *Pongo*, *Pan*, *Gorilla*, *Alouatta*, *Cercopithecus neglectus*, *Macaca fascicularis* (curiously) and *M. sinica* are close to the caeco-colic fermenting folivores.

Considering the effects of dietary change, as evidenced by comparisons of wild and captive populations of the same species, data are available for the squirrel monkey (*Saimiri sciureus*). In two (admittedly) different populations the stomach was 25% shorter, the small intestine 16% shorter and the colon 48% shorter (caecum no change) in captive specimens (Beischer and Furry, 1964; Fooden, 1964). Similarly, in a captive toque macaque (*Macaca sinica*) in Sri Lanka the surface areas of the stomach, small intestine and large intestine relative to body size were 42, 85 and 74%, respectively, of those for a wild specimen (Amerasinghe *et al.*, 1971).

Another indicator of the lability of gut dimensions is the intraspecific variation of measurements found by Martin *et al.* (1985), with the overall coefficient of variation approximating 25%, much greater than that found for any other body system. In other words, the 95% range of variation for the surface area of a given gut compartment is approximately 50%, and the 99% range of variation is 75%. Average coefficients of variation for 65 individuals of nine primate species (with five or more specimens) were 29, 22, 25 and 24% for stomach, small intestine, caecum and colon, respectively; that is, variation was greatest in the stomach and least in the small intestine.

Chivers and Hladik (1980) derived gut specialization indices for absorptive area and for fermenting volume. When compared with the dietary index (also varying between −100 for exclusive faunivory and +100 for exclusive folivory), there are differences, even with reliable data from directly comparable populations, that indicate the flexible relationship between gut structure (and function) and diet (Table 1.1). Thus, the talapoin monkey (*Miopithecus talapoin*) appears to consume less animal matter than its anatomy suggests; the toque macaque (*Macaca sinica*) and the howler monkey (*Alouatta palliata*) show close correspondence between gut structure and diet; for the Barbary macaque (*Macaca sylvanus*) the volume index gives clearer indication of the folivorous dietary content than the area index, which is to be expected

Table 1.1 Examples from primates of morphological and dietary indices (from Chivers and Hladik, 1980)

	Gut specialization index		*Dietary index*
	By area (A)	*By volume (V)*	*X* (see Figure 1.11b)
Talapoin monkey (*Miopithecus talapoin*)	−85		−40 (estimate from Figure 1.11a)
Moustached monkey (*Cercopithecus cephus*)	−27	−2	−10 (estimate from Figure 1.11a)
Mandrill (*Papio sphinx*)	+9	0	
White-faced capuchin (*Cebus capucinus*)	+11		−5
Toque macaque (*Macaca sinica*)	+16	+22	+15
Howler monkey (*Alouatta palliata*)	+40	+31	+40
Barbary macaque (*Macaca sylvanus*)	+45	+85	
Banded langur (*Presbytis melalophos*)	+82		+53

(the area index corresponding more closely to the predominance of animal matter in the diet).

It is through such approaches that one can hope to specify the range of diets that can be coped with by a particular anatomy and, conversely, the range of gut structure that can cope with a particular diet (i.e. with a particular blend of foliage, animal matter and fruit).

A confounding variable is the passage time of digesta; it is difficult to determine this and little information is available. Gut compartments can be smaller if passage time is longer, and vice versa, without difference in digestive ability. According to Milton (1984), among Neotropical primates the animal-fruit diet of capuchin monkeys (*Cebus*) takes about 3.5 h from ingestion to time of first appearance of markers in the faeces; the fruit diets of uakaris (*Cacajao*) and bearded sakis (*Chiropotes*), with seeds predominating, take 5 h and that of woolly monkeys (*Lagothrix*), with a leaf component, takes 7 h; and the much more folivorous diet of howler monkeys (*Alouatta*) takes as much as 20 h. More detailed investigations for more subtle differences in diet would clearly be very productive. Langer (1987) points out that forestomach fermenting systems are much less able to vary passage time than caeco-colic fermenting ones; this clearly imposes greater restrictions on dietary variation in the former.

1.6 PRIMATES: DIETS AND CHANGE

It has been shown in the preceding sections how flexibility in positional and feeding behaviour and, to a lesser extent in the special senses, have encouraged primate radiations, initially from insectivory and subsequently, having crossed the body-weight threshold of 1 kg, from frugivory among the haplorhine primates (Ripley, 1979, 1984). Such dietary radiations among strepsirhine and haplorhine primates (Figure 1.10) correspond with changes in positional behaviour, especially in relation to contrasts between above-branch and below-branch niches.

This approach is reflected in Rosenberger's (1981) reclassification of the Ceboidea, which involves divergence from a frugivorous ancestor into the more frugivorous atelids (including the seed-eating pithecines and leaf-eating *Alouatta*) and the more faunivorous cebids (including the much smaller callitrichids). As pointed out by Ford (1986), the position of *Aotus, Callicebus, Cebus* and *Saimiri* remains equivocal, in relation to each other and to the more divergent forms, which may reflect their closer proximity to the ancestral form (Rosenberger separates the first two genera into Atelidae and the latter two into Cebidae). Reference has already been made to the uniqueness of ceboids, with unparalleled opportunities for study of their radiation within tropical rainforest, in contrast to the cercopithecoid radiation into other habitats (as well as forest) with less morphological change. It is here, in particular, that we see flexibility within genera, and even within species, in adaptive responses to different habitats.

This adaptive flexibility is suggested to be based on frugivory – a dietary strategy that distinguishes primates from most other mammals. First, they could exploit unoccupied niches; then some veered towards either faunivory or folivory, but not both (for anatomical and physiological reasons), hence the uselessness of the commonly used term 'omnivory'. The chosen direction has been based on body size, which has subsequently changed in some cases; for example, callitrichids, other ceboids and cercopithecoids such as the talapoin (*Miopithecus)* have become secondarily small, whereas other ceboids and some cercopithecoids and hominoids (great apes in particular) have become secondarily large. This has resulted in corresponding changes in positional behaviour (and anatomy) and in dentitions and gastrointestinal tracts.

Thus, plotting mean annual diets of as many primate species for

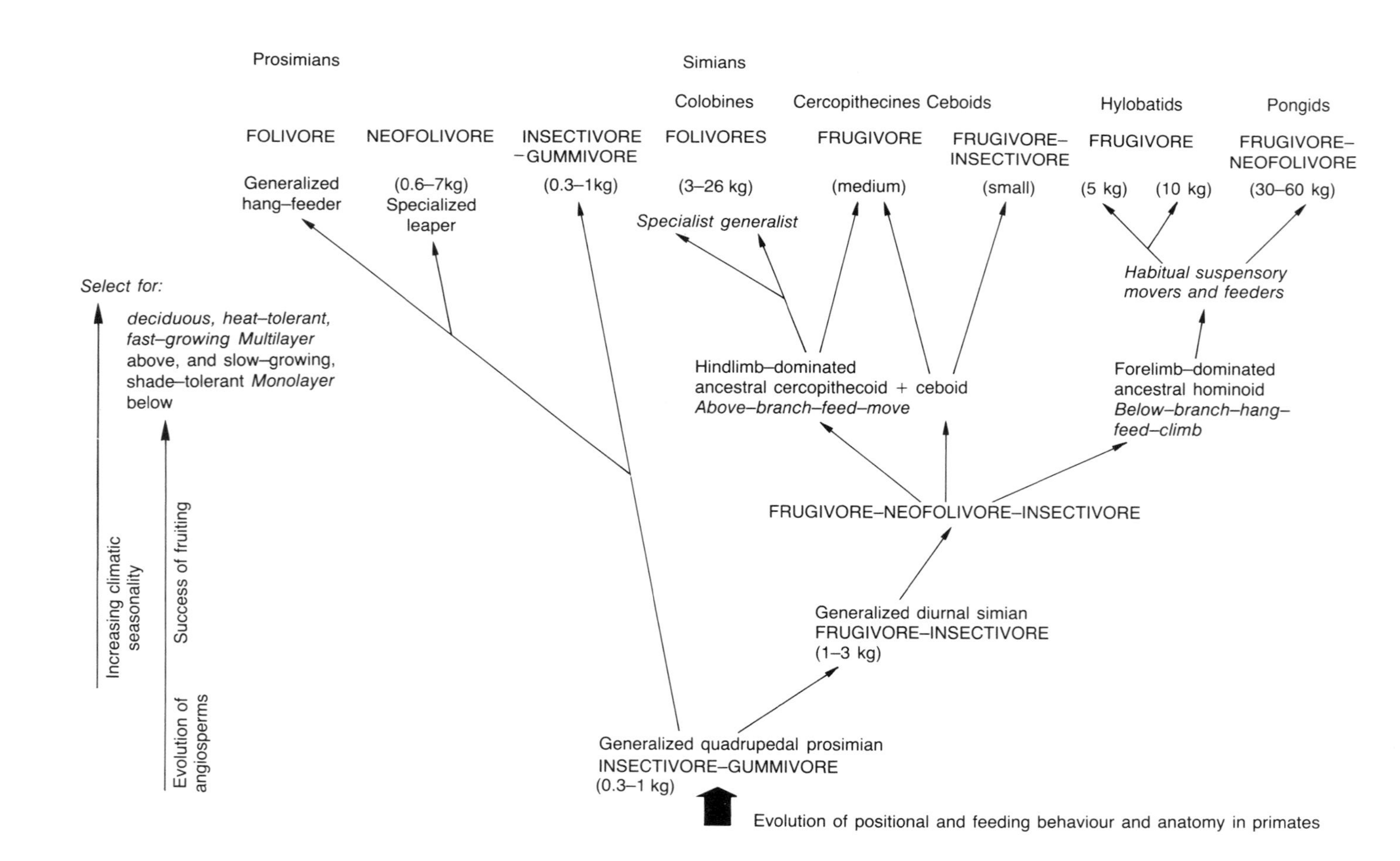

Evolution of positional and feeding behaviour and anatomy in primates

which data are available (Figure 1.11a) shows the strong clustering towards the frugivorous pole, with divergence back towards the faunivorous ancestry or on towards the folivorous pole – the latter two poles being monopolized by other, more specialized mammals. Hladik's data (Chivers and Hladik, 1980) on certain New and Old World primates (Figure 1.11b) show the extent of annual variations about such means, which are much greater than found in other mammals; comparable data are available for Malayan forest primates over just 6 months of the same year (MacKinnon and MacKinnon, 1980b).

Inspection of data on species' diets shows that strepsirhine primates span the greatest range (having occupied haplorhine and nonprimate niches in isolation in Madagascar), hominoids (with fewest species) the least, and with cercopithecoids spanning a greater range than ceboids (Figure 1.11a). Increased folivory is associated with increased biomass density, more than 4 kg/ha, and increased faunivory with a decrease, less than 0.1 kg/ha (Hladik and Chivers, 1978).

While there may be no change in body size within species in different habitats, such changes in biomass do occur, as evidenced for the grey langur, *Presbytis entellus* (Hrdy, 1977). Data from 15 study sites were collated – from Simla in the hills of northern India to Polonnaruwa in Sri Lanka in the south – encompassing a great variety of habitats from forested to arid, from moist to dry. While population densities and social structures have been fully compared, there seems to be less information on dietary contrasts (e.g. A.F. Richard, 1985). The genus *Macaca* offers opportunities comparable to *Presbytis* for investigating differences within and between closely related species according to differences in habitat – from open arid or woodland and montane habitats through semideciduous forest to evergreen rain forest (e.g. Lindburg, 1980; Davies and Oates, in press). In the more seasonal habitats, especially montane ones, dietary shifts from reproductive to vegetative parts of plants are quite marked; the ability to cope with such dietary change must be preadaptive to surviving more permanent dietary shifts through environmental change.

Reference has already been made to the ability of folivorous colobine monkeys to shift to seed-eating when the more nourishing leaves are scarce (McKey, 1978; Bennett, 1984; Davies, 1984), but their digestive systems cannot tolerate sugary fruit pulp. Apart from the

Figure 1.10 Evolution of positional and feeding behaviour and anatomy in primates, in relation to generalist and specialist frugivores, folivores and faunivores and to above-branch or below-branch movement and feeding. (Constructed from text by Ripley, 1979.)

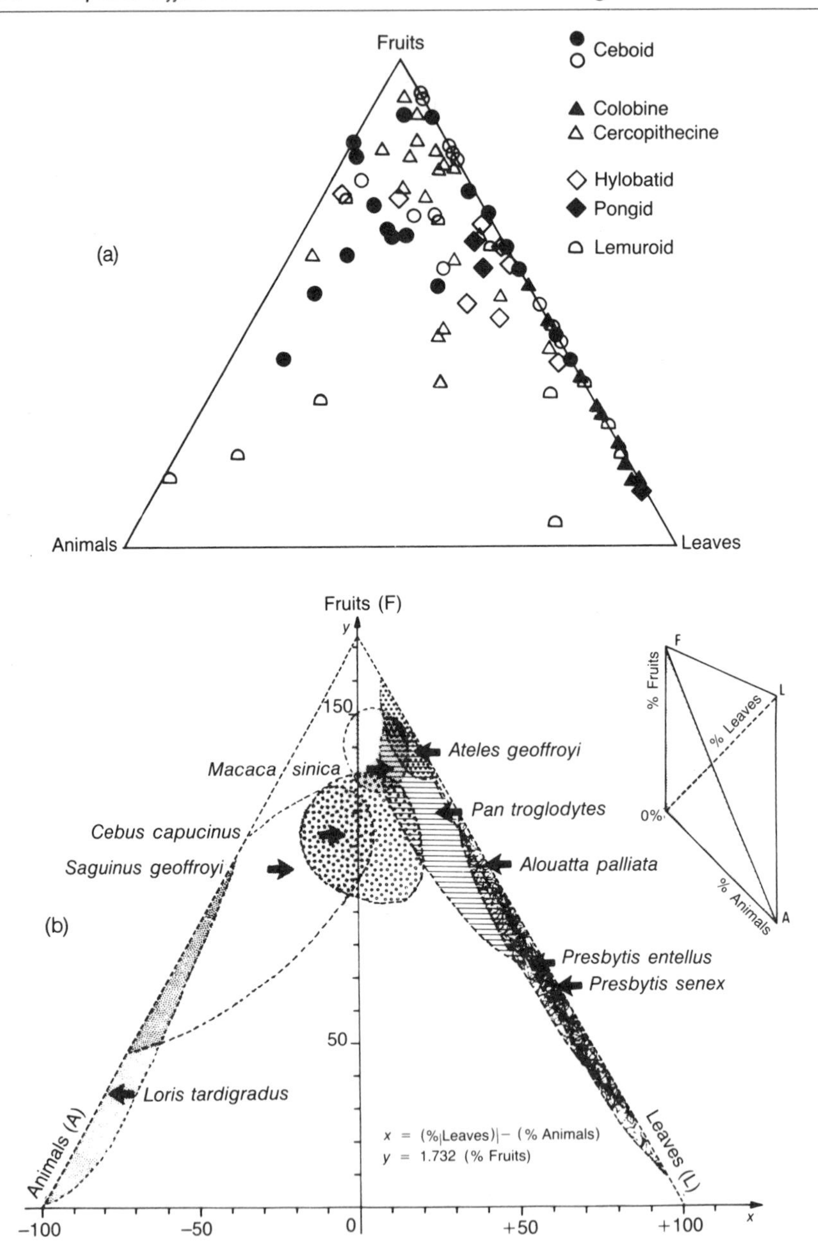

Figure 1.11 (a) Plots of mean annual diets of different primate taxa, according to the proportions of fruit, leaves and animals. (b) Plots of mean annual diets and extent of variation over a year for nine primate species in Panama, Sri Lanka and Gabon, indicating calculation of dietary index, from −100 (100% faunivory) to +100 (100% folivory). (From Chivers and Hladik, 1980.)

seasonal and annual variation shown by frugivore-folivores, according to the availability of food resources, persistent daily patterns of variation in diet occur. Fruits are eaten earlier in the day to replenish depleted levels of blood sugar after the overnight fasting, and then to harvest such foods, which are energetically more costly so to do; leaves are consumed later in the day when the primates are fatigued, to load up the digestive system for slower digestion through the night (e.g. Chivers, 1980).

Alternation in the intake of different food types may also be related to reducing any negative effects on the digestive system of plant secondary compounds, which may be harmless in small quantities in terms of toxicity or digestion inhibition and worth consuming for energetic or nutritional reasons (e.g. Oates *et al.*, 1980). It is too idealistic to expect the diet to be balanced each day, but one would expect the diet to be more or less balanced over several days.

There is clear evidence, however, even in tropical forest where seasonality is depressed, that primates may be relatively deficient in energy and nutrients for quite long periods, and in marked surplus at others. This is reflected by marked contrasts in the amount of travelling, calling, playing and mating from one time of year to another (e.g. the siamang, *Hylobates syndactylus*; Chivers, 1974). Thus, this not only affects activity budgets, which vary accordingly, but social activities, including reproductive behaviour. Hence the cyclicity of breeding – restricted to seasons in open habitats with marked wet and dry seasons, and producing peaks in evergreen forests that have very short dry spells (Chivers, 1980; Hrdy and Whitten, 1986). It is the varying effect on food supply that underlies these behavioural changes. The facility of hominoids, in particular, to store food surpluses as fat, is crucial to helping them through such fruit shortages – in serving to iron out the effects of such environmental fluctuations.

1.7 CONCLUDING REMARKS

It is this flexibility in feeding strategies that enables primates to cope not only with short-term and long-term environmental changes, but to respond to the more drastic human-induced changes in habitat (Johns, 1983; see Chapter 6). Fruit is the resource most markedly depleted by logging, and so survival depends – until the forest regenerates sufficiently – on abilities to move through the changed habitat (through their diverse locomotor repertoires) and to consume either more animal matter or more foliage (according to their diverse postural and feeding abilities).

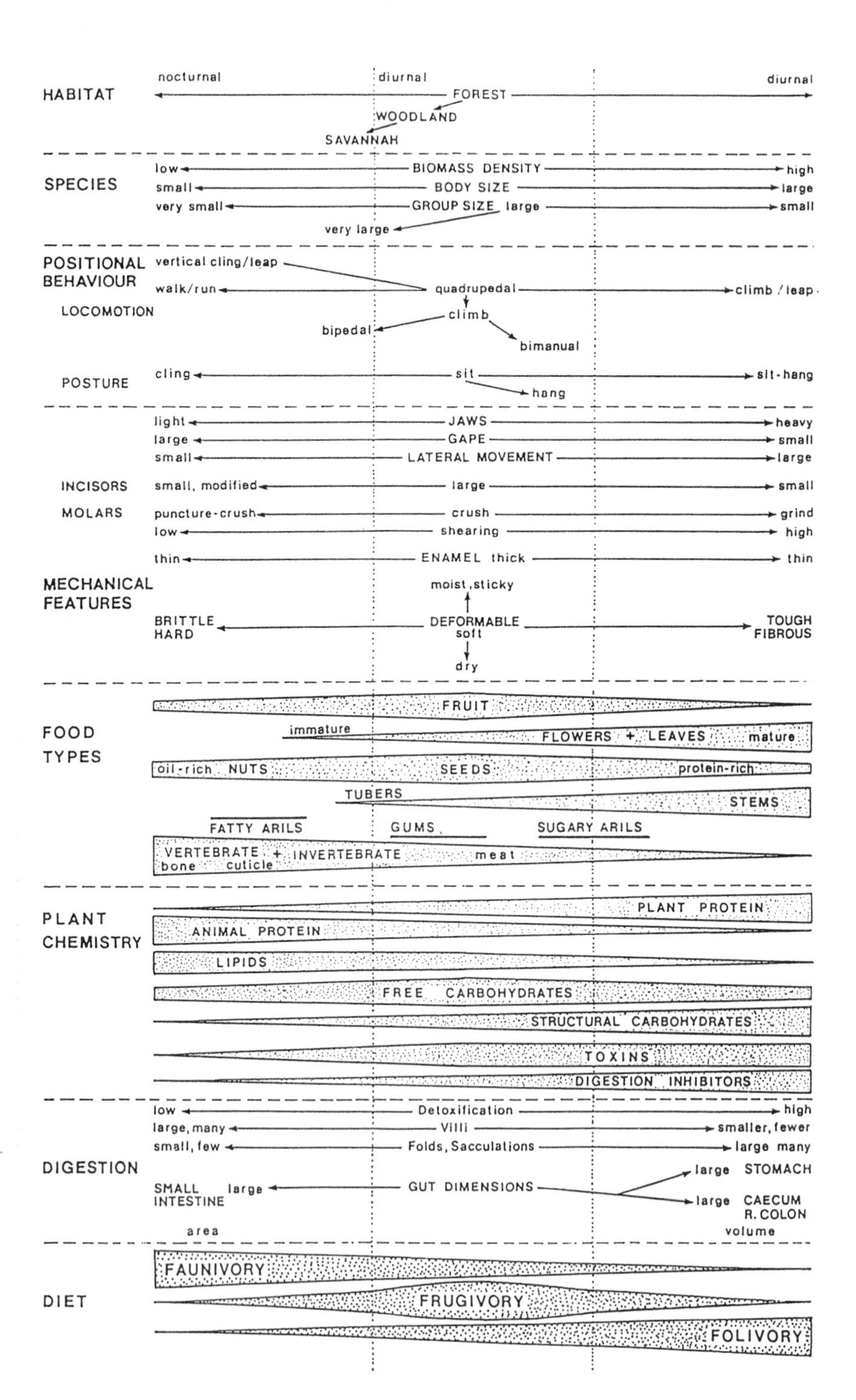

HABITAT
nocturnal
diurnal
diurnal
FOREST
WOODLAND
SAVANNAH
SPECIES
low
BIOMASS DENSITY
high
small
BODY SIZE
large
very small
GROUP SIZE large
small
very large
POSITIONAL BEHAVIOUR
LOCOMOTION
vertical cling/leap
walk/run
quadrupedal
climb / leap
climb
bipedal
bimanual
POSTURE
cling
sit
sit-hang
hang
light
JAWS
heavy
large
GAPE
small
small
LATERAL MOVEMENT
large
INCISORS
small, modified
large
small
MOLARS
puncture-crush
crush
grind
low
shearing
high
thin
ENAMEL thick
thin
MECHANICAL FEATURES
moist, sticky
BRITTLE HARD
DEFORMABLE soft
TOUGH FIBROUS
dry
FOOD TYPES
FRUIT
immature
FLOWERS + LEAVES
mature
oil-rich NUTS
SEEDS
protein-rich
TUBERS
STEMS
FATTY ARILS
GUMS
SUGARY ARILS
VERTEBRATE + INVERTEBRATE
meat
bone
cuticle
PLANT CHEMISTRY
PLANT PROTEIN
ANIMAL PROTEIN
LIPIDS
FREE CARBOHYDRATES
STRUCTURAL CARBOHYDRATES
TOXINS
DIGESTION INHIBITORS
DIGESTION
low
Detoxification
high
large, many
Villi
smaller, fewer
small, few
Folds, Sacculations
large many
large STOMACH
SMALL INTESTINE
large
GUT DIMENSIONS
large CAECUM R. COLON
area
volume
DIET
FAUNIVORY
FRUGIVORY
FOLIVORY

Other animals may depend for their survival on moving to undisturbed habitats, but the complex societies depend on their knowledge of a known area and, thus, dietary rather than range shifts are of prime importance. More faunivorous or folivorous species are clearly less affected by such disturbance. The year-round capacity for breeding, as afforded by the menstrual cycle (emancipated from the rigid hormonal controls of the oestrous cycle), clearly aids greatly the readjustment of population levels more efficiently and rapidly after the mild levels of disturbance resulting from carefully controlled selective logging.

The spectrum of variation among primates in (1) habitat, (2) species characteristics of body size, group size and biomass, (3) positional behaviour, (4) jaw, teeth and gut features have been correlated here with (5) dietary categories and (6) the chemical and mechanical properties of different foods (Chivers *et al.*, 1984). Compared with other mammals, primates can shift more easily along this spectrum (Figure 1.12) because of their lack of restrictive specializations. Thus, they can survive more readily in undisturbed or disturbed habitats and can respond successfully to both short-term and long-term environmental changes.

Figure 1.12 Correlation between various aspects of feeding. (From Chivers *et al.*, 1984.)

2
Adaptations to environmental change: an evolutionary perspective

PHYLLIS C. LEE

2.1 INTRODUCTION

Investigations of the ways in which mammalian species respond to environmental change are of particular interest since they illuminate questions of evolution, of behavioural adaptation, and of the proximate solutions to problems posed by the environment. The past 20 years of intensive research on primates in their natural environment and in captivity have provided us with an almost bewildering demonstration of variability in morphological adaptation, in social organization and behaviour.

The focus of this work in behavioural ecology has been adaptation to the environment. What is of interest here is whether this observed behavioural flexibility, and the principles underlying it, can be extended to the more specific problem of adaptation to environmental change through time. Environmental change and the reactions of animals to it can be considered at scales that range from the immediate responses to short-term environmental change, through the longer-term changes in reproductive potential, in the nature and expression of relationships and, ultimately, to evolutionary change across generations and geological time. All these levels are interrelated and build upon changes from one to another. Variation in behaviour in the context of environmental change must therefore underly genetical adaptation to the environment.

Under the Darwinian paradigm, large-scale evolutionary changes are the outcome of variation in individual survival and reproductive success. To relate changes in individual behaviour to concepts such as adaptation, we need to examine the ascending order of interactions between the organism and its environment – starting with the effects of environmental change on individual reproductive parameters and behaviour, which are linked to changes in the social and demographic context of behaviour, and which in turn affect the long-term processes of speciation and extinction. Of interest here is the scale of the change, from short-term to long-term, and the degree of accuracy and detail with which we can assess behavioural changes.

The problem of resolution in assessment of change is highlighted by the failure of simple ecologically deterministic relations to explain variation in social organization in even such a well-studied genus as baboons *Papio* (e.g. Popp, 1983). From the more sophisticated eco-correlates approach have come some suggestive causal ecological relations in comparisons across species (e.g. Clutton-Brock and Harvey, 1977; Sailer *et al.*, 1985; Dunbar, 1988), but our ability to define, and indeed to measure, the critical ecological variables and to incorporate a wide range of behavioural flexibility into these approaches still appears to be limited. Detailed information about behavioural and habitat characteristics is still restricted to a relatively small number of species.

Despite these limitations, defining and explaining behavioural variability and adaptation can be approached in several ways. The first is to focus on the cognitive capacities of different species, their abilities to solve problems, to learn, to innovate and to manipulate their environment (see Chapter 3, section 3.2). The second general approach is an understanding of life history parameters – energetics, growth, reproductive rates and longevity. A third is to examine the physiological and morphological specializations linked with the behaviour of different species, in their digestive and reproductive physiology, locomotor behaviour, and so on (see Chapter 1). A fourth approach is to examine how changes in the environment are related directly to behavioural flexibility. All of these approaches are aspects of the biology of the organism, but the last is the main focus of this chapter. Behavioural flexibility can be viewed as the changes in individual behaviour resulting from variation in social or ecological conditions, and which consists of a complex set of possible responses. Environmental variation is defined through alterations in levels of energy and nutrients available for maintenance, growth and reproduction. It has the indirect effect of changing amounts of energy on life history parameters and the expression of cognitive capacities.

The environment is also a set of interrelated variables. It is not merely a refrigerator containing more or less food upon opening its door. It is a complex system of refuges and places of shelter from heat or cold stress, or from potential predators. These predators may be more or less numerous, skilled, or hungry through time. It offers similar opportunities for its exploitation to individuals of other species, who thus may be competitors of the primates in question. The numbers, distribution and exploitative skills of interspecific competitors in the community also vary through time. Individuals may not have access to critical resources at the necessary times since not all are equally skilled at opening the fridge door. All of these variables, which can improve, deteriorate or change cyclically through time, have consequences for the survivorship and reproduction of individuals, and thus pose problems of adaptation.

A synthesis of all modes of response to observed environmental change is beyond the scope of this chapter, and is probably premature. Of specific interest here is the categorization of variability in behaviour, first, in the form of direct changes in behaviour mediated by energy availability, and, second, in the consequences of these changes in terms of demographic parameters and social systems. A system of interactions relating the potential range of behavioural responses to the nature and types of environmental variation is proposed, using a small number of specific examples. Both the degree of resolution, when assessing behavioural flexibility, and the scale of environmental change are emphasized. The specific examples will be used to establish predictive principles of adaptive behavioural responses, as part of a theoretical framework that allows us to make sense of primate behavioural flexibility.

2.2 SOURCES OF VARIATION

From a primate's perspective environments, and hence the resources available for exploitation, are not stable entities with fixed parameters. They change through time, on a scale that can vary from a few hours (as in changes in temperature) to millions of years (as in changes in the continental distribution of forests). Environmental variation thus forms a continuum from short-term changes that are unique and unpredictable, to cyclical changes that may be somewhat longer term (diurnal, seasonal), through long-term changes such as woodland regeneration or loss over decades, and, finally, to changes over geological time. The scale of environmental change is presented in Table 2.1, along with a scale of adaptation that again is part of a continuum of behavioural flexibility. Short-term behavioural changes can be a direct response to changes in the environment, or may

Table 2.1

	Scale of environmental change			
	Short-term unpredictable	*Cyclical predictable*	*Long term*	*Geological*
Scale of adaptation				
Behavioural	Social flux	Seasonal changes in rates of interaction	Changes in social structure	
	Change in relationships	Altered patterns of energy optimization		
	Cognitive and behavioural innovation			
Demographic		Seasonal migrations	Altered reproductive rates	Expansion into new habitats
		Seasonal breeding	Changed life history parameters	
Genetic			Altered kin structures	Speciation
				Extinction

themselves alter the environment further. Short-term and long-term behavioural flexibility can influence demographic and life history variables, which subsequently can have consequences for behaviour at a population level, leading to genetical and adaptive changes. Both scales of environmental and behavioural change require a second scale relative to lifespan and generation time in order for comparisons between species to be valid.

What, then, causes environments to change and elicit behavioural flexibility? Floods, droughts, superabundances or 'plagues' of army worms or locusts, are examples of short-term disruptions to prevailing or 'average' environmental conditions. A single hurricane that eliminates feeding and sleeping trees can have dramatic and long-term effects on behaviour (e.g. Dittus, 1986). Such events are probably rare in the life of an individual (or underreported in studies of mammalian ecology; see Weatherhead, 1986), but they may occur regularly through geological time. While the effects of rare events tend to be localized to particular groups or regional populations, they provide information on the nature and extent of behavioural responses, and on the duration of their effects.

Most environmental variation is probably cyclical since most changes take place on a seasonal basis. Such seasonal changes, associated with predictable diurnal and monthly rhythms, are common in most of the tropical environments of primates. The effects of predictable seasonal variation on behaviour have been documented in numerous studies (see also Lee, 1988). However, observed changes in surface structure (i.e. interactions and relationships that can be observed to vary in response to environmental cycles) have yet to be related to processes of change in the deep structure (i.e. the organizing principles or solutions to environmental problems (see Seyfarth *et al.*, 1978; Dunbar, 1988)). Alteration in demography and life history variables linked with variability in surface structure may provide the process for changes in the long term which, over generations, lead to genetic change. Alteration of habitats or community dynamics over the medium term, such as several generations, are poorly documented for the obvious reason that few studies continue for the time required to detect and observe such responses. We can approach very long-term, or geological, changes with the aid of genetical studies on rates of species divergence and the use of fossils, in order to assess the evidence for adaptive responses. We may now take some examples of behavioural responses.

2.3 SOCIAL FLUX IN CHIMPANZEES

One striking example of short-term changes in the environment leading to radical changes in behaviour (what I have called 'social flux' in Table 2.1) has been observed in wild common chimpanzees (*Pan troglodytes*). Chimpanzees live in relatively large communities of 20–100 animals within a well-defined home range (reviewed by Nishida and Hiraiwa-Hasegawa, 1987). Long-term studies at Gombe Stream National Park and the Mahale Mountains in Tanzania have found that relationships between neighbouring communities tend to be overtly aggressive (Goodall, 1986; Nishida, 1979). Males jointly patrol and defend boundary areas and attack strangers found within these areas. However, in one community at Gombe these attacks escalated to the point where a smaller adjacent community was completely annihilated through the death of its males and many older females as a result of the concerted and extremely violent attacks from the neighbouring community's males. Goodall (1986) provides a complete description of the attacks and their effects on the two communities.

This example is of interest in that it can be related to a gradual change in the demographic composition of the two communities; in the familiarity between community members which took place shortly after a period of relatively rapid ecological change. Over the 10 years before the intensification of confrontations between the communities, members of both communities appeared to mingle readily at the banana feeding station established to assist with habituation (Wrangham, 1974). During these contacts at the feeding station, subgroups could be distinguished (Bygott, 1979) and aggression between individuals was frequent (Wrangham, 1974). Interactions at the feeding area did not appear to alter behaviour in other parts of the home range (Goodall, 1983). However, when banana feeding was reduced in 1972, the distinction between the subgroups became more apparent, with infrequent contacts between individuals, and a gradual separation of geographic ranges. Between 1974 and 1975, the smaller community was exterminated and their range was appropriated by the larger community. Goodall (1986, p. 501) is careful to point out that an increase, through gradual changes in demographic parameters, in the number of adult males in the successful community initiated these unusual levels of aggression, and not the banana feeding *per se*. The banana feeding, or its cessation, did not precipitate the intercommunity aggression. But I would suggest that the reduction in the feeding may have played a role in the timing of the subsequent events, and indeed accelerated the trends already underway. When bananas were abundant and unrestricted, members of the two communities had good energetic and nutritional reasons for aggregating at this superabundant resource. High levels of contact and familiarity, expressed through complex greeting and appeasement behaviour, are essential to the maintenance of chimpanzee societies (de Waal, 1982; Goodall, 1986). The two communities, which may have been in the process of fragmenting (e.g. Bygott, 1979), were kept in close social contact through aggregating at the feeding area.

With the restriction of ad lib feeding, the resource not only became less abundant but also less predictable in its abundance. Individuals spent less time associating, were less familiar and interacted less frequently. Ultimately, they had few reasons to tolerate rather than to compete with each other. Such changes in patterns of association and tolerance related to resource quality, abundance and distribution, combined with a demographic change (i.e. the increase in the numbers of adult males) resulted in 'warfare' between the communities.

Social flux, as in the above example, is a short-term behavioural response with long-term effects – the extinction of a community. It resulted from a series of connected and cumulative changes, in demography, in the maintenance of relationships between familiar and unfamiliar individuals, in the expression of power politics (e.g. de Waal and Hoekstra, 1980), and was accelerated by a rapid shift in resource base with the termination of artificial feeding. Both the scales of the ecological change of the behavioural response were short term, and unpredictable. These behavioural responses were not in themselves novel. Aggressive behaviour and displays seen during intercommunity interactions (e.g. Nishida, 1979) may be basic to the processes of conflict resolution and appeasement (e.g. de Waal, 1982) between individuals and they act to structure relationships between communities as well as within them. A propensity to manipulate social situations and relationships, and well-established tendencies for neighbouring communities to interact in a hostile manner, were pushed over a threshold into previously unseen and unsuspected levels of violence within the context of a relatively short-term change in a specific ecological parameter.

This example points to two problems in relating short-term and unpredictable behavioural changes to ecological changes. Firstly, such social responses seldom arise *de novo* (see below). The basis for the response must be present as in the propensity for chimpanzees to engage in power politics, and a pattern of antagonistic interactions existing between strangers. Secondly, the same ecological change might not produce any effect should the baseline conditions be different. If the number of males in the two communities had been equal, then no community would have been able to escalate hostilities as effectively (see also Nishida, 1979). Can we then argue that the ecological change alone was the cause of the social response? The answer, clearly, is no. The demographic change was probably the main causal variable, while the ecological change was the precipitating factor. The aim of establishing a framework for describing the relations between ecological changes and social changes is to allow us to explore which factors are necessary to effect change, and in what sequences, how these are interrelated, and how effects can be varied when the parameters change.

2.4 INNOVATION, TRADITION AND 'CULTURE'

In the chimpanzee example above, an existing behaviour pattern was given new intensities within an established context. For the expression of a completely novel behaviour, some form of innovation is required, where innovation refers to the introduction of a novel mode of coping with the environment. The behaviour can be either one not previously known, or it can be an existing behaviour, or a combination of behaviours that are applied in a novel context. The transmission of innovations to other group members, and across generations, often is considered to be 'cultural' (McGrew and Tutin, 1978; Nishida, 1987). (Debates over the definition of the term 'culture' and its application to nonhuman primates are continual (Hinde, 1987). Here, I have used the term in the sense of McGrew and Tutin (1978): innovation, standardization, durability, transmission to others through diffusion and dissemination, as well as across generations in the form of traditions.)

In relation to ecological variation, two important factors can be considered. Firstly, innovation may originate as a solution to a specific ecological problem. The classic examples of innovation during the extraction and preparation of food – the washing of wheat and potatoes by Japanese macaques (*Macaca fuscata*) (see Nishida, 1987 for a review) and tool-use in chimpanzees (Goodall, 1963; McGrew and Tutin, 1978; McGrew, 1989) and some other primates (e.g. *Cebus*: Visalberghi and Antinucci, 1986) – demonstrate how the intake of food is either enhanced or made energetically more efficient when that food is exploited in a novel way. The second factor is the rate of spread of the novel behaviour to other group members. The ability to innovate with respect to the exploitation of food can spread rapidly to other group members, and Nishida (1987) has recently argued that once transmission has occurred, a new ecological niche has become available.

Experimental studies of the responses of chacma baboons (*Papio ursinus*), mandrills (*Mandrillus sphinx*) and vervet monkeys (*Cercopithecus aethiops*) to new food types and to novel unpalatable foods found differences between the species in the form of learning and in the mode and speed of transmission to other group members (Cambefort, 1981). Among the baboons, juveniles were more likely to discover, and to take advantage of, novel foods. Transmission to other group members was rapid. Among vervets, animals from any age-gender class were equally likely to discover the novel food, but transmission was slow and took place through a pivotal individual. Only among mandrills did observations of other animals' success at avoiding unpalatable foods, rather than direct individual experience with the foods, lead to wider transmission of knowledge about the foods.

Species differences, both in the propensity to experiment with novel foods, and in the transmission of information to others, reflected feeding strategies and social interactions (Cambefort, 1981). Opportunistic feeders may be more likely to incorporate novel foods and, as in the Japanese macaques, juveniles who have yet to experience all components of their diet may experiment more frequently. However, the transmission of information about these foods to others is possibly only when animals feed in associations, when the possibilities for observation of the success or failure of others are present. As Kummer (1971) has noted, a mode of learning that avoids the costs and risks associated with failure may be especially important for long-lived animals where experience of rare events can be accumulated over a lifespan and transmitted to animals who have not yet had such an experience (see also Chapter 3, section 3.2).

The role of innovation in creating new ecological opportunities, which in turn influence social responses (see also Kummer and Goodall, 1985; Lee, 1988), is an example of how ecological problems can be solved by a short-term behavioural response (i.e. innovation). These new opportunities thus can have intergenerational consequences when transmitted as traditions. Since much of adaptation results from the exploitation of a new habitat or resources in the face of competition with other species or from environmental change, such an evolutionary process is probably continual and many examples exist.

The point here is that a small behavioural change, not a genetical, morphological or physiological change, is creating the new opportunities. However, several factors are involved in complex forms of innovation. The materials necessary to exploit a new resource or to create a new response must exist in the surrounding environment. For example, an appropriate substrate for tool-making must be present (McGrew, 1989). The new resource may offer an energetic or nutritional return that makes its exploitation cost-effective. Among the nut-cracking chimpanzees of the Taï Forest, Ivory Coast, foods of a high nutritional value are available only through the use of a hammerstone, which is a marked innovation for apes (Boesch and Boesch, 1983). When chimpanzees were given opportunities to use hammerstones in a semi-experimental context, young females were more likely to acquire the novel behaviour than were older males (Hannah and McGrew, 1987). The use of tools to extract high-quality resources such as invertebrates or nuts tends to be more frequent among females and young chimpanzees (McGrew *et al.*, 1979; Boesch and Boesch, 1981; see also Chapter 9, section 9.4), which have higher energetic requirements. Finally, the opportunities to experiment in terms of time and to incur the costs or risks of failure must be available (Torrence, 1983). Social factors, such as the types of relationships affecting proximity while feeding or the emotional state of the group, may also be critical to the probability of an innovation being generalized within a group.

In the case of innovation, an environmental change is effectively created through the behaviour of an individual. As a consequence, the efficiency of exploitation or access to resources is enhanced in the short term, and longer-term reductions in the risk of foraging are achieved with the cross-generation transmission of innovations in the form of traditions.

2.5 LIFE HISTORY VARIABLES

How can longer-term ecological changes be related to behavioural flexibility over time? A change in the reproductive and life history parameters of a group can have marked effects on the demography of the group, which in turn influences the social outcomes that we observe (e.g. S.A. Altmann and J. Altmann, 1979; Dunbar, 1986; Lee, 1988; Datta, 1989). Furthermore, differences between species in life histories are presumed to be adaptive; to result from different selective pressures (see Martin, 1984a; Harvey *et al.*, 1987). Thus attention has been focused on life history variables in a number of primate studies (reviewed by Richard, 1985; see also Dunbar, 1988). How do such behavioural and evolutionary changes arise? A direct link between nutrition and fertility among primates has been well established in comparisons between food enhanced and wild populations (e.g. Mori, 1979; Lee, 1987a; Fa and Southwick, 1988). A consistent pattern emerges from these comparisons. Low quality of food or of nutrient intake is directly related to low fertility and often to poor survival of infants and of adults (see also Cheney *et al.*, 1988). These reproductive constraints may arise through variability in individual behaviour, such as differences in intake as a result of dominance rank (reviewed by

Harcourt, 1987a), or through different modes of parenting (e.g. J. Altmann, 1980; Hauser and Fairbanks, 1988). Ultimately, changes in reproductive parameters can be translated into long-term effects on groups and populations.

One well-documented example of altered reproductive parameters resulting from environmental changes leading to variability in social behaviour is that of yellow baboons (*Papio cynocephalus*) in Amboseli, Kenya. With a 20-year decline in habitat quality and food availability, the population of baboons decreased dramatically (J. Altmann *et al.*, 1985). Among the Amboseli baboons, low nutrient availability has been related to a late age of menarche, low fertility, long interbirth intervals, and low infant survival (see J. Altmann *et al.*, 1977; J. Altmann, 1980). Growth rates are slow relative to captive baboons (J. Altmann and Alberts, 1987). Gestation length increased under conditions of low nutrient availability (Silk, 1986), while the energetic costs of lactation and carrying dependent infants were difficult to sustain (J. Altmann, 1980, 1983).

These changes in reproductive parameters, in turn, affected the social behaviour and social structure of the Amboseli baboons. Small family size was related to variability in the nature of mother–infant relationships (J. Altmann, 1980) and juvenile social development (J. Altmann, 1979; Walters, 1980; Walters, 1987b). Play partners, alliance structures and sibling relationships are influenced by the availability of other individuals for interactions (see also Lee, 1986, 1987b). Changes in the size and composition of the family also alter the number and types of potential competitors, affecting the transmission of maternal status (e.g. Walters, 1980; Shopland, 1987) and the stability and persistence of female hierarchies (Hausfater *et al.*, 1982; Samuels *et al.*, 1987). All such changes influence the types of relationships observed as part of the surface social structure (Lee, 1988).

How changes in group composition could potentially affect social structure is revealed in studies of the acquisition of dominance among rhesus macaques (Datta, 1983, 1986, 1989). Datta (1989) has shown that a general phenomenon, the transmission of maternal rank to daughters, is a function of several processes, the outcome of which depends on the composition of the group, namely, the availability of allies and close kin, and the relative size and age of the interacting animals. Using models approximating to populations under different nutritional regimes, she has shown that demographic processes affect the stability of matriline ranks and indeed the nature of the hierarchy within the groups.

Given that demographic processes can be directly related to variation in patterns of relationships either within groups over time or between groups, can they influence the deep or underlying structure of a group? Examination of further consequences of such demographic changes may help to answer this question. Changes in group composition and the size and numbers of groups can affect opportunities for movement of males between groups (Manzolillo, 1986), influencing the age at transfer and strategies of transfer or maintenance of male ranks within groups (e.g. Noe, 1986). Thus, patterns of both male and female kinship and relationships are affected, and, with opportunities for novel relationships, variability in deep social structure can be produced. Examples of apparent changes in underlying structure do exist. Harem types of male/female relationships have been observed among the Amboseli baboons, at least over a short period (S.A. Altmann and J. Altmann, 1979). Other examples are the facultative switches of social organization among the callitrichids and the development of female kin lineages in the typically nonkin-bonded mountain gorillas, *Gorilla g. beringei* (reviewed by Lee, 1988). While such changes in group structure may be temporary, they suggest that transition between a finite number of possible types of social systems can occur (Foley and Lee, 1989). Underlying these shifts in deep structure may be the processes of change in surface structure that create new ecological opportunities. When competitively effective, the new form of deep structure will become established as the rule rather than the exception.

For the Amboseli baboons, declines in food availability over several generations have led to behavioural flexibility, mediated by altered demographic parameters and consequent changes in group composition affecting the nature of relationships. Interestingly, long-term changes in habitat quality can also be related to community dynamics, with similar effects on fertility and ultimately on social structure. Strum and Western (1982) have documented variations in fertility of female baboons as a function of year-to-year productivity of food plants. Furthermore, they suggested that the biomass of other herbivores is an important factor affecting that fertility, since the competing herbivores reduce the food available to baboons. A similar effect of livestock biomass directly influencing caloric intakes among primates was found for Barbary macaques (Drucker, 1984). Other species of primates (Terborgh, 1983) can also act as agents producing long-term changes in the environment, thus affecting life history variables.

Environmental changes, as a result of extrinsic habitat changes or from the presence and density of competitors, can strongly influence female reproductive parameters. Variation in reproductive parameters produces changes in demographic composition of groups, affects the nature of relationships, and leads to changes in both surface and deep social structure. In addition, such changes place strong selection on life history variables and body size (Demment, 1983), resulting in long-term evolutionary changes.

2.6 INVASIONS AND RADIATIONS

Throughout the evolutionary history of primates, opportunities for expansion into previously uninhabited geographical areas have arisen. In many cases, these opportunities have been taken, either accidentally, as perhaps in the dispersal of primates to the New World (Fleagle, 1986), or through the gradual expansion of ranges, as in the spread of apes into Europe and Asia (Bernor, 1983). While behavioural flexibility may have little to do with accidental colonization, it may radically affect the survival chances of the animals under new conditions. Furthermore, the capacity for behavioural flexibility may allow the same basic type of animal to remain relatively similar, despite the occupation of new regions or a variety of habitats. The subsequent diversity may be one of minor degree rather than major differentiation.

One example of a successful invasion is that of the macaques. Their dispersal from Africa and subsequent radiations can demonstrate how behavioural flexibility in the face of changing environmental circumstances leads to evolutionary changes. The evolutionary history of the macaques has been reasonably well documented (see Delson 1980; Eudey, 1980). The northern African group of cercopithecoids, the macaques, moved into Europe at the end of the Miocene around 5 million years ago as sea levels lowered with the onset of drier conditions. Subsequent invasions of India and Asia led to radiations when global temperature changes resulted in alternations in sea levels, combined with regional vulcanism. These changes produced forest and island habitats which expanded and contracted through geological time, and high levels of endemism resulted. Such a process of ecological separation within expanding and contracting forest refuges during periods of climatic change has also been proposed as underlying the radiations of the *Cercopithecus* monkeys (Kingdon, 1980).

What behavioural characteristics, then, can be related to the successful occupation of new habitats and radiations within these habitats? In common with most mammals, altering the resource base affects female reproductive parameters and life history variables. However, as the baboon example above suggests, among primates, adaptation also takes place within the context of social flexibility. Changes in life history variables influence group composition, the types and nature of relationships within groups and social structure.

Thus, individuals confront new environments, new ecological opportunities, within a social context, which is itself subject to change.

Macaques show the typical cercopithecoid pattern of relations between nutrition, fertility, group composition and flexibility in social behaviour. Furthermore, most species reproduce seasonally and, under conditions of yearly births, are capable of rapid population growth (see Melnick and Pearl, 1987). When populations are growing and group size increases, new groups are formed through fission along genealogical lines (Chepko-Sade and Olivier, 1979). Food limitations appear to be less important than social considerations in determining the nature of the fission. The new group maintains as its structural base kin relations between females.

That new groups can be formed through fissioning, and expand their range to incorporate new or abundant resources while maintaining the basic organization around female kin, facilitates the successful invasion of new habitats. Lability in group size and group composition between the species of macaques, and differences in patterns of resource exploitation (e.g. Teas *et al.*, 1981; Melnick and Pearl, 1987), are related to species differences in fertility (e.g. Small, 1983; Wolfe, 1986) and in the degree of paternal care and relationships between and within the genders (Caldecott, 1986).

The capacity for behaviour to vary with circumstances is also evident in responses to resource constraints. Small groups among some species appear to have the option of fusion, a rare phenomenon among the female kin-bonded (*sensu* Wrangham, 1980) cercopithecines (see also Hauser *et al.*, 1986). Group fusion allows for increase in range size and enhanced competitive potential in contests between groups. Both toque (*Macaca sinica*: Dittus, 1986) and rhesus (*M. mulatta*: Malik, 1986) macaques have been observed to form fused groups under conditions of food limitation. Among the rhesus, plentiful supplies of food and water resulted in high fertility, low mortality and rapid (23% per year) population growth (Malik *et al.*, 1985). As group size increased, small splinter groups formed by separation between lineages, as has been observed on Cayo Santiago island, Puerto Rico. With subsequent reductions in food availability, two splinter groups fused and the combined group was subsequently able to supplant a previously dominant equal-sized group at feeding sites (Malik, 1986). Group fusion in the toque macaques also resulted from scarcity of foods (Dittus, 1986). In this case, however, females in the smaller and previously subordinate group were more likely to die after the fusion than were females from the previously dominant group. Differential mortality as a result of food scarcity and competitive relations between females within groups (Dittus, 1979; Cheney *et al.*, 1981; Wrangham,

1981) may be the processes underlying the structuring of female groups (Datta, 1989), especially in the context of resource limitation.

Thus we have resource constraints and opportunities interacting with principles of female fertility and kinship that affect group composition and the formation of relationships, all of which result in genetic differentiation (e.g. Silk and Boyd, 1983; Melnick and Pearl, 1987) between populations. A high degree of behavioural flexibility as a response to environmental uncertainty through time and over geographic areas may have led to the successful occupation of a wide range of new habitats during the expansion into new regions, and subsequently to speciation.

2.7 BEHAVIOURAL FLEXIBILITY AND EVOLUTIONARY FAILURE

Not all primates have been as successful as the macaques; indeed, from a perspective of the fossil record, most primate species are extinct (Martin, 1986). Is behavioural flexibility the trademark of the successful, or extant, species? During the Pliocene and Pleistocene, baboons of the genus *Theropithecus* were widely distributed in Africa, from North Africa to the Cape (Jolly, 1972). At least four and possibly as many as six species have been recognized (Jolly, 1972; Jablonski, 1986). Today, only a single species, the gelada *Theropithecus gelada*, remains in the remote high plateaux of Ethiopia (Dunbar, 1984b). In comparison to modern primates, some of the extinct species were extremely large in size. A male *T. oswaldi* may have weighed over 70 kg (Jolly, 1972), and most of the species appear to have been specialized grass-eaters (Jolly, 1972; Dunbar, 1984b; Jablonski, 1986).

The modern gelada is behaviourally highly flexible; indeed it has often been used to exemplify variability in behavioural strategies (Dunbar, 1984a, 1986, 1988). Social groups are complexly structured one-male, female kin-bonded units which are embedded in larger interactive groups or bands, and which vary in size as a function of the probability of predation rather than resource availability (see Dunbar, 1984a, 1986, 1988). However, the extant gelada baboons live in areas of higher rainfall and are smaller in body size than were the extinct giant species (Jolly, 1972; Dunbar, 1984b). When food availability changes temporally and spatially, and is low in quality, as is grass, then large body size should be advantageous as long as intake rates can sustain energetic requirements. Large body size allows for the processing and digestion of low-quality foods in bulk (Demment and Van Soest, 1983; Demment, 1983), and for enhanced efficiency of locomotion, thermoregulation and reproduction (Peters, 1983). The problem facing the giant gelada was that of maintaining a sufficient intake of low-quality foods to sustain energetic requirements for growth, maintenance and reproduction (Lee and Foley, in press). With the onset of drier and more seasonal weather and corresponding seasonal patterns of grass productivity (see Foley, 1987), and the radiation of the more digestively efficient grazing ungulates (Vrba, 1985), reproductive rates would have been difficult to sustain. Ultimately, extinction of most of the lowland species resulted. Only the species living in the highly productive, less-seasonal plateaux, survived.

A long-term change in climate – increasing seasonality – could not be compensated for by behavioural flexibility. The links between body size, reproductive rates, life history parameters and diet cannot be overcome by changes in social or foraging behaviour when ecological constraints are severe. The presence of more efficient competitors may also have played a role in the extinctions. It should perhaps be borne in mind that the ecological and behavioural principles leading to success as opposed to extinction may differ, and we have as yet only begun to explore those for the extant, or recently successful, species.

2.8 CONCLUDING REMARKS

I have presented above a few brief examples of primate behavioural variability within a context of environmental change from the very short to the longer term. Does the base of this behavioural variability lie in the changes in social dynamics (primates as social opportunists) or in the environmental change itself (primates as ecological opportunists)? I suspect that primates are social opportunists and ecologically relatively conservative. For many species, maintaining complex relationships often takes priority over short-term resource maximization (see Cheney *et al.*, 1986; Dunbar, 1988) and this may limit their ability to take advantage of all ecological opportunities. However, when an opportunity is exploited, it tends to occur in the context of social flexibility. The dramatic escalation of hostile intercommunity interactions among the chimpanzees at Gombe was a result of changes in community composition and patterns of familiarity between individuals, associated with an ecological change. It produced an ecological opportunity for the larger community – the occupation of the other's home range.

Foraging innovations are examples of novel responses to ecological problems and may represent primates as ecological opportunists. They allow individuals to exploit new foods, or make for more efficient exploitation, and they give the innovators a competitive (i.e. social) advantage. Innovators can have long-term effects on behaviour when the innovation is transmitted between individuals and across generations as cultural traditions. Again, the social context for maintaining the ecological opportunity is important.

Individual responses to long-term environmental changes tend to be found at the level of alterations in reproductive and life history parameters, as documented for Amboseli baboons. The consequences of these changes for behaviour were indirect in that they were mediated through changes in relationships, in group composition, in patterns of kinship – in social dynamics. Changes in social dynamics and in social structure can have consequences at a population level, resulting in genetic variability, which leads to radiations or the occupation of new habitats, as in the macaques. It should be pointed out that the more ecologically conservative leaf-digesting colobines have also radiated in India and Asia, suggesting that social rather than ecological opportunism again lies at the root of successful primate adaptations. The foraging specializations and constraints of large size on reproductive rates and life history parameters among the giant gelada have led to their extinction and replacement on the savannahs of Africa by vast herds of wildebeests and gazelles. Perhaps, had they not had to devote time and energy to maintaining social relationships in the typical primate pattern, a solution to the ecological and energetic problems of increasing seasonality and growing numbers of competitors might have been possible.

The causal relations between ecological changes and behavioural flexibility have, for the most part, been ignored here in order to concentrate on general patterns of observed responses. In only a few cases are these causal relations understood for primates, and tend to focus on large-scale or relatively easily defined or measured differences between species or groups in food quality and social organization (e.g. Wrangham, 1986). Examination of the broad patterns outlined in Table 2.1 may suggest predictive relations between ecological variation and behavioural flexibility by focusing at an individual level and building up towards social responses. The context of adaptation for the primates is both social and ecological; the two are linked through complex dialectics which have as their root the individual imbedded in the social and physical environment.

ACKNOWLEDGEMENTS

Hilary Box, Saroj Datta and Robin Dunbar have made many very useful comments on the ideas developed in this paper. Rob Foley has critically assisted in all ways, and has provided the theoretical basis for branching from the known primates to the fossil record. Bill McGrew stimulated many of the ideas on innovation. And to Jane Goodall and the Gombe Chimpanzees, I owe a debt of starting my interest in primate behaviour. I thank Robert Hinde and Pat Bateson for discussion and facilities while writing, and Dr J. Garlick for further facilities.

3

Responsiveness to environmental change: interrelationships among parameters

HILARY O. BOX

3.1 INTRODUCTION

The aim of this chapter is to consider some of the biological parameters that underpin behavioural responsiveness to environmental change. The discussion is necessarily selective, but has been chosen to highlight recent areas of research that indicate innovative steps for further investigation. For example, a key area, to which little attention has been paid until recently though it is now attracting a rapid expansion of interest, is that relating to the relative cognitive capacities of simian primates as expressed in their natural patterns of behaviour and their responses to environmental change.

3.2 COGNITIVE CAPACITIES

The term 'cognition' generally relates to the coding and manipulation of information; it includes the ability to be flexible in responsiveness, to anticipate and withhold actions, to predict future actions, and to misinform a potential recipient. Hence, cognitive capacities include a variety of processes which, as Beer (1986) in his discussion of comparative intelligence has emphasized, are irreducible to measurement on a single quantitative scale.

Cognitive capacities and responsiveness to environmental change are of topical importance in various ways. In contrast to traditional ideas about 'intelligence', by giving individuals of different species abstract problem-solving tasks and observing their differential successes, it becomes realistic to consider the potential role of cognitive processes in the expression of life strategies. This assumes special importance for comparatively big-brained, long-lived species such as simian primates that have long periods of development, during which they accumulate much information about their social and physical environments.

There is evidence for sophisticated manipulations of the physical environment by simians, as in the inventive use of tools to extract embedded sources of food; these may be considered to indicate complex cognitive capacities. The frequency and variety with which simian primates use tools are greater than in other taxa (Beck, 1980; Essock-Vitale and Seyfarth, 1987). Varying degrees of dexterity, in combination with cognitive capacities within the suborder, present opportunities to respond to environmental conditions in novel ways, which may further create environmental change.

However, a greater emphasis on the role of cognition in behaviour is necessary because of the extensive complexities of primate social life. It is commonly emphasized, for example, that nonhuman primates show a greater variety and complexity of social relationships and interactions than any other group of animals (Wrangham, 1983; Dasser, 1985). Recent observations have done much to extend our perspectives in this respect. There is especially impressive information on complex cooperation and reciprocity among individuals of various species (Silk, 1987; Smuts *et al.*, 1987). Field studies yield a substantial array of examples which indicate purposive behaviour and social knowledge, often involving long-term strategies of behaviour. Good examples include baboons, which, when they have been treated aggressively by individuals of higher social status than themselves, subsequently show aggression to animals that are socially close, or related to, their aggressors (Smuts, 1985). Again, Cheney *et al.* (1986) note that vervet monkeys are more aggressive to individuals with whose close matrilineal kin they have recently been in aggressive conflict.

Stimulating discussions of general theoretical importance relating to the complexity of primate social life include those on reciprocal altruism, in which the trade-off for a cooperative action appears to be gained at some future date (e.g. Trivers, 1971; Packer, 1977; Bercovitch, 1988); coalitional behaviour, which involves complex associations among individuals (Smuts, 1985; Cheney *et al.*, 1986; Smuts *et al.*, 1987; Harcourt, 1988), and tactical deception, in which individuals may misrepresent a situation for some behavioural gain (Byrne and Whiten, 1988). The complexities of such social life present a constant milieu of environmental change.

There has been continued discussion of the influence of the demands of social life as a primary selection pressure in the development of marked mental flexibility among higher primates, since the idea was first advanced by Chance and Mead in 1953, and subsequently taken up by Jolly (1966), Humphrey (1976) and Kummer (1982).

In captivity, also, simian primates show hitherto unsuspected sophistications in behaviour, which observers have taken to indicate complex cognitive abilities. Since Menzel (1974) showed that common chimpanzees (*Pan troglodytes*) could withhold information within their social group as to the whereabouts of hidden food in a large enclosure, there have been other similar examples, such as that reported by Woodruff and Premack (1979) for chimpanzees in a more formal laboratory setting. Moreover, in recent years chimpanzees living in larger social groups than have normally been kept in captivity have shown behaviour that has provided yet more impressive, if somewhat controversial, information. de Waal's (1982) *Chimpanzee Politics*, based on observations of a group of chimpanzees at Arnhem Zoo in The Netherlands, is an outstanding example (see Chapter 19). Whatever the interpretive detail involved, social tactics of behaviour have been recorded at Arnhem that have not been observed in nature, and this very flexibility of responsiveness further implicates the role of cognitive processes. In fact, captivity may provide excellent opportunities for observing how social and technical problems that are not present in nature may be solved (see Figure 3.1). Goodall (1985), for instance, has emphasized that conditions in which material needs such as food are provided for, but where the individuals are forced to live closely together, provide new environmental challenges that may lead to novel solutions.

The interest of such examples for cognitive ethology relates to the nature of, and evidence for, cognitive processes that are inferred from all such behaviour. For example, complex social interactions involve the ability to store and readily retrieve a large amount of information about individuals as individuals, as well as in categorized forms and relational contexts.

At this point, some intriguing questions need to be asked. One of these concerns differences in cognitive capacities between species with broadly different social organizational predispositions. Fragaszy *et al.* (1989) considered the possibility that memory capacity may be more highly developed in species that naturally live in large bisexual groups, with much more varied social environmental change, than in monogamous species. There is also the question as to whether '*specialized* cognitive capacities are involved in social behaviours, or alternatively that the general processes, such as memory, are elaborated specially

Figure 3.1 Cooperation and tool-use between two male common chimpanzees to obtain fresh leaves from trees that were protected by electric fencing. A suitable length of branch had been selected, carried to the site and placed against the tree with a fork end upwards to give more support. Finally, the branch was held firmly by one male while the other climbed up into the tree. (From de Waal *Chimpanzee Politics* 1982.)

with regard to social information' (Fragaszy *et al.*, 1989).

The position at present, however, is that there have been few empirical studies (Fragaszy and Box, 1986), and there remains a longstanding self-consciousness about the potential pitfalls of anthropomorphism. However, the depth and the breadth of the observations such as those indicated above now force us to consider both the complexity of the behaviour and its implications for the involvement of cognitive processes, as well as other biological propensities. Whatever our perspectives, it is certainly healthy and timely to open up new lines of enquiry on this topic.

The major point to clarify is that relationships between cognition and social behaviour are ill-defined and poorly understood. We have neither objective measures of the complexity of social interactions nor of cognitive performance compared with nonsocial activities within and across species (Essock-Vitale and Seyfarth, 1987). Moreover, Dasser (1985) and several other researchers have emphasized that although sophisticated social behavioural interactions clearly presuppose 'excellent memory and learning capacities, it is less obvious what higher cognitive abilities over and above those basic capacities are needed to explain the behaviour of non-human primates' (Dassar, 1985, p. 21).

At present, cognitive processes are commonly and intuitively inferred from behaviour without detailed information, by asking questions about what is required to do particular acts in particular ways. Subjective judgements are also sometimes made about the relative levels of complexity of the processes inferred. Itani and Nishimura (1973), for example, assumed, without any critical evaluation, that the wheat-washing behaviour of Japanese macaques (*Macaca fuscata*), in which the separation of wheat from sand is achieved by floating it in water, is more cognitively complex than the activity of washing provisioned sweet potatoes in the sea. Cognitive inferences are intuitively and empirically reasonable first steps to delineate the problem and its solution, but they need to be followed through by identifying the processes by which classes of information are acquired and handled in different situations, and in different species.

Hence, clear evidence is still required in order to delineate the role of cognitive processes in many aspects of behaviour. Moreover, although cognitive abilities are implied by the degree of flexibility of behavioural responsiveness that different taxa make to their environments, it is also the case that cognitive processes interact with a variety of other biological propensities. It is with some specific aspects of this general perspective that I shall now be concerned.

3.3 TEMPERAMENT AND RESPONSE STYLES

A central pivot of the interfacing between biological propensities and the responses that individuals of different species make to environmental change is what we may call **temperament**. The term refers to the ways in which an individual characteristically reacts to the environment and its challenges, physiologically and behaviourally (see Mendoza Chapter 17, section 17.3). A series of examples will indicate some of the principal issues.

First, we may continue and amplify some of the themes of the previous section by drawing attention to the concept of **response style**. There are few comparative data but a study by Visalberghi and Mason (1983) provides a good case in point. It concerns squirrel monkeys (*Saimiri sciureus*) and titi monkeys (*Callicebus moloch*), that were given a series of problem-solving tests of increasing complexity. The tasks involved obtaining a food reward that all the monkeys (as shown by preliminary tests) were eager to have. Although individuals of both species were less proficient with increasing complexity of the task, they did improve over the various phases. However, squirrel monkeys performed significantly better at all levels of difficulty of the tasks; they actually solved the problems more regularly, as well as doing them in significantly shorter periods of time.

The question then arose as to the principal influences on the differences in performance. Differences in motivation were ruled out by the preliminary food tests. Again, cognitive abilities did not appear to be involved because both species made similar patterns of error during learning, and improved their performances similarly across the levels of complexity of the problems. The point at issue is that differences in the style of responsiveness between the two species most reasonably accounted for differences in their problem-solving success. Style of responsiveness refers to species-specific characteristics that include the time taken to contact the task, the persistence of responding, the tempo of activity and the variety and vigour of the responses made. Squirrel and titi monkeys were significantly different on all of these measures. Squirrel monkeys contacted the tasks more swiftly, were more persistent, stayed for longer periods at the tasks, were more frequently in contact or proximity with them, and were generally more vigorous in their handling of the problem. By contrast, titis explored the tasks more by looking at them and contacted them less (see Figure 3.2).

Comparative work on these species by Mason and his colleagues over many years (see Chapter 17 for a detailed discussion) provides excellent examples of the interaction of social behaviour, cognition and physiological responsiveness. Both species are members of the same taxonomic family, the Cebidae. They are of similar size, eat similar foods and are frequently sympatric. However, they show striking and consistent contrasts in their modal patterns of social organization. Titi monkeys are monogamous, adult heterosexual pairs behave with close reference to one another, they stay close physically and monitor each other's behaviour; squirrel monkeys live in large heterosexual groups, and males and females maintained as heterosexual pairs behave independently. The species also differ in their physiological responsiveness to environmental change under pre-change and baseline conditions. Levels of cardiac activity and adrenocortical output are significantly higher in squirrel monkeys. In fact, a major difference between these species, which makes them particularly interesting for comparative studies, concerns differences between them at a physiological level that stem from basic differences in the organization of their adrenocortical systems (Mendoza and Moberg, 1985). Differences in both physiological responsiveness and social organizational propensities are believed to influence the ways in which each of these species responds to changes in their social and physical environments.

The concept of response style is, in fact, extremely important in considerations of species differences in responsiveness to environmental change, and much more comparative work is needed to widen our perspective on the functional significances of such differences among different taxa. This concept is a good example of the value of considering the interfacing of biological propensities in studies of responsiveness with reference to different strategies of coping with environmental change. It is also important, however, that responsiveness is not a unitary concept (Moberg, 1985); that 'The particular measures that will differentiate species most clearly on any given occasion are likely to depend on the specific features of the situation' (Clarke *et al.*, 1988a, p. 48). It is especially important that studies such as that by Visalberghi and Mason also draw attention to the fact that although current environmental circumstances are undoubtedly important in influencing behavioural responsiveness at any one time, we also need to understand the ways in which behaviour is organized in different species (Mason, 1984). This is of theoretical interest because, among other things, it lends a different perspective, and redresses the balance to the pervasive inclination of many biologists to emphasize the role of proximate ecological factors as principal influences upon responsiveness.

We may take another example to illustrate aspects of the interfacing of biological parameters underpinning differences in responsiveness to environmental change among species. The case is also drawn from the

(a)

Figure 3.2 (a) A squirrel monkey obtaining food by vigorously tackling the problem of extricating it from a 'closed' container contrasts with (b) the gentler and more visually acquisitive response style of a titi monkey, which is also in close social proximity to its pairmate. (Visalberghi and Mason, 1983; photos: E. Visalberghi.)

work of Mason and colleagues, and was based on three species from the Old World monkey genus, *Macaca*. A.S. Clarke and associates (Clarke, 1985; Clarke and Mason, 1988; Clarke *et al.*, 1988a,b) studied juvenile female cynomolgus (*Macaca fascicularis*), rhesus (*M. mulatta*) and bonnet macaques (*M. radiata*) born and reared under similar conditions. Behavioural and physiological responses of the three species were compared over a variety of conditions. For example, when

(b)

responses indicating habituation to confinement in the transport cage were measured, it was found that both behavioural and psychophysiological reactions to this mild stressor differed among the three species. Although rhesus and cynomolgus monkeys performed similarly on measures of habituation training, cynomolgus monkeys showed the least habituation to confinement, as measured by adrenocortical response, which were derived from assays of blood samples (Clarke *et al.*, 1988b). In other experiments, monkeys of the three species were exposed to various handling stressors (capture, restraint for venipuncture and wearing telemetry equipment) and to exposure to a large but unfamiliar cage environment.

Under all of these conditions, cynomolgus monkeys showed the greatest signs of behavioural disturbance and physiological responses (heart rate and adrenocortical response), whereas rhesus monkeys were lowest on these measures. Responses to an observer (who was also directly involved in all other experimental procedures in the series) were measured while the monkeys were in their home cages. Cynomolgus monkeys were primarily fearful in their reactions to the observer, rhesus were primarily aggressive, and bonnet monkeys showed evidence of submissive affiliation with the observer (Clarke and Mason, 1988). Interestingly, bonnet monkeys showed intermediate (between the rhesus and cynomolgus) cardiac and adrenocortical responses to the variety of manipulations and environmental events in this series of experiments. Unlike the differences between the squirrel and titi monkeys described above, there is no evidence that these three species differ in any aspect of basal adrenocortical function (Clarke *et al.*, 1988a). Rather, the evidence from these investigations supports the hypothesis that differences in physiological responsiveness to environmental events are 'primarily mediated' by differences among the species in their characteristic ways of coping with environmental change, or temperamental dispositions. The results agree with a variety of other observations from different laboratories that cynomolgus monkeys, for example, are fearful of their caretakers, and that rhesus monkeys are relatively aggressive in both intraspecific and interspecific contexts (cf. Clarke and Mason, 1988). The findings are also in accord with data on social interactions from other age–gender classes of these species housed in more natural conditions (Hawkes, 1970; Shively *et al.*, 1982).

Studies such as these raise questions about the comparative data base both across and within species for different levels of responsiveness in various environmental situations. They also draw attention to the fact that although closely related species (such as these macaques) may show similarities in life history patterns (e.g. gender ratios, mating strategies and degree of mating seasonality), variations in the quality of their intraspecific social relationships appear to be especially important in interpreting species differences in responsiveness. For example, and with specific reference to the studies mentioned above, both cynomolgus and rhesus macaques are generally more socially intolerant than bonnet monkeys. Again, interindividual differences in social status are considerably more influential in behaviour among rhesus macaques than among bonnet monkeys. The burden of the point is that differences between species in their life modes and social organization are believed to be at least partially underpinned by differences in psychophysiological characteristics.

Consideration of species differences, in behavioural and physiological responsiveness in these ways, leads not only to an appreciation of theoretically interesting questions but has practical implications. The choice and appropriateness of particular species may be influenced by such knowledge, as Clarke and her colleagues emphasize. The fact that rhesus monkeys are generally aggressive to observers, as well as to conspecifics, and are difficult to handle (Hawkes, 1970; Clarke and Mason, 1988), may obviously cause various problems. By contrast, they are relatively easy to train in laboratory routines, are high in motor activity, and take well to tasks such as operant conditioning. In addition, and very importantly from a welfare point of view, rhesus monkeys habituate relatively rapidly in terms of adrenocortical responsiveness; they are less likely to suffer prolonged periods of stress. They do, however, react adversely to being housed as individuals and are prone to develop behavioural stereotypes. It was also noted that bonnet monkeys are the least disturbed of the three species to changes in their social and physical environments. They are the most tractable to human caretakers and are tolerant of each other, which is clearly an advantage in terms of animal management. Moreover, their relative tractability makes them more suitable than rhesus monkeys to conditions involving physical restraint. On the other hand, bonnet macaques are less easy to train and are slow in tasks that call for active responses. By contrast again, cynomolgus monkeys show that although they are quick to learn routine procedures, their high levels of physiological responsiveness, as measured peripherally by heart rate and by adrenocortical output, put them at risk from stress-related diseases and possible reproductive failure.

With reference to the suitability of cynomolgus monkeys as experimental animals in a wider context, high levels of physiological responsiveness may influence a variety of other physiological measures of responsiveness such as blood pressure, which may, in turn, influence their responsiveness as subjects in biomedical experiments (Clarke *et al.*, 1988b).

One may go further and advocate that the responsiveness of a species to various conditions of captivity, and its relative responses to stressful conditions, should be considered as uncontrolled variables that may influence experimental results in unsuspected ways (see also Chapter 19). In this context, the procedures of workers such as Wrenshall and Gilbert (1986), in providing social and physical enrichment for a colony of cynomolgus monkeys with regular monitoring of such parameters as blood variables and behaviour, together with the long-term toxicological testing, is an instructive paradigm, beneficial both to the animals and to the efficacy of the research programme.

3.4 PHYSIOLOGICAL INDICES – ADDITIONAL CONSIDERATIONS

Questions about physiological indices of responsiveness raise additional points of methodological and conceptual interest. For example, there is considerable scope for the development of a wider range of measures of physiological responsiveness. Moreover, an obvious methodological advantage in studies of responsiveness to environmental change involves noninvasive methods of data collection. We may note here the work of Fuchs (1987), who is concerned with developing wide-ranging indices of biochemical change in spontaneously eliminated urine from animals that experience different social interactions.

There is also a need to consider a wide range of physiological parameters at different ages in the lifespan. There is relatively little information, for example, on patterns of physiological (or behavioural) responsiveness among ageing primates, and how these may relate to stress, distress and health. The two most commonly used measures of physiological responsiveness – namely, heart rate and adrenocortical output – complement one another well. They reflect activity in the autonomic and pituitary-adrenal systems. It is also obviously the case, however, that the responsiveness of a whole individual involves a variety of physiological subsystems and their interrelationships. Critical examples involve those that influence the immune system (Moberg, 1985) and reproductive success (see Chapter 4), both of which may have critical effects on fitness. The influence of environmental change upon health deserves special attention from all points of view. As yet, however, there are few studies (see Chapter 18). Similarly, the impact of change upon reproduction is obviously crucial to the adequate maintenance of animal species and conservation programmes.

From a different standpoint, physiological and behavioural indices do not always track each other consistently. We may note here good evidence to show that, taken alone, behavioural indices may be limited to indicate continued reactivity to environmental change over time. M.B. Hennessy (1986) examined the cortisol and behavioural responses of infant squirrel monkeys in a novel situation when they were separated from their mothers for 80 sessions, each lasting 2 h. Compared to resting levels, cortisol output was much elevated from sessions 2 to 28, with a significant decline thereafter. There was no further decline from session 28 until the final one, although all the results for this period remained significantly greater than those of the resting, baseline period. He also found, however, that as the study progressed the young monkeys ceased to show signs of behavioural (distress) responsiveness to the separations. Hence, monkeys that were 9 months old showed normal signs of independence from their mothers, and yet remained physiologically 'stressed'. As Hennessy noted, several questions remain open from this study. The possibility of conditioning of pituitary-adrenal responsiveness, and the age at which young monkeys are separated, present good examples. Nevertheless, as he pointed out, young monkeys may appear to be behaviourally calm but continue to show a significant neuroendocrine responsiveness indicative of stress. Implications for past and future studies, which have concentrated on long-term behavioural responsiveness, are clear. The lesson from this kind of study is that, depending on the questions we want to ask of course, changes in behaviour alone are insufficient indices of responsiveness to environmental change.

Furthermore, species vary considerably in the range of behavioural indices that are readily available for reliable quantification. Anyone who has kept different species of tamarin (Callitrichidae), for instance, is quickly aware of the differences in their degrees and ways of responding to the routines of daily life, as in their responsiveness to human caretakers and to changes in their physical and social environments. A few experimenters have observed that different species react differently to changes in their physical environments, as with the addition of unfamiliar objects and the provision of foraging tasks (Box and Rohrhuber, 1987; Box, 1988). It is important in a more general sense, however, and in the context of long-term husbandry and welfare that, compared with so many simian species, callitrichids have a limited repertoire of behavioural responsiveness, as in facial and body postures.

Moreover, information from veterinary records including relative rates of mortality for different species maintained under similar conditions in captivity, as well as the comparative difficulties of establishing viable breeding colonies, for example (Box, unpublished data), tend to support substantial differences among species in their responsiveness to captive conditions. Information about such differences has important implications for the establishment of breeding programmes for species that are endangered in nature, and/or much used in biomedical research. There is a need for studies of physiological responsiveness to supplement behavioural observations.

It is also important that an individual's responses to environmental change normally take place in the context of a social group (Hall, 1963). The influence of this contextual milieu has been shown to be a fruitful area of research.

3.5 SOCIAL CONTEXT

The influence of social context is, then, important for individual responsiveness, and there are excellent discussions which interrelate both behavioural and physiological parameters (see Chapters 4, 15, 16 and 17). In a noteworthy example, Mendoza and Mason (1984) examined the responsiveness of male/female pairs of the monogamous titi monkey when they were temporarily separated from each other, compared with situations in which each adult was left with its offspring and when the infant was temporarily removed from the pair. Interestingly, the highest levels of physiological responsiveness occurred when the mother was left alone with her offspring. This finding is in accord with the natural condition for titi monkeys in which the young initiate much of the care-giving which they receive, and that this comes predominantly from the father.

Other examples of the interrelationship of the social context of responsiveness with physiological parameters include the fact that familiar individuals, with which close bonds have been established, are potent in reducing the physiological consequences of stress. In one instance, when infant squirrel monkeys were caught and removed from their mothers or mother substitutes, both infant and 'mother' showed considerable elevation in adrenocortical responsiveness after half an hour of separation. The influence of separation was substantiated by the fact that when infants were captured, but immediately returned to their 'mothers', there were no such increases in plasma cortisol in either offspring or mother (Mendoza *et al.*, 1978). Further, when infants were returned to a female with which no such close bond had developed, reunion was not effective in ameliorating physiological stress (Coe *et al.*, 1978).

There is a variety of behavioural demonstrations that strong social bonds are beneficial to an individual in its adjustment to a new and potentially stressful situation. There are well-known experiments in which fearfulness in young primates may be at least moderated, if not eliminated, by the presence of an attachment figure. Ainsworth and Bell's (1970) studies of young children and that by Miller *et al.* (1986) of young chimpanzees are cases in point.

The study of Miller *et al.* (1986) is one of a series on the development of responsiveness of young common chimpanzees to different kinds of unfamiliar social and nonsocial conditions. The presence of a familiar caretaker was included to eliminate the possibility that responsiveness to novelty would be confounded by social separation. There were quantitative differences in behaviour among chimpanzees of different ages but, overall, the data were generally similar to those from human children. Hence, the presence of an attachment figure influenced behavioural responsiveness by attenuating distress in the young individuals when they were faced with the novelty of an unfamiliar room. A secure base was also thereby provided, from which explorations could be made. Such explorations are, of course, a necessary part of adaptation to a new environment. Young chimpanzees and children tested under such circumstances, without familiar attachment figures, show very marked signs of stress.

From a somewhat different point of view, the type of relationship and its quality is known to be instrumental in inhibiting or facilitating what an animal learns about its environment. There is little detailed information for nonhuman primates, but research on young children has shown deficits in the acquisition of practical skills, and of mental performance, under conditions of poor social relationships. For example, Cohler and Grunbau (1977) found that children of mothers who suffered depression were considerably lower in their attainment, and in their spans of attention, than were children from more normal social situations.

Moreover, the emotional climate within a social group, both in nature and in captivity, may vary considerably and it may interact with the particular conditions of the physical environment to inhibit or facilitate the possibility to learn (Box, 1984; Box and Fragaszy, 1986). A good example here is Kawamura's (1965) description of the different responses of various groups of Japanese macaques to being initially provisioned with food. Acceptance of the food depended substantially upon the social behaviour of the leader in different troops. Food was taken readily in cases where the leader was socially relaxed and relatively permissive. Conversely, it took much longer to accept the new food in groups in which the leader was aggressive and restrictive. The quality of social responsiveness by some key members of social units is important in determining the extents to which other individuals (and perhaps especially young ones) are prepared to respond to environmental change.

As a follow-on from this, an important aspect of the influence of social context upon responsiveness concerns differences among individuals in their social status. In Sigg's (1980) experiments with wild hamadryas baboons (*Papio hamadryas*), for example, the positions of buried food and water were cued with the aid of markers of different colours in the large field enclosures. Females on the periphery of the group learnt the task readily, whereas the central females did not. The hypothesis is that the central, higher status monkeys were much more preoccupied socially in the maintenance of their status to attend to and thereby learn the task. The influence of social status upon responsiveness has also been shown with reference to new objects. High-ranking animals are frequently less responsive than lower-ranking ones (E.W. Menzel, 1971; Chamove, 1983; Bunnell *et al.*, 1980). Joubert and Vauclair (1986) have considered various hypotheses as to why this might be so. Once again, greater social vigilance of high-status individuals in the control of their social interactions was considered. These authors also examined the alternative hypothesis that it may be more hazardous for high-ranking animals to risk predation (e.g. Katzir, 1982, 1983). Of course, neither hypothesis precludes the other. Comparative studies with species of different social organizational predispositions would illuminate these issues, and contribute to the development of functional hypotheses. We have already observed, from other points of view, that differences among species in the quality of their social relationships may be substantially influential.

3.6 INDIVIDUAL DIFFERENCES

It is also important to our general perspectives that individuals may vary consistently in their responsiveness. Moreover, although individual differences are often ignored as irrelevant biological data in many areas of interest (see Chapter 5), there is a variety of evidence to show that individuals differ early on in their lives with respect to physiological, constitutional and temperamental factors (Suomi, 1985). As it turns out, such differences can be noted to identify those animals that will be socially competent later in their lives and to predict those that may make inadequate adjustment to environmental change.

Central to the whole topic of responsiveness to environmental events is that anxiety or fearfulness may vary considerably among individuals. Suomi (1985), for example, has emphasized the role that such aspects of responsiveness play in the social development of young monkeys. Thus, young, relatively anxious animals need more security; their development may be restricted in terms of what they learn about their environments. They play and explore less than bolder monkeys, which, in turn, influences their potential acquisition and use of information. It is important generally that there is a close inter-relationship between emotional development and the development of cognitive abilities as expressed in social and technical skills (Box and Fragaszy, 1986).

Suomi (1985) describes experiments in which young rhesus monkeys showed substantial individual differences in response to mild stress. The monkeys were temporarily separated from their familiar conspecifics in an area with unfamiliar objects. Under these conditions, some infants showed behavioural and physiological anxiety, whereas other infants explored the environment. These differences were evident physiologically by cortisol indices when the monkeys were 30 days old, and behaviourally by the second month.

It is noteworthy that these differences became stable and persistent, so that they were described as characteristic responses. Persistent differences in behaviour were evident into adolescence and, for females, into young adulthood. Differences in physiological responsiveness for both genders also persisted into adolescence. Further, if the test situation was made more stressful, as when young animals were separated from their social environment for several hours, the differences in behavioural and physiological responsiveness among anxious and relatively nonanxious monkeys diverged further. The nonanxious monkeys showed coping behaviour, while the anxious individuals became depressed (see also Chapter 18). Indirect evidence for genetic influences, as well as good evidence for experiential factors, such as inadequate mothering (e.g. Suomi, 1981), support the case for heritable characteristics, as well as environmental influences, in the development of individual differences in responsiveness, which has important implications for health, welfare and reproductive success. This is an important area of research that will repay much greater effort and substantially aid our understanding of the nature of responsiveness to environmental change in different primate species (see Chapter 5 for detailed discussion).

3.7 CONCLUDING REMARKS

There is a variety of good quality evidence to show that simian primates vary considerably within and between species in their responsiveness to environmental change. Such variations are associated with, first, cognitive, physiological and social propensities that reflect the natural life styles of different species, and, second, ontogenetic variance.

Until fairly recently, the role of cognition in natural behaviour was rarely considered. However, we are now faced with a variety of complex and flexible behavioural responsiveness to environmental events, especially those within social domains that press for a consideration of the cognitive processes involved. At present, our knowledge about social behaviour is far in advance of what we know about cognitive abilities in the context of behavioural interactions. The questions are hardly formulated, but at least they are beginning to be so and a new field of empirical endeavour will undoubtedly develop.

We are in a more fortunate position in our understanding about interactions between physiological and social organizational propensities of responsiveness. In some well-documented cases, variations between species of the same taxonomic family in, for example, their modal patterns of social organization as well as in basic differences in the organization of their adrenocortical systems, are implicated in the contrasting ways in which each responds to changes in the social and physical environment. The concepts of temperament and response style assume particular importance in these respects. Further, recent work has shown that species within the same genus may show significant behavioural and physiological differences in respect to environmental stressors in ways that are not attributable, either to any aspect of their basal adrenocortical function, or to any of a variety of their life history patterns, such as their mating strategies. Interestingly, differences among such species (as with some macaques) are readily consistent with variations among them in the quality of their intra-species social relationships.

There are also robust data to indicate the importance of the social context in which individual responsiveness to change takes place. The important general point is that the quality of social interactions among individuals of different species may directly influence both their behavioural and physiological responsiveness to environmental events.

It is also central to our perspectives about responsiveness that individuals differ with respect to physiological, constitutional and temperamental factors. As with all the areas that have been mentioned, there are few comparative data as yet, but questions of individual variation relate to the identification and prediction of competence to environmental change.

Finally, there are both theoretical and practical implications from all these lines of enquiry. The time is now right to develop as many indices of responsiveness as possible and, most importantly, to address specifically the ways in which they are interrelated. Thereby, we shall increase our knowledge of the ways in which responsiveness is organized in different taxa.

ACKNOWLEDGEMENTS

It is my pleasure to thank Susan Clarke, whose work is referred to in this chapter, for many critical thoughts and detailed advice. I also thank Dorothy Fragaszy and Donald Broom for their useful comments, and, as always, Joan M. Morris for her help and excellent word-processing skills.

4 The social control of fertility

DAVID H. ABBOTT

4.1 INTRODUCTION

Social factors are now well recognized as having profound effects on reproduction in many mammals, from rodents (McClintock, 1983) to humans (Preti *et al.*, 1986). Given that a high degree of social complexity is synonymous with the primate order, it is not surprising that the social environment of primate groups has become a predominant environmental factor influencing reproduction at both behavioural and physiological levels (Wasser and Barash, 1983; Abbott, 1987; Keverne, 1987; Dunbar, 1988). The ability of social factors to exert such control over reproduction in primates is probably related to the development of the neocortex in the brain. The neocortex influences the neuroendocrine limbic brain, including the hypothalamus. Such increased cognitive input into the neural centres that control internal physiological events might explain the diminished effects of environmental cues, such as photoperiod on primate reproduction (Ruiz de Elvira *et al.*, 1982; Wehrenberg and Dyrenfurth, 1983; Herndon *et al.*, 1987), and the diminished effects of fluctuations in gonadal hormones on the expression of primate sexual behaviour (Dixson, 1983; Kendrick and Dixson, 1984; Keverne, 1985).

One of the more exciting and novel aspects of primate social organization being investigated at present is the direct social control of fertility that operates on many, but by no means all, primate species. These studies have been fuelled by reports showing that social factors can cause specific alterations in primate reproductive physiology, and can thus dramatically alter fertility (Abbott and Hearn, 1978; Bowman *et al.*, 1978; Wilson *et al.*, 1983; French *et al.*, 1984; Schilling *et al.*, 1984; Adams *et al.*, 1985). Two excellent examples of this appear later in this volume. They will examine the salient role of socially dominant female marmosets (see Chapter 16) and tamarins (see Chapter 15) in the complete suppression of reproduction of their female subordinates.

Dominance status is the social factor most commonly linked to the control of fertility in primate studies. Dominance status or high social rank is not just passively attained and maintained by individuals of both genders in primate societies. It is an active process, which is vigorously pursued (Keverne, 1979; Abbott, 1984; Zumpe and Michael, 1986). The attainment of high social rank can be rather transient and volatile as, for example, in cynomolgus monkeys (*Macaca fascicularis*: van Noordwijk and van Schaik, 1985, 1988), Barbary macaques (*M. sylvanus*: Witt *et al.*, 1981), and baboons (*Papio*: Bercovitch, 1986; *Theropithecus gelada*: Dunbar, 1988), or it can be stably maintained between the same individuals over long periods of time, such as in Japanese macaques (*M. fuscata*: Kawai, 1958), rhesus monkeys (*M. mulatta*: Sade, 1967), baboons (*Papio*: Hausfater *et al.*, 1982) and vervet monkeys (*Cercopithecus aethiops*: Lee, 1983). Certainly, female primates will compete for high social status in the absence of any apparent limited resource (Dunbar, 1980; Walters, 1980; Datta, 1983), and females that leave their social groups tend to be subordinate (Chepko-Sade and Sade, 1979; Crockett, 1984).

The differential effects on fertility of low and high social ranks are manifest in a more obvious, tangible and succinct manner in females than in males. For example, in females, ovulation may be suppressed or no offspring are born (Abbott, 1987; Harcourt, 1987a). In males, there may be a lack of (presumed) fertile matings or of offspring sired (Vessey and Meikle, 1987). This chapter concentrates on recent studies of the social control of fertility in female primates. It takes a comparative approach and examines particular findings from captive, experimental and field studies to demonstrate the variety of ways in which physiological and behavioural factors are manipulated so effectively by the social environment that fertility is altered directly.

4.2 PHYSIOLOGICAL STUDIES OF FEMALE REPRODUCTIVE SUPPRESSION

Lesser mouse lemur

In this representative of the ancestral primate stock (Dutrillaux, 1979), social suppression of female fertility appears to operate only as a density-dependent mechanism. Increasing the housing density of female lesser mouse lemurs (*Microcebus murinus*) in captivity has led to a lengthening of the interval between oestrous bouts (Perret, 1986), which was due to an elongation of the luteal phase of the ovarian cycle, and possibly suppression of ovulatory cycles. Social stimulation of the hypothalamo-adrenal axis, resulting in elevated levels of blood cortisol, was suggested as the physiological causative factor impairing ovarian function, but other possible neuroendocrine mechanisms have still to be explored.

As mouse lemurs breed seasonally, in captivity as well as in the wild (Martin, 1972; Petter-Rousseaux, 1980), and are normally encountered singly in the wild (Martin, 1973), density-dependent suppression of female reproduction may serve to limit population growth and density, and to maintain established female territories. In the wild, a group of female territories overlap with that of a single adult male in the form of an extended harem (Martin, 1984b). Density-dependent reproductive inhibition has been well documented in female rodents (McClintock, 1983; Marchlewska-Koj, 1984), and involves inhibitory urinary chemical signals or pheromones (Novotny *et al.*, 1986). Lesser mouse lemurs possess a well-developed olfactory system and exhibit numerous behavioural patterns associated with olfactory signals (Schilling, 1979). Such signals may mediate the socially induced reproductive suppression in females. This is certainly the case in male lesser mouse lemurs, where the seasonally induced rise in blood testosterone levels and appearance of sexual behaviour is inhibited by exposing isolated males to the urinary odour of dominant males (Schilling *et al.*, 1984).

Marmosets and tamarins

Marmosets and tamarin monkeys (Callitrichidae) probably provide the best examples of the social control of fertility in female primates. The suppression mechanism is not density-dependent, as in the female lesser mouse lemur, but is completely determined by social status. In the wild, marmosets and tamarins live in one-female breeding groups (Kleiman *et al.*, 1988; Snowdon and Soini, 1988; Stevenson and Rylands, 1988). In captive studies, the breeding female is the dominant female and is the only female in the group to produce offspring (Epple, 1967; Rothe, 1975; Abbott and Hearn, 1978). Subordinate females, including both daughters and unrelated females, rarely copulate or ovulate (e.g. common marmosets, *Callithrix jacchus*: see Chapter 16; cotton-top tamarins, *Saguinus oedipus*: French *et al.*, 1984; Ziegler *et al.*, 1987b; saddle-back tamarins, *S. fuscicollis*: Epple and Katz, 1984), but there are exceptions to the anovulatory rule, as in golden lion tamarins (*Leontopithecus rosalia*: see Chapter 15). Recent evidence also suggests that in Goeldi's monkeys (*Callimico goeldii*), a close relative of marmosets and tamarins, a combination of behavioural and physiological controls may affect nonbreeding females (Carroll *et al.*, 1990).

Reproductively suppressed female marmosets and tamarins rapidly attain the fully fertile condition when they are removed from subordinate social status. In marmosets, the fertile, ovulatory condition is equally rapidly reversed into the anovulatory condition within a few days, by introducing ovulating females to groups where they attain only subordinate status (see Chapter 16). Such precise social control of ovulation is without parallel in anthropoid primates.

Ovarian suppression in subordinate female marmosets goes beyond suppression of the ovulatory rupture of pre-ovulatory follicles from the ovary. In ovarian dissection (Harlow *et al.*, 1986), ultrasound (Abbott *et al.*, 1989a) and histological studies (Abbott *et al.*, 1981), ovarian follicular development in subordinate female ovaries is arrested at an antral stage with follicles approximately 1 mm in diameter. This contrasts with the 2–4 mm diameter of pre-ovulatory follicles seen prior to ovulation in dominant or ovulatory female marmosets (Hillier *et al.*, 1987; Abbott *et al.*, 1989a).

The immediate physiological cause of this ovarian suppression is suppressed secretion of pituitary gonadotrophins. There is direct evidence for the suppressed secretion of luteinizing hormone (LH) from findings, such as significantly reduced plasma LH concentrations, in subordinate females (Abbott *et al.*, 1988). The evidence for suppressed secretion of follicle-stimulating hormone (FSH) is indirect, because of a lack of an available assay for marmoset FSH. However, when small antral follicles are removed from the ovaries of subordinate females, they are highly responsive to FSH treatment when cultured *in vitro* (Harlow *et al.*, 1986), suggesting that blood FSH levels are low *in vivo*.

The physiological origin of socially suppressed ovulation lies within the brain of subordinate female marmosets. Suppressed hypothalamic secretion of gonadotrophin-releasing hormone (GnRH) is apparently responsible for suppressed pituitary gonadotrophin secretion and ovulatory failure, because subordinate females treated with GnRH experienced increases in circulating LH concentrations (Abbott, 1987), and increased ovarian activity, which led to one short-term pregnancy (Abbott, 1987, 1989). The neuroendocrine mechanisms that activate and maintain GnRH suppression in subordinate female marmosets are not fully understood. However, they do not involve elevated plasma concentrations of prolactin (Abbott *et al.*, 1981) or of cortisol (Abbott *et al.*, 1981; L.M. George and D.H. Abbott, in preparation), low body weight (Abbott, 1988) or alterations in the pattern of plasma melatonin secretion (Webley *et al.*, 1989), all of which have been previously implicated in mediating the suppression of hypothalamic GnRH secretion in other mammalian species.

Two components of the inhibitory neuroendocrine mechanisms have been identified so far in subordinate female marmosets. Firstly, an increased sensitivity to oestradiol negative feedback was shown when plasma LH concentrations were suppressed to intact levels in ovariectomized subordinate females implanted with a small oestradiol capsule, but not in similarly treated ovariectomized dominant females (Abbott, 1988). However, because plasma LH concentrations in long-term ovariectomized subordinate females failed to reach those found in ovariectomized dominant females (Abbott, 1988), a further inhibitory neuroendocrine mechanism is thought to be operating. So when plasma LH concentrations of ovariectomized subordinates were elevated to those of ovariectomized dominants, following the injection of an opiate antagonist (naloxone), the endogenous opioid peptides were also implicated in suppressing hypothalamic GnRH secretion in subordinate female marmosets (Abbott *et al.*, 1989b). Nevertheless, as naloxone failed to elevate plasma LH concentrations in intact subordinate female marmosets, the endogenous opioid peptides may not be the only inhibitory neurotransmitter systems involved in this social suppression of GnRH secretion.

Attention has also been given to the nature of the external cues from dominant females which activate and maintain the physiological suppression mechanism in subordinate female marmosets and tamarins. Epple and Katz (1984) were the first to show that pheromones from dominant females could play an important part in suppressing ovarian function in subordinate female saddle-back tamarins. Savage and her colleagues (1988) took this further, by showing that scent from dominant female cotton-top tamarins significantly delays the onset of ovulation in subordinate daughters, when isolated from their families and paired with males, and it also prevents pregnancy. Barrett and Abbott (1989) also found that the dominant female's scent delays the onset of ovulation in isolated subordinate female marmosets. They showed that pheromones may not provide the only inhibitory cue. Subordinate females remaining in their social groups received bilateral ablations to both the main olfactory epithelium and vomeronasal organ (housing the accessory olfactory epithelium), leaving the subordinates effectively without the sense of smell. Despite this olfactory blockade, none of the subordinates started to ovulate (J. Barrett and D.H. Abbott, unpublished results). Other inhibitory cues that possibly involve visual or physical contact with dominant females apparently play an overriding part in suppressing ovulation in comparison to pheromonal cues. Auditory cues are unimportant (Abbott *et al.*, 1989c).

The golden lion tamarin is the only callitrichid studied to date which fails to demonstrate suppression of ovulation among subordinate females (see Chapter 15). However, as the ovarian cycles of females living in a group appear to be synchronized (French and Stribley, 1987), these monkeys may have developed the next best social manipulation to ovulation suppression. Hence, as all mature females in a group ovulate at approximately the same time, the dominant female would therefore have the best opportunity to monopolize the breeding male or males during this most fertile period for all the females.

Talapoin monkey

In the wild, talapoin monkeys (*Miopithecus talapoin*) live in large social troops, where more than one female breeds (Rowell and Dixson, 1975). Like marmoset and tamarin monkeys, socially dominant female talapoin monkeys suppress reproduction in their subordinates, but reproductive suppression is far less extreme than in groups of marmosets and tamarins because more than one female breeds (Rowell and Dixson, 1975; Abbott *et al.*, 1986; Abbott, 1987). In laboratory studies of groups of talapoin monkeys, low-ranking females participated less frequently in sexual behaviour than high-ranking females (Bowman *et al.*, 1978; Keverne, 1979), and had a poorer reproductive output than high-ranking females (Abbott, 1987).

The first evidence of a physiological impairment in the reproduction of subordinate female talapoins came in 1978 when Bowman and colleagues showed that the lowest ranking females in their groups had reduced LH responses to an oestrogenic treatment, i.e. an oestrogen positive feedback test. As all the females tested were ovariectomized, the results were taken to imply that intact subordinate females would have a reduced capacity to ovulate, because they would have a reduced ability to generate an ovarian oestrogen-induced ovulatory LH surge. This was certainly the case for some intact subordinates, but not all: out of three social groups examined only two out of five subordinate females were anovulatory (Abbott *et al.*, 1986). Nevertheless, ovulatory subordinate females still failed to conceive in most cases, because of a lack of ejaculatory mounts from males.

Plasma concentrations of both prolactin and cortisol were elevated in low-ranking female talapoin monkeys (Keverne *et al.*, 1982, 1984), but it was the elevated prolactin concentrations that were implicated in the impaired LH responses of subordinates (Bowman *et al.*, 1978; Keverne, 1979). Treating ovariectomized (oestrogen-implanted) subordinate females with the dopamine agonist, bromocriptine, reduced plasma prolactin concentrations and reinstated their ability to respond to an oestrogenic positive feedback test with an LH surge. Treating ovariectomized (oestrogen-implanted) dominant females with a dopamine antagonist, haloperidol, elevated plasma prolactin concentrations and removed their LH responses to a positive feedback test. The results suggest that hyperprolactinaemia in female talapoin monkeys impairs hypothalamic GnRH or pituitary LH secretion, or both. Hyperprolactinaemia certainly suppresses GnRH and gonadotrophin secretion in women (Yen, 1986).

It is unclear whether or not endogenous opioid peptides are involved in mediating this gonadotrophin suppression, because naltrexone, a long-acting opiate receptor antagonist, did not differentially stimulate LH secretion in subordinate females in comparison to dominant females (Meller *et al.*, 1980), whilst in subordinate male talapoin monkeys there was clear evidence of increased concentrations of beta-endorphin in cerebrospinal fluid in comparison to dominant males (Martensz *et al.*, 1986).

Elevations in plasma prolactin concentrations are correlated with aggression received, and low social rank in female talapoin monkeys (Keverne *et al.*, 1984). Impairments in the reproductive physiology of subordinate females are therefore the direct result of agonistic interactions with higher ranking females.

Cynomolgus and rhesus monkeys

Social subordination is also associated with a diminished number of ovulatory cycles in both cynomolgus (*Macaca fascicularis*) and rhesus monkeys (*M. mulatta*) (Walker *et al.*, 1983; Adams *et al.*, 1985). Subordinate females not only ovulated less frequently than dominant females, but they also had more progesterone-deficient cycles (luteal insufficiency). In subordinate cynomolgus monkeys, anovulatory and deficient cycles accounted for over 50% of subordinates' menstrual cycles (Adams *et al.*, 1985). The ovarian failure was linked to altered gonadotrophin secretion in subordinate female rhesus monkeys (Walker *et al.*, 1983).

Kaplan *et al.* (1986) showed that subordinate female cynomolgus monkeys had greater cortisol responses to an injection of adrenocorticotrophic hormone (ACTH) than dominant females. The suggestion was therefore made that social subordination activated the hypothalamo-pituitary-adrenal axis to suppress GnRH secretion. Certainly, both cortisol (Dubey and Plant, 1985) and corticotrophin-releasing factor (CRF) (Olster and Ferin, 1987) have been implicated in suppressing GnRH secretion in rhesus monkeys. In rhesus monkeys, at least, these results would help explain the poor birth rates of low-ranking females in captive groups (Wilson *et al.*, 1978). A lack of sexual interactions with males is not a contributing factor to the poor reproductive success of low-ranking female rhesus monkeys, because low-ranking females engage in copulatory activity with males as frequently as high-ranking females (Wilson, 1981).

4.3 IS THERE A COMMON PHYSIOLOGICAL CAUSE OF REPRODUCTIVE SUPPRESSION?

Studies of the social inhibition of ovarian function in primates have suggested three apparently different physiological suppression mechanisms:

1. activation of the hypothalamo-pituitary-adrenal axis in the lesser mouse lemur and cynomolgus monkeys (Perret, 1986; Kaplan *et al.*, 1986);
2. increased sensitivity to oestradiol negative feedback and inhibitory endogenous opioid peptides in the marmoset monkey (Abbott *et al.*, 1989c); and
3. hyperprolactinaemia in the talapoin monkey (Bowman *et al.*, 1978; Keverne, 1979).

Nevertheless, all three mechanisms are accompanied by either circumstantial evidence or suggestions for suppressed hypothalamic GnRH secretion as the main physiological cause of the ovarian failure. This therefore provides the possibility for a common mechanism of GnRH suppression operating in all three types of fertility inhibition.

Activation of the hypothalamo-pituitary-adrenal axis will occur with increased hypothalamic CRF release; in rats and monkeys, CRF suppresses GnRH or LH secretion via inhibitory opioid pathways (Gindoff and Ferin, 1987; Almeida *et al.*, 1988). In women, suppression of LH secretion in hyperprolactinaemic patients is reversed with naloxone, an opiate antagonist (Grossman *et al.*, 1982). Consequently, all three inhibitory mechanisms may employ endogenous opioid peptides as common neuroendocrine suppressors of hypothalamic GnRH secretion. As endogenous opioid peptides are activated in times of stress (Grossman, 1988), and are known to inhibit GnRH secretion (Almeida *et al.*, 1988; Kalra *et al.*, 1989), such a final common inhibitory pathway in the social suppression of ovulation in female primates would not be surprising. The behavioural and physiological ways in which the inhibitory opioid pathways are activated probably vary because of the different social environments of the various primate species involved.

4.4 FIELD STUDIES OF FEMALE REPRODUCTIVE SUPPRESSION

Social factors clearly have pronounced effects on the reproductive physiology of female primates in captivity, especially on low-ranking, subordinate females. The logical extension of this is to find evidence that high-ranking, dominant females produce more offspring than low-ranking females in nature. This seems to be the case with marmoset and tamarin monkeys, where (except for rare, transient occasions: Terborgh and Goldizen, 1985; Dietz and Kleiman, 1986) only one female produces offspring in a social group: the remaining adult females do not (common marmoset: Hubrecht, 1984; Scanlon *et al.*, 1988; Stevenson and Rylands, 1988; pygmy marmoset: Soini, 1988; saddle-back tamarin: Terborgh and Goldizen, 1985; moustached tamarin (*Saguinus mystax*): Garber *et al.*, 1984; Sussman and Garber, 1987; golden lion tamarin: Dietz and Kleiman, 1987). Only in the golden lion tamarin is there positive evidence that the breeding female is the dominant female in free-living groups (Baker, 1987). In this species, aggressive expulsions of subordinate daughters from groups can be so violent as to result in daughters' deaths (A.J. Baker, personal communication).

Such tight social control of fertility in marmosets and tamarins may have developed because cooperation within a group may benefit individuals more than living outside groups. MacDonald and Carr (1989) have suggested that this may occur because, in certain ecological niches, food sources and other essential resources for survival, are dispersed in such a manner that feeding territories have to be maintained, but that each territory can support more than two, and maybe several, individuals. In such situations, dominant females could then monopolize the help of others in their groups to rear their young in order to maximize the chances of their offsprings' survival, provided that female 'helpers' were nonreproductive (Sussman and Garber, 1987; MacDonald and Carr, 1989). Nonreproductive, subordinate females may also benefit from such a system, by delaying their departure from their group because of difficulties in establishing a separate feeding territory and breeding group. They might benefit further by gaining experience in infant caretaking, which would be useful for the rearing of their own offspring (Sussman and Garber, 1987). Why such niche specialization occurs in the first place is still not well understood.

However, in further studies of other primate species, the benefits of high social rank to female fertility have not always been found to be so straightforward. Part of this problem lies in the nature of the reproductive suppression itself. In marmoset and tamarin monkeys, reproduction in all subordinate females is completely suppressed. By contrast, reproductive suppression of subordinates in other primates is incomplete, such as in talapoin (Keverne, 1979; Abbott, 1987) or cynomolgus monkeys (Adams *et al.*, 1985; Kaplan *et al.*, 1986), where subordinate females do ovulate to some extent. As a result, the consequences of social suppression of reproduction among low-ranking females in such primate species are more difficult to detect in the wild. Captive conditions may also exaggerate the normal social factors operating to limit subordinate female reproduction, so that field studies may not necessarily reflect the same intensity of reproductive suppression.

Studies of free-ranging Indian rhesus monkeys introduced to islands in the Caribbean showed that high-ranking females began reproducing at earlier ages, had higher birth rates and had a greater proportion of their offspring survive than low-ranking females (Drickamer, 1974; Meikle *et al.*, 1984). In the same populations of rhesus monkeys, Sade and colleagues (1977) attributed the faster rate of increase in the numbers of individuals in high-ranking maternal lineages, to high-ranking females reproducing for the first time at an earlier age than low-ranking females. A similar delay in the age at first reproduction was found in daughters of low-ranking female Barbary macaques (Paul and Thommen, 1984), Japanese macaques (Sugiyama and Ohsawa, 1982) and yellow baboons (*Papio cynocephalus*: J. Altmann *et al.*, 1988).

Three further studies have shown that adult dominant females have significantly higher birth rates than subordinate females: toque macaques (*Macaca sinica*: Dittus, 1986), vervet monkeys (Whitten, 1984) and gelada baboons (Dunbar, 1989). However, in the case of the vervet monkeys, no such rank-related fertility difference was found by Cheney *et al.* (1981), suggesting, perhaps, that even within a species, some populations may face different environmental conditions than others, and so may behave differently (see also Dunbar, 1989, p. 85).

Perhaps the most well-known field study illustrating the benefits of high social rank on female primate reproductive success has been carried out on gelada baboons (Dunbar and Dunbar, 1977; Dunbar, 1980, 1984a, 1988, 1989). Subordinate female reproduction is impaired by harassment from higher-ranking females, especially around the time of presumed ovulation (Dunbar, 1980). Consequently, this 'oestrous' phase of the menstrual cycle (Dunbar, 1989) is prolonged in subordinates, and they undergo more menstrual cycles prior to conception than dominant females (Dunbar, 1980). As low-ranking females copulate with males as frequently as high-ranking females, Dunbar (1980, 1988, 1989) has argued that these findings provide strong evidence for stress-induced social suppression of ovulation in low-ranking females. In fact, Dunbar (1989) has calculated that each loss of one unit of social rank (e.g. a decline from rank 3 to rank 4) reduces a female's potential contribution to the species' gene pool by 10%, which 'is clearly a very significant selection pressure'.

Social stress is generally invoked as the causative behavioural suppressor of reproduction in subordinate female primates (Keverne, 1979; Abbott, 1987; Harcourt, 1987a; Dunbar, 1989), but in some species, such as vervet monkeys (Whitten, 1983), brown capuchin monkeys (*Cebus apella*: Janson, 1985) and Japanese macaques (Sugiyama and Ohsawa, 1982), but not gelada baboons (Dunbar, 1989), female competition for food and water might introduce a confounding or exaggerating inhibitory influence on subordinate female reproduction. Reduced body weight or reduced food intake might be the cause of suppressed female reproduction (Sadleir, 1969a; Dubey *et al.*, 1986; Yen, 1986). However, as the links between fertility and female social rank are found in groups provisioned with food, as well as in wild groups of primates (Harcourt, 1987a), it is difficult to establish to what extent restricted access to resources plays in the reproductive impairments of low-ranking females (except, of course, to breeding males).

Dunbar (1989) also raises a further complication in the study of social factors controlling female primate fertility. Low-ranking female gelada baboons can buffer themselves from the reproductive consequences of their low social status by forming a coalition with another female. The birth rates of both members of a coalition are improved in contrast to comparable females outside coalitions. The buffering effect of being in a coalition appears to be the result of reduced harassment from other females (Dunbar, 1989).

In yellow baboons, formation of coalitions by females is an integral part of the social mechanism operating to suppress reproduction, particularly among low-ranking females (Wasser and Starling, 1988). Coalitions of attack involve at least two females attacking a third. Recipients of coalition attacks are always of lower social status than their attackers, and are most likely to be undergoing ovarian cycles. Attackers significantly concentrate on menstruating recipients or recipients in the follicular phase of the ovarian cycle (i.e. during the pre-ovulatory phase of the cycle), which probably directly results in recipients experiencing increased numbers of ovarian cycles prior to conception, and longer interbirth intervals (Wasser and Starling, 1988). Pregnant females that receive frequent coalition attacks suffer from spontaneous abortion, premature delivery or prolonged gestation. Female offspring born to low-ranking mothers have particularly poor

Table 4.1 Summary of the extent of suppression of reproduction in low-ranking female primates in the wild.

Species	*Maximum extent of observed suppression of reproduction*	*Field study*
(a) 'Helpers' required to rear the offspring of one female*		
Common marmoset (*Callithrix jacchus*)	No offspring produced	Hubrecht, 1984; Scanlon *et al.*, 1988
Saddle-back tamarin (*Saguinus fuscicollis*)	No offspring produced	Terborgh and Goldizen, 1985
Black-chested moustached tamarin (*Saguinus mystax*)	No offspring produced	Garber *et al.*, 1984; Sussman and Kinzey, 1984
Golden lion tamarin (*Leontopithecus rosalia*)	No offspring produced	Dietz and Kleiman, 1986
(b) Females can raise infants unaided		
Vervet monkey (*Cercopithecus aethiops*)	Fewer offspring produced	Whitten, 1984
Toque macaque (*Macaca sinica*)	Fewer offspring produced	Dittus, 1986
Rhesus macaque (*Macaca mulatta*)	Fewer offspring produced; first successful conception delayed	Drickamer, 1974; Sade *et al.*, 1977; Meikle *et al.*, 1984
Japanese macaque (*Macaca fuscata*)	First successful conception delayed	Sugiyama and Ohsawa, 1982
Barbary macaque (*Macaca sylvanus*)	First successful conception delayed	Paul and Thommen, 1984
Gelada baboon (*Theropithecus gelada*)	Fewer offspring produced	Dunbar, 1989
	Delays in successful conception	Dunbar, 1980
Common baboon (*Papio cynocephalus*)	Delays in successful conception	Wasser and Starling, 1988
	First successful conception delayed	Altmann *et al.*, 1988

* In these species, breeding females appear to require help with rearing their young if the young are to have the maximum chance of surviving (Sussman and Garber, 1987; Garber *et al.*, 1984); hence, the total suppression of fertility in subordinate 'helpers' (MacDonald and Carr, 1989; Wasser and Barash, 1983).

chances of surviving, if they are born around the time of the seasonal birth peak, because of attacks from other adult females. In this primate, a complex and rather brutal social mechanism has developed, whereby females aggressively attempt to suppress the reproduction of other females so as to reduce the competition their own infants face following weaning. Table 4.1 summarizes the social suppression of female reproduction in free-living primates.

Primates may be more adapted to social factors affecting fertility than is immediately apparent. Rhesus monkeys certainly seem to have taken adaptation to social control of fertility to an extreme. In a 20-year study of a captive colony, Simpson and Simpson (1982) found that dominant females produce significantly more daughters than sons, whereas subordinate females produce significantly more sons. In the matrilineal rhesus monkey, dominant females would probably benefit more from daughters, because the latter inherit their mothers' ranks (Sade, 1967), while sons do not (Koford, 1963). Low-ranking females would probably benefit most from sons because the latter often leave their natal group at puberty (Drickamer and Vessey, 1973; Packer, 1979), and stand some chance of achieving higher rank in another group. Female rhesus monkeys may be able to adjust the gender ratios of their infants by controlling the timing of mating in relation to ovulation (Simpson and Simpson, 1982), which in itself may occur as a result of differences in female social rank.

4.5 OVARIAN/MENSTRUAL SYNCHRONY AND IMPLICATIONS FOR HUMANS

Suppression of female primate reproduction demonstrates the profound effects of social factors on female fertility. Synchronization of ovarian or menstrual cycles has also attracted much recent attention, especially because synchronization of menstrual cycles has been found in women (Graham and McGrew, 1980; McClintock, 1983). Synchronization of ovarian cycles has been found in several other primate species; for example, captive golden lion tamarins (synchrony of ovarian hormone cycles: French and Stribley, 1987), wild vervet monkeys (synchrony of mating: Whitten, 1983), captive chimpanzees (synchrony of sex skin swelling: Wallis, 1985) and humans (synchrony of menstrual cycles: McClintock, 1971; Skandkhan *et al.*, 1979; Graham and McGrew, 1980; Quadagno *et al.*, 1981; Preti *et al.*, 1986).

Such reproductive synchrony may be due to some form of social facilitation. For instance, in gelada baboons, females return to oestrus prematurely if their unit is taken over by a new male (Dunbar, 1980; Mori and Dunbar, 1985). In patas monkeys (*Erythrocebus patas*), Rowell and Hartwell (1978) found that one female returning to oestrus following lactational amenorrhea triggered the other females in her group to do the same. Furthermore, in the same species, challengers for the harem-holding position of a group male are likely to stimulate the onset of oestrus among the challenged male's harem females (Rowell, 1978). Social facilitation of ovarian function by new breeding males in a group may accelerate the initial conception rates. Amongst females, social facilitation of ovarian or menstrual synchrony might enable high-ranking females to monopolize males during times when most females in their group are ovulatory and thus contribute to a conception delay in their subordinates (Whitten, 1983).

In an interesting development, pheromones have been implicated in synchronizing ovarian function in female rodents (McClintock, 1983) and women (Russell *et al.*, 1980; Preti *et al.*, 1986). In the study by Preti and colleagues (1986), solvent extracts of axillary secretions from women donors significantly synchronized the menstrual cycles of other women receiving the scent extract to the cycle pattern of the donors. In this double-blind study, the donors and the recipients had not met, unlike in previous human studies (e.g. Graham and McGrew, 1980), which suggests that specific chemical signals activate olfactory pathways to alter human female reproductive physiology directly. In a parallel study, Cutler and colleagues (1986) found that exposing women with either overly short (< 26 days) or long (> 32 days) menstrual cycles (considered aberrant) to the solvent extracts of axillary secretions from men, significantly changed the menstrual cycle length of the women recipients to a regular 29 days (approximately). This again suggests a specific olfactory effect on human female reproductive physiology.

If these findings came from a nonhuman primate species, one could be forgiven for speculating that in groups of women with synchronized cycles, the more socially dominant might try to monopolize the available men around the time of ovulation, to obtain a conception rate advantage over subordinates. This might be possible in the light of recent studies showing a peak in human female sexuality around the time of ovulation (Adams *et al.*, 1978; Harvey, 1987). Secondly, it may be that the close presence of a novel man stimulates human female reproductive physiology so that the chances of rapid conception are increased. While the latter possibility might fit well with the predominantly monogamous human family culture, the former would not. The actual significance of these social and olfactory effects on human reproduction is not well understood.

4.6 CONCLUDING REMARKS

Social factors clearly have strong controlling effects on female primate fertility. For example, ovulation is suppressed by the stress of low social status (e.g. marmosets), or by increased population density (e.g. lesser mouse lemurs). Ovarian cycles can be synchronized between females (e.g. golden lion tamarin), and oestrus can be stimulated by events such as the presence of a new breeding male (e.g. wild gelada baboons). It is not surprising, then, that changes in a female primate's social environment will have profound consequences for her fertility. Moreover, as primate social systems cover almost the full array of social possibilities, our understanding of how the social environment controls female fertility may be only in its infancy.

ACKNOWLEDGEMENTS

I thank my colleagues J. Barrett, L.M. George and C.G. Faulkes for their collaboration in some of the work reported here, and for their and Professor A.P.F. Flint's criticism of the manuscript, and Kathy O'Sullivan for typing the manuscript. This work, and the preparation of this chapter, were supported by an MRC/AFRC programme grant, and grants from the Wellcome Trust, SERC, Nuffield Foundation, Association for the Study of Animal Behaviour and the University of London.

5

Individual variation in responsiveness to environmental change

ANNE B. CLARK

5.1 INTRODUCTION

Responsiveness to environmental change can vary not only between species but within species. A variety of studies in evolutionary and behavioural ecology, developmental psychobiology and neuropsychology deal directly or indirectly with this individual variation in responses to change. The aim of this chapter is to review and integrate some of the diverse theoretical and empirical results in a functional and evolutionary framework. While the emphasis is clearly on primates, data on other species serve to augment the primate literature, and also to demonstrate the generality of some results and ideas.

The discussion will proceed from theoretical ideas from evolutionary and behavioural ecology on the selective effects of environmental change and uncertainty on animal behaviour. These ideas lead us to expect not only intraspecific variability in behaviour, but also differences between species in the origins and degree of individual variability. Empirically, there appears to be widespread individual variation, specifically in the response to change. The psychological literature on human and nonhuman primates provides evidence for both congenital and experiential influences on individual responses.

The empirical results often offer little or no functional explanation. But, as a major goal of this discussion, some possible functional approaches to variation are considered. It is clear that consideration of the natural selection pressures and naturally existing variation in developmental experience are necessary to any useful interpretations of responses to change in the laboratory or in the field. As a final step, both empirical results and functional interpretations are related to the potential for captive breeding to alter or limit adaptive variation, a problem which conservation programmes for endangered species may well face. Thus, a functional and evolutionary approach could provide a useful framework to studies on individual responsiveness to change.

5.2 INTERSPECIFIC AND INTRASPECIFIC VARIATION IN BEHAVIOUR: THEORETICAL EXPECTATIONS

Clark and Ehlinger (1987) emphasized the close relationship between behavioural variation at the level of the individual and at the level of the species. They pointed out that it would be difficult to build broad theories at one level which would not predict or necessitate theories of variation at the other. Furthermore, environmental change plays a key part in selection, resulting in behavioural variability. This is clear in theoretical treatments of the effects of fluctuating as opposed to stable environments on the evolution of life history characteristics in both plants and animals (e.g. Levins, 1968; Stearns, 1977; Caswell, 1983; Smith-Gill, 1983). Organisms in spatially and temporally homogeneous environments are expected to evolve those suites of characteristics which best allow them to exploit their resources and compete with other species. Organisms faced with environmental heterogeneity have several potential responses. They could ignore or wait out the changes, making no fundamental alterations in behaviour, physiology or morphology. Any adaptation to environmental shifts would then be seen at the population level after selection among genotypes occurs. Alternatively, they could track the environment phenotypically, altering in ways which adapt them to new conditions. The ability of a given genotype to generate more than a single phenotype is termed phenotypic plasticity. Plasticity can range from reversible physiological adaptation to physical changes (e.g. temperature), to fundamental, irreversible changes of phenotype during development. Developmental plasticity can further be divided into continuous, graded 'modulation' of a phenotype versus 'conversion' to a discrete distinct phenotype with its own integrated suite of behavioural, physiological and morphological characteristics (Smith-Gill, 1983). We understand less about modulation and how it might evolve adaptively (Caswell, 1983). This problem is considered later in the context of functional interpretations of variation (section 5.4).

There are broad expectations for the necessary and sufficient characteristics of environmental variability, under which adaptive phenotypic plasticity of any kind can evolve (Fagen, 1987). Cues for change must be available, detectable and sufficiently predictive of the change. The time required for phenotypic response must be brief, relative to the time scale of the environmental variation, so that the organism will not find that the environment has changed yet again during its response. Thus, the effects of environmental fluctuations are to be scaled to the expected lifetime and developmental schedule of a given species: mayflies emerging to breed, with life expectancies of a few days, can have little opportunity to profit from plastic responses to cues predicting a warm month to follow. The characteristics of the environmental variation (e.g. the timescale of a change or the stability of the new conditions), should dictate the kind of plasticity that will be adaptive (see Bateson, 1963). We know little about the costs of plasticity, but maintenance of the 'machinery' for such a switch, the energetic costs of the response itself, and perhaps increased risk or decreased efficiency during a switch all seem likely problems. These must, of course, be outweighed by the expected gains in survival and eproduction due to the new phenotype (Fagen, 1987) if plasticity is to evolve.

Thus, the temporal characteristics of environmental fluctuations, to which members of a species are subjected, should select for species-characteristic abilities to adjust and modify the phenotype. On time-scales approximating to the life of most individuals, regular fluctuation in the environment can produce alternating episodes of selection among several existing genotypes. These episodes may produce intra-specific variation in the form of genetic polymorphisms, as has been suggested for clutch size variation in tits (*Paru* spp.: Nur, 1987). The individuals themselves, however, are not detecting, tracking or adjusting to change, and the variants might be described as discrete types. In quite constant environments, but also in environments which are alternating more rapidly than an animal can produce and utilize a new phenotype, animals may be expected to develop and behave conservatively. They should not alter their characteristics from those which are, on average, most profitable. One would thus encounter little plasticity and relatively little intraspecific variation in phenotype.

Frequent novelty and rapid change may select for another attribute, 'flexibility' (Fagen, 1982). Flexibility is usefully distinguished from plasticity (Fagen, 1982; Clark and Ehlinger, 1987) in that plasticity allows a single genotype to take on more than one distinct phenotype through alternative courses of development. Flexibility, however, allows animals to vary short-term behaviour in the face of changes or new information, largely through processes of learning. All individuals may be equally flexible with identical learning rules and, having encountered different events and information, may be acting differently. An animal which is developmentally plastic may not be at all flexible in its behaviour once a developmental 'choice' of phenotype has been made.

Primates and other relatively long-lived animals will be exposed to change on more timescales than organisms living a very short period of time. Each individual can expect to experience enough repetitions of very short changes that finer adjustments become worthwhile. For instance, while many shorter-lived animals must make one-time seasonal adjustments in behaviour and physiology, a primate such as a baboon (*Papio*) or a Japanese macaque (*Macaca fuscata*) also experiences within its lifetime the variation in availability of specific foods among repeated dry, wet or cold periods. Learning about alternative ways to forage, given the severity of that particular year's season, can be the key to survival. Flexibility as fine-tuned, rapid, adaptive modification of behaviour requires learning from individual experience about predictors, ranges of variation and outcomes of responses. This information-processing itself entails behaviours which add to the experience: exploration, manipulation, active search for change and learning to detect it. The costs of flexibility as a means of adapting to immediate conditions may lie both in the information-processing machinery and in the information-gathering behaviours themselves. For instance, sampling and exploration of previously unproductive food sources can lower the current rate of feeding, but may still be selectively advantageous (Lea, 1979). Indeed, as one might expect if gathering information for the future is critical, such sampling can be quite resistant to 'extinction', even if punished (Shettleworth, 1978).

There is one interesting point of disagreement on how extreme novelty, or the occurrence of one-time events, should affect rules of behaviour and the observed degree of variability. P.B. Slater (unpublished) suggests that when past events shed no light on how to behave now (i.e. when animals are left with too little information), 'gambling' on a course of action may result. With no predictable best approach, nonadaptive variation in choices of action and resulting behaviour can result. On the other hand, Bookstaber and Langsam (1985) argue that any rules which allow animals to respond appropriately to changes presuppose an internal model of the world which specifies the nature and probability of change. In the face of truly novel events outside the species' experience (e.g. volcanoes and oil spills), that internal model is challenged, while no new information of future probabilities of events or results of responses to events are added. This engenders what Bookstaber and Langsam term 'extended uncertainty'. Optimal behaviour rules under this new possibility of novel and incomprehensible changes are, they argue, likely to be simpler and less finely tuned to the environment, because it is now less likely that the environment is fully 'understood'. Thus, environments which are more prone to radical, unprecedented change may actually produce coarser, more inflexible, behaviour. In contrast to P.B. Slater's suggestion (unpublished), animals may therefore be less flexible and appear less variable at both ends of the environmental novelty spectrum than in the mid-range of moderate and regular amounts of change. (This result is obviously similar to the prediction on plasticity in the face of rapidly alternating environments, but not identical. Animals might well have some upper limit on the degree of novelty which will result in learning or 'trying to adapt'. Therefore, this is a prediction on the rules for using their flexibility, not necessarily on species differences in flexibility. Empirical results of responses to novelty and change as stressors tend to support Bookstaber and Langsam's idea, as discussed below (section 5.4).

The social environment

As most behavioural biologists are acutely aware, the application of game theory to rules of social behaviour (Maynard Smith, 1982) has led us to expect that the relative success of what one individual does depends upon what associates or competitors are doing. For example, behavioural ecologists now routinely seek and find intraspecific variation in the form of two or more coexisting alternative strategies (West Eberhard, 1979; Gross and Charnov, 1980; Dunbar, 1982; Thornhill and Alcock, 1983 (for mating strategies); Ehlinger, 1986, 1989 (for foraging strategies). As with morphological polymorphisms, behavioural variants such as alternative mating strategies may be based on genetic polymorphisms maintained by frequency-dependent or density-dependent selection, the developmental plasticity of single genotypes, or a combination by which several more or less environmentally sensitive genotypes coexist (Cade, 1981).

The coexistence of phenotypes with alternative strategies adds one more form of diversity to the environment of social animals whose milieu is already enormously variable with respect to number of conspecifics, the relative frequencies of age and gender classes among them, their degree of kinship and any individual behavioural traits. Such variability exists between populations and through time in any one population. This high rate of change and unpredictability within many social groups may select for flexibility (Humphrey, 1976). If so, Fagen (1982) points out, the effect of social living is amplified. Having conspecifics who routinely modify their behaviour through individual experiences and learning will further increase the variety and unpredictability of the behaviour of associates. Based on a theoretical model, Fagen (1987) suggests that social environments may provide the necessary uncertainty to select for a trait of 'phenotypic plasticity', or the ability to adopt the phenotype of nonplastic conspecifics. Extrapolating further, the ability to adopt any of a range of social characteristics might be an adaptive trait because it allows an animal to decide with whom it will most directly compete. In other words, social animals might find a rather different, less discrete form of plasticity advantageous in adjusting intragroup competition and cooperation.

The amount of social change experienced by members of different species varies with life history, particularly with longevity and social organization. Individuals may typically associate with one age cohort, reproduce and die or, more typically for the primates, they may associate with several generations, move through several social roles and/or change social groups (Periera and Altmann, 1985; Clark and Ehlinger, 1987). These interspecific differences in social complexity and social change should be expected to result in interspecific contrasts in the degree of plasticity, flexibility and observable intraspecific behavioural variation. An excellent discussion of complex primate social trajectories and differences in social 'strategies' among life stages of different species is provided by Periera and Altmann (1985).

5.3 RESPONSIVENESS TO CHANGE AS AN AXIS OF INDIVIDUAL VARIATION

The foregoing sections explore the relationship between interspecific and intraspecific variation in behaviour, and identify the scales of environmental and social change as key elements in the amount of observable intraspecific variation. This variation represents the degree to which individuals of a given species typically alter their behaviour and physiology, contingent on specific characteristics of their social or physical environment. It is not necessarily implied that conspecifics will differ in how they respond to a given change. There is, however, considerable evidence for 'responsiveness to change' being a major axis of variation shown by many and diverse species.

Our empirical knowledge of intraspecific variation in responses to change exists largely because change in the form of unfamiliar environments, novelty, uncertainty and moderate social disruption are all stimuli which psychologists use to produce distinct individual differences in response behaviour of a wide variety of vertebrates, from fish to humans. For most of these studies, functional interpretation of the individual variation is lacking (Cowan, 1983). The emphasis is on understanding what attributes sum to a competent, optimal or successful organism. Much of the variation is thus treated as nonadaptive, or at least as a less-desirable deviation from an adaptable or socially successful norm. Despite the range of species studied and experimental routines used, many responses seem to vary on a continuum at least analogous to the ranges 'Neophilia ↔ Neophobia' and 'bold ↔ shy' (Suomi, 1987; Kagan *et al.*, 1988). Furthermore, these studies do suggest answers to the questions on the underlying basis of variation in responses to change. Are mammals, especially primates, completely flexible? Does variation often seem to develop at particular stages, like 'developmental conversions'? What does the genotype bring to the environmental stage, and how? We will briefly mention some of the diversity of studies, emphasizing long-term results on primates. Possible directions in the functional interpretation of intraspecific variation are then discussed below.

The kinds of change offered experimentally tend to differ with the species studied, and the nature of change offered affects the precise type of response elicited (Russell, 1983). The responses studied may be those which are immediately elicited in one species, or longer-term effects on behaviour in another. We have learned about different aspects of the response to change in different species. Thus, it is striking that a common theme links the responses. For example, rodents are often faced with unfamiliar or novel environments (cages and mazes), as well as brief introductions to novel social partners or rivals. Such short-term tests yield variation on dimensions, such as Bold ↔ Timid, Active ↔ Passive, or Aggressive ↔ Submissive. In experimental ecological studies, fish faced with predators vary intraspecifically in 'boldness' in terms of their propensity to return to foraging (Fraser and Gilliam, 1987). Boldness in antipredator behaviour is reported to co-vary with boldness in conspecific aggression in sticklebacks (Huntingford, 1976).

Tests of primate characteristics often include disruptions of stable social situations, as in mother-infant separations. Individuals are judged, not only on their behaviour during the disruption but also on their return to normalcy, and sometimes on their later social behaviour. It is primarily in studies of primates and a few other mammals that the relationship of early responses to later social characteristics are specifically studied. The measured responses thus reflect both immediate and longer term effects of change. The assumption is that, during the change, individuals reveal aspects of their social abilities or relationships which are continually present but hidden in familiar circumstances. As a case in point, the 'strange situation' of Ainsworth and colleagues (Ainsworth *et al.*, 1971), tests the responses of young children to an unfamiliar environment, a stranger and a brief separation from the mother. This now-classic test procedure can produce a tripartite classification of youngsters on the basis of behaviour during reunion with the mother. The reunion behaviour is thought to reflect characteristics of the mother-child relationship, which are important because they follow and influence the child during ensuing development. The classification of children is based on simultaneous variation along familiar axes, such as Bold ↔ Timid, Outgoing ↔ Withdrawn, Confident ↔ Anxious, and also on relationships such as Attached ↔ Insecure. In summary, a wide variety of vertebrates, including primates, have been found to vary intraspecifically on relative neophilia as opposed to neophobia, but primates are probably our best source of information on the early development and continuation of variation in social behaviour and relationships.

Two long-term studies, Jerome Kagan's of shyness in young children (Kagan *et al.*, 1988, 1989) and Stephen Suomi's of susceptibility to behaviour disruption in young rhesus (Suomi *et al.*, 1981; Suomi, 1983), are now providing evidence of intraspecific physiological variation in responses to stress which underlie such behavioural contrasts. Both studies show that the physiological variation may be relatively stable from very early in life, and may consistently produce sharply contrasting responses to stress, notably in the form of environmental and social change (see also Chapter 3).

Kagan *et al.* (1988) followed several cohorts of children longitudinally, from less than 2 years to 7 years. At several ages, they characterized the children's behaviour in novel social situations, and also tested sympathetic and limbic system physiological responses in the face of mild to moderate stress. Two of these cohorts consisted of children chosen from a large sample aged 1 and 2 years old as consistently extremely inhibited (clinging, nonexploratory) or uninhibited across novel situations. An unselected group of about 80 children followed from 14 months to 4 years comprised a third cohort.

The inhibited groups of children continued to differ behaviourally from uninhibited groups in testing at 5.5 and 7.5 years of age. Inhibited children continued to be slow to interact and to investigate novel toys or situations. Physiologically, inhibited groups had higher, less-variable heart rates, greater acceleration of heart rate and decreased variation in pitch of vocalizations under stress. Moreover, individual differences in heart rate and variability were maintained from less than 3 years to 5.5 and 7.5 years old. Urinary norepinephrine was moderately positively correlated with inhibited behaviour at both

4 and 5.5 years of age, as was salivary cortisol level at 5.5 years. Perhaps more importantly, a composite index of physiological arousal showed a significant positive correlation with behavioural inhibition scores at all ages.

In contrast, among the broad range of children in cohort 3, the degree of inhibition at 14 and 20 months did not predict the inhibition observed at 4 years. If, however, only those children consistently in the top and bottom 20% of the distribution at 14 and 20 months are considered (13 per group), the two groups differ significantly in behaviour at 4 years, as was found for cohorts 1 and 2. Kagan *et al.* (1988) draw several conclusions: the physiological data suggest that extremely shy children 'belong to a qualitatively distinct category of children who were born with a lower threshold for limbic-hypothalamic arousal to unexpected changes in environment or novel events that cannot be assimilated easily' (p. 171). Second, the behavioural profile of shy/quiet children may emerge and endure when such a temperament (*sensu* Thomas and Chess, 1977) is exposed to chronic environmental stress. Thus, a few of the initially extremely shy children did not continue to act so at later ages. Exactly what and how much stress was involved remains a matter of conjecture. Kagan *et al.* (1988) point out that most of the shy 1- and 2-year-olds were later-born children, in contrast to many first-borns among uninhibited children, and suggest that certain kinds of sibling interactions act as stressors. Obviously, when we are dealing with behavioural differences realized as a function of a variable physiological threshold in the context of variable developmental environments, we should not expect strong determinism. Kagan *et al.* (1988, 1989) argue, however, that the much greater consistency of 'behavioral style' among children at the extremes of the bold–shy continuum make it worthwhile to treat these children as qualitatively distinct. We should note that, although the emphasis is on the etiology of shyness, extreme boldness might be treated similarly. We can ask whether it also represents a problematic extreme or a socially successful behavioural type, a point to consider when hypothesizing functional explanations.

Suomi's studies of behaviour and physiology in young rhesus macaques, *Macaca mulatta* (e.g. Suomi *et al.*, 1981; Suomi, 1983, 1987) have revealed a similar pattern of marked individual differences in both responses to stress and susceptibility to anxiety and depression, which can be at least partially predicted by early physiological and developmental measures. The precipitating situations have been both short-term and longer-term social changes, largely removal from a mother or 'attachment object' (mother substitute for motherless monkeys), and introduction into a group of strangers. Some monkeys show stable elevations of heart rate and avoidance in the face of novelty (mild stress), or extreme adrenocortical responses and general behavioural disruption in the face of separations from mothers or mother substitutes. These are characterized as 'highly reactive' individuals. Those at the opposite end of the spectrum, which show brief heart rate responses, exploratory behaviour as well as recovery from separations, are characterized as 'low reactive' monkeys. A genetic component to these suites of characters has been suggested by the similarity among siblings in levels of cortisol and ACTH and in several behavioural responses (Suomi, 1981; Scanlon *et al.*, 1982).

One potential modifier of 'reactivity' during development is obviously the mother's treatment and her own behavioural responses. In the first report of a recent cross-fostering study (Suomi, 1987), young monkeys with a pedigree of high or moderate reactivity were paired with adoptive mothers, which were themselves either high or moderately reactive (a). Mothers were also classed as nurturant (b) or punitive (c), within the normal range of mothering shown by multiparous colony females. Neonatal tests during the first month showed that young with high reactivity pedigrees did differ from others, as expected. Maternal differences did not seem to affect these results. By contrast, in ensuing day-to-day observations, the behaviour of young monkeys differed mostly in accord with the kind of mothering they were receiving. However, in a series of separations at 6 months, the differences were primarily between infants who were of genetically 'low' or 'high' reactivity. The foster mother's reactivity contributed primarily to the degree of behavioural disturbance during reunions when the highly reactive mothers themselves showed self-directed behaviour, and probably affected that of their foster young. Finally, and intriguingly, continued longitudinal studies of these young monkeys in social groups suggest that social success ('dominance') can come to highly reactive young, but this has been true only for the three (genetically) highly reactive monkeys with nurturant foster mothers. Suomi (1987) stresses that while biological or genetic tendencies may be influential during marked environmental challenges, or changes, those tendencies may be irrelevant to behaviour under familiar normal conditions, and may even be overridden by the behaviour of important conspecifics, such as mothers. Both congenital differences and developmental experience apparently contribute to the behavioural outcome.

Both Suomi *et al.* (1981) and Kagan *et al.* (1988) report evidence that physiological contrasts between extreme types of monkeys and children are apparent as early as measurements can be made. ('Extreme' is not defined across species. For children, it was taken as consistency of scores in the upper or lower 20% of the range (Kagan *et al.*, 1988).) In rhesus, sibling and paternal half-sibling comparisons suggest a familial component, apart from maternal rearing effects (Suomi *et al.*, 1981). In both rhesus and humans, however, the physiologically measurable differences do not become behaviourally obvious until a disruption of the family occurs. In children, the differences may continue into later life only under certain social conditions, but more physiologically extreme individuals seem more likely to maintain their early characteristics (Kagan *et al.*, 1988). In rhesus, the differences may apparently disappear in a stable environment, only to reappear in exactly the same way if social change and stress are reintroduced.

Several other primate studies, notably by J. Altmann (1980), Simpson (1985) and Fairbanks and McGuire (1988), report that mothering styles which encourage independence and allow young monkeys more contact with novelty may increase their interest in, exploration of, and ability to deal with novelty and change. Conversely, restrictive or very protective mothering decreased these characteristics. However, early independence and any attendant advantages in the face of later change may normally come with a cost; in the wild, such young were more likely to be predated (J. Altmann, 1980).

Play is another social experience which may increase the individual's ability to be flexible and to respond adaptively to change or new environments (Geist, 1978; Bekoff and Byers, 1985; Fagen, 1987). There is little real data on the relationship of play experience to an individual's ability to adapt to change, but we do know that the amount of play during the development of young monkeys varies for a number of reasons – for example, number of available peers or siblings (Cheney, 1978), or time or energy limitations due to food shortfalls (Baldwin and Baldwin, 1974; Lee, 1981). Thus, there is natural variation in early peer experience which might contribute to differences in responses to change, particularly in concert with any physiological predispositions.

The overall picture from these developmental studies of young primates is that responses to novelty and change form a strong axis of intraspecific variation in both physiology and behaviour, one which has congenital components and can yield highly contrasting 'social styles', especially in the face of moderate stress. What these studies do not generally provide is any functional explanation for the existence and maintenance of such variation, particularly for the responses of extreme phenotypes. Furthermore, with interest centred on the extremes, we are left with few insights into the vast mid-range between them. Physiological measures do not map onto behavioural measures over the larger part of the range, nor do the individuals' ranks on these measure remain relatively stable, except at the extremes in these studies. While emphasis on the extreme individuals is pragmatically justified by their greater stability, as well as their greater potential for 'pathology', I believe we will have to offer specific hypotheses for the mid-range of variation before we can even understand why these extremes occasionally occur.

5.4 FUNCTIONAL HYPOTHESES FOR INDIVIDUAL VARIATION IN RESPONSE TO CHANGE

Limitations from our current research methods

Comparative studies of intraspecific variation are, of course, no different than other cross-species comparison. The same stimuli will not be appropriate for all species. Investigations of exploration and curiosity (Archer and Birke, 1983), as well as individual differences in behaviour, utilize a variety of environmental and social change as stimuli, but many studies neglect to justify their choices. Fragaszy and Adams-Curtis (see Chapter 13, section 13.1) discuss the choice of environmental challenge in captivity for the purposes of understanding species-normal characteristics, pointing out that both the stimulus and available responses should be relevant to aspects of the natural environment. While not necessarily identical, tests should at least be comparable to types of changes which are critical to survival and success in the evolutionary history of a given species. In the case of rhesus monkeys, manipulations such as separation from mothers, which 'reliably elevate, in a highly predictable fashion, the incidence of anxiety in most individuals' are arguably similar to natural, highly dangerous situations; in this case, maternal death (Suomi *et al.*, 1981). It also goes without saying that experimenters need to be sensitive to what the responses tell them about the stimuli. Some forms of change may seem equivalent to scientists, yet be processed very differently by their subjects. For instance, unfamiliar environments (cages or enclosures) and novel objects can elicit quite independent responses (Russell, 1983). When young rats encountered both unfamiliar environments and a specific novel event simultaneously, they did not orient to or explore a novel auditory stimulus until some learning about the new area had occurred (Richardson *et al.*, 1988). Apparently the rats were not fearful, but simply could not process the two levels of new information, general and specific, at one time. This problem of absolute novelty (surroundings unfamiliar) differing from relative novelty (unfamiliar object in familiar surroundings) (Hennessy and

Levine, 1979) is the subject of considerable discussion in psychological studies of exploration (see Archer and Birke, 1983).

Interspecific comparisons of responses to the same stimuli can reveal differences in the ways species deal with their environments (e.g. Fragaszy and Mason, 1978), in their adaptability (see Chapter 13) and, by inference, may also suggest differences in the relevant ecological problems that typically face them. For instance, wild *Rattus novegicus* and *R. rattus* both avoid strange objects in familiar environments, possibly as a result of selection due to our attempts to exterminate these two with various poisons, traps, etc., which may exploit their curiosity (Cowan, 1977). Two other species, *R. fuscipes* and *R. villosissimus*, as well as laboratory rats, do not show this avoidance (Cowan, 1977, 1983).

Very similar problems of comparability face developmental studies of responsiveness within species. The abilities of the organism and the relevant variables in its environment change with age. Fundamental and continuing aspects of responsiveness may not be measurable with similar stimuli or by the same responses across ages. For instance, what induces fear will very likely change with age and increased motor abilities. This problem of continuity and stability arises in developmental studies of humans and other animals. In many cases we know too little about the developmental process itself to make assumptions about which responses are related across ages. (For critical discussion, see Hinde and Bateson, 1984; Kagan, 1980, 1984.) Functional or evolutionary interpretations will often require our knowing how the range of individual variation changes as animals age and gain experience, or as selection acts on an initial distribution.

Suomi (1983) has made the point that the degree of change or stress which is useful in highlighting individual responsiveness is also critical. Mid-range to moderate stressors or novelty are most useful because extreme conditions can exceed the thresholds of all individuals and produce rather uniform extreme responses. Similarly, increasing 'stimulus complexity' in rodents increases exploration only up to a point, which may be related to the level of complexity in the species-normal environment (Russell, 1983). These empirical results may shed light on two contradictory predictions of Slater (unpublished) and Bookstaber and Langsam (1985). If extremely unfamiliar, novel and unprecedented events do generally produce fairly uniform responses, or little response at all, Bookstaber and Langsam's suggestion of 'coarse behaviour rules' appearing as a function of extreme uncertainty would be supported. P.B. Slater (unpublished) would predict that behavioural responses under these conditions might be wildly variable, though often maladaptive. Specific studies looking at the relationship between the degree of intraspecific variation in responses and the degree of novelty or change are obviously needed.

In summary, we face the problem of choosing both an appropriate kind and degree of change for the species and for the class of individuals under study, particularly if we are to elucidate adaptive behavioural differences in responsiveness to change. Interpretation of the typical responses, or intraspecific variation in responses of species, must be based on some understanding of environmental regimes encountered by a species, and selection in the wild.

Some specific hypotheses

Looking for the advantages or disadvantages of particular types of physiological/behavioural responsiveness may be taking the wrong approach. Especially in a highly social species, where an individual's relative fitness is a result of both what it does and what others are doing, the important factor in success may be the differences between it and other individuals, rather than its exact physiological make-up or its absolute level of response to a given amount of change (Clark and Ehlinger, 1987). For instance, as a competitive strategy, it may pay to do what others are not doing, whatever that is. Particularly where close kin are involved, behaviour which differs from theirs may reduce the costs of direct, confrontational competition and work to the net benefit of both (Fagen, 1987). Underlying covert differences (Clark and Ehlinger, 1987) in thresholds to social challenges or stressors could predispose one individual to do something different than another; to take on a contrasting behavioural phenotype or personality, and to make different decisions at those critical points in the life history.

To the extent that costs and benefits of those decisions are dependent on the decisions of others, they are not simply good or bad, adaptive or maladaptive. They become so primarily in the context of others' behaviour. Thus, the function of continuous variation in an underlying physiological trait or threshold would not be to determine a particular behavioural phenotype, but rather to provide a direction or proximate decision rule on how to differ from another individual. (Note that this is the converse of phenotype matching through plasticity, as suggested by Fagen (1987), but the functional effects in terms of controlling competition are similar.) This interpretation is in keeping with the results of Kagan *et al.* (1989), for example, in that early physiological measures were not correlated with later behavioural measures for children whose responses fell in the middle range of inhibition. Under the foregoing hypothesis, one might make predictions of the trajectory of such a child on the basis of both his/her early scores and the behaviour and physiology of individuals with whom he/she had interacted during intervening development.

From a maternal point of view, reduction of competition among a female's offspring would be particularly desirable (Bekoff, 1977; Armitage, 1986) because it raises the chances of success within her offspring set. If she reduced competition by raising a diversity of social strategies which were collectively suited to exploit a range of social or physical environments, a mother would have achieved another theoretical advantage of diversity among offspring. She would be prepared for a range of future environments with an adapted offspring. The mother's expected fitness in an unpredictable future would therefore increase, even if some of her offspring were maladapted at any one time (e.g. Armitage, 1986). If social experience works on continuously variable physiological characters, minor differences in the treatment of her young might be more than sufficient to reinforce long-term phenotypic differentiation.

Dispersal is one arena where these differences among siblings and among peers might be particularly important to realizing an individual's advantage to 'being different'. Dispersal may remove an individual from competition with siblings, or from an environment where success is unlikely, to one where it is more likely. The effect of 'arranging competitive interactions as far as possible between unlike genotypes' can select for dispersal even in stable habitats where the chance of success in going elsewhere is unchanged (Hamilton and May, 1977, p. 579). Proximate causes of dispersal in primates is controversial (Periera and Altmann, 1985), but at least in some species dispersal is often not a result of direct aggression, and particularly not sibling aggression (Bekoff, 1977; Cheney, 1978; Packer, 1979; Harcourt and Stewart, 1981; Strum, 1982). Furthermore, if all young males were following the same rules, and responding similarly to their own changing social status and the increasing aggression they are receiving from a set of adults, one might expect that many young males in a troop would leave at the same age. I know of no reports of great similarity within cohorts. This is admittedly a weak argument for the idea that differences in sensitivity to changing social stress tend to remove close-aged siblings and half-siblings from competition through resulting variation in age to dispersal. On the other hand, a few studies offer suggestive, if limited, evidence for such within-litter differences affecting the decisions to disperse (Fox, 1972; Bekoff, 1977 (for wolves); Clark, 1982 (for galagos); see Clark and Ehlinger, 1987).

Gender differences: a testing ground?

It has been suggested elsewhere (Clark and Ehlinger, 1987), that the life histories of male and female primates are often markedly different in such parameters as the stability of associations with other individuals, the ability to predict which particular conspecifics will be critical in reproductive competition, the timing and probability of dispersal and possibly also the reproductive benefits and costs of investigation, sampling and engaging in novel behaviour. Such intraspecific differences in life history, together with an existing database on gender differences in such aspects as neural plasticity (Juraska, 1986), variability in response to change (see discussion in Juraska, 1986), tendencies to explore (Archer, 1975 for comparative rodent data), and the nature, frequency and duration of play during development, offer a testing ground for predictions on the fine tuning of responses to typical and relevant patterns of social and environmental change.

For example, we can consider the lives of baboons and macaques. In a simplified but not atypical scenario, a female remains in the natal troop, most closely associated with members of her matriline, and her expected reproductive success is most affected by her dominance status within her matriline, as well as the relative status of the matriline itself. Her social environment is determined early in her life and she may be able to influence her success more by how she behaves within her social role than by changing her status, her close associates or her troop. Young males, on the other hand, grow up among many other males who will ultimately not affect their lives because the young males will have dispersed, or the dominance order will have changed radically by the time they are of reproductive age. If they do disperse, they will be adapting to a whole new set of individuals with their own idiosyncrasies. Thus, we might predict, however naively, that young females would be more stable in their response characteristics, that apparent roles and dominance orders among playmates might be less changeable, and that, if play functions do increase behavioural flexibility, that female play would at least have a different quality from male play and perhaps duration during development. We might also expect that female play and choice of partners would reflect the building of long-term bonds rather than the testing of social partners and the behaviour appropriate to different roles. Adaptability for females might mean the ability to put up with one of a range of possible social positions and make the best of it.

In contrast, young males will be called upon to make repeated adjustments to new associates into maturity, and we might expect that they would be more changeable, less stable in apparent behavioural characteristics for a longer time. Their social strategy might be either to delay 'choosing' a social phenotype until later in development, or actively to seek out a social environment which best fits them. Certainly, they should not simply polish a set of behavioural responses which serve to make the best of the social position in which they find themselves when young. Thus, male play might function in providing experience with the behavioural requirements of social roles, and allow juveniles to discover how to change relative status within play groups. In an abstract sense, males might be expected to be very socially sensitive, able to detect social situations in which they could succeed or fail, and prone to move out of situations in which they are subordinate.

How might this translate into differences in responsiveness to change? Realizing that this is simply an attempt to build an example of the logic, we can consider some of the data on gender differences in environmental sensitivity and neural development. Male behaviour is generally more strongly affected by environmental conditions than female behaviour in rodents and primates (see Juraska, 1986). Male rats respond more than females to enrichment (Joseph, 1979) or isolation (Freeman and Ray, 1972). Male rhesus monkeys react with more behavioural abnormality and reduced exploration of novelty to isolation than do females (Sackett, 1974). Changes in the brains of rats show that for some areas (e.g. visual cortex) males have responded more to isolated as opposed to enriched environments (Juraska, 1984). In the hippocampal regions, females appear more plastic (Juraska *et al.*, 1985). But the gender differences are dependent on the environmental challenge, and thus gender differences in the brains of rats at least are not absolute features, but are influenced themselves by the kind of environments that individuals experience. This meets our very general expectations that adaptive responses to the same apparent social environment should differ for males and females for many species, but to go farther would mean understanding much more about the socioecology of rats.

Such studies suggest that female adaptability in the sense of maintaining normal behaviour in the face of less than adequate environments is greater than that of males. Males may change behaviourally more with the environment. For primates, more abnormal responses to isolation could reflect a greater input of social experience with conspecifics to development (e.g. through play), or could be a pleiotropic effect of being more behaviourally responsive to conspecifics over a longer period of time. Perhaps males would be more prone to leave abnormal or undesirable social situations than females, given that they normally have that option. Situations in which there is no way to alter their social circumstances may be more abnormal for males than females. Thus, greater responsiveness of males is in the direction of differences we might have predicted from the social picture. The life history differences in pattern and effects of social change between genders may offer some alternative interpretations for the apparent susceptibility of male primates and other mammals to social stress.

5.5 EFFECTS OF CAPTIVITY ON VARIATION IN RESPONSES TO CHANGE

Hediger (1950) was one of the first to call attention to the adverse effects of captivity on behaviour of wild animals. Concern since then has been largely focused on damage due to inadequate or abnormal conditions, especially during development, rather than on the genetic changes due to inadvertent selection (Stevenson, 1983). The process of bringing animals into captivity and then breeding from a small initial stock – or at least from that subset which both survives and agrees to reproduce successfully – provides a nice bottleneck for selection. This has been recognized (Hediger, 1950), especially with respect to responses to novelty (Glickman and Sroges, 1966; Stevenson, 1983), and certainly modern zoos do a great deal to try to provide naturalistic environments with some degree of complexity, if not daily change. If, however, the goal is to establish a captive population of animals capable of returning to the wild, as some of the endangered species programmes hope to do, then the problem of genetic change or genetic loss due to selection looms larger. Then, beyond the initial problems of survival of animals with less than adequate experience of predators, foods, and/or environmental complexity, one has to ask what subset of initial individual variants are no longer represented in the population.

Let us suppose, for instance, that there is a continuous heritable variation along an axis which tends to produce animals varying behaviourally in their willingness to explore, take risks and also disperse. When a sample of animals is first brought in, one might find that the boldest, least fearful, perhaps the greater explorers and risk takers, are overly represented among those which adapt rapidly and breed most readily. However, these animals are also very prone to lose all predator shyness and would require many negative experiences with predators while young, to develop the necessary caution. Furthermore, these are also among the most aggressive individuals. Obviously, as a largely made-up scenario, there are several ways for the story to end when the now captive-bred population is to be returned to a new park. But some social as well as survival problems might ensue.

The conclusion from the possibilities outlined above, is that we need some research directly on selection during the early phases of captive breeding in a sample of (nonendangered) species which themselves differ ecologically and in their species-normal experience of and responses to environmental change.

5.6 CONCLUDING REMARKS

The conclusions of this brief review of several distinct areas of investigation are several-fold. First, both theoretically and empirically, individuals within a species vary most often on axes potentially related to responses to change. This variation appears often continuous, although it may be argued that some extremes differ sufficiently in their developmental trajectories to be dealt with as 'categories' or distinct phenotypes. Second, functional hypotheses for the existence of this variation are rare, and centre either on the extremes, or are constructed only when clear types are apparent. The least-studied problem seems to be the significance of the often wide range of variation between the extremes. It is suggested here that variation in underlying physiological characteristics might allow social animals to develop behavioural characteristics which differ from those of conspecifics, especially kin. Further, the adaptive significance of this differentiation is that direct competition is reduced and a wider range of routes to social success is open to exploitation. Thus, it would be the differences between the behavioural characteristics between one animal and those of its conspecifics, not the exact value of their characteristics on the continuum, that are adaptive. Under this hypothesis, selection acts on underlying characteristics in terms of their utility in making a more or less fine-tuned differentiation from relevant conspecifics possible.

We are, of course, hampered when integrating results from diverse studies of a wide range of vertebrate species, but an unexpected uniformity of behavioural variation seems to indicate that we might be on the right track. If a socially important degree of variation is selectively maintained in many animals, one would expect it to be most critical in long-lived social animals, such as primate species, in which a great many associates are together over a long period of time. In such circumstances, there is ample time to modify phenotypes with respect to others; and associates are a major factor in reproductive success. An empirical knowledge of intraspecific behavioural variation may be essential to understand both how behaviour develops adaptively in highly social animals and why individuals achieve their relative reproductive success. It also will be critical that captive breeding programmes for endangered species consider what captivity as a selective force may do to this form of adaptive individual variation in responsiveness to social and physical change.

Part Two
Environmental Change in Nature

An adolescent female chimpanzee watching another adolescent female crack open oil-palm nuts using a hammerstone and a concrete slab as an anvil at an island rehabilitation site in Liberia. (Photo: Alison Hannah.)

The most extensive worldwide and pervasive environmental change in nature concerns changes in the patterns of land use, as in the extensive loss of animal and plant life that may result from the logging of forest areas. Hence, it is critically important to assess the responsiveness of different species to the heavy destruction of their habitats; to be able to pinpoint their chances of survival by studying, for example, differences among them in their behaviour and feeding ecology, which makes them more or less at risk. Andrew Johns' study in Amazonia is valuable in this context (see Chapter 6). Moreover, it is important because much of our information on the influence of logging comes from forested areas in Africa and Asia. The vast and relatively untouched forests of Amazonia are still largely unknown to us, but data are urgently needed towards the preservation of indigenous species.

Simon Bearder (Chapter 11) considers issues of conservation and wildlife management. His concern is not only to indicate the scope of the problems involved, but to discuss in wider comparative perspective the core of these concerns; namely, an appreciation of the ways in which different taxa respond to the environmental changes that we inflict upon them.

A central thrust of conservation aims is, of course, to re-establish particular species in nature. Environmental changes, as with logging, may require the relocation of animals from one area to another. Such situations would seem to be ideal in the sense that individuals with the appropriate skills to live in nature continue to do so without interruption and stress of captivity. Relocation exercises are, however, complex and have to be undertaken with much care (see Strum and Southwick, 1986).

Under a variety of circumstances, however, the candidates for re-establishment in nature emanate from captive environments, even if they were not actually born there. The rehabilitation of species such as chimpanzees that have been used in medical research is a realistic task as far as many people are concerned. Alison Hannah and Bill McGrew (Chapter 9) provide a sound and careful description of what is involved. There are few good empirical studies in this area, and this one is an obvious exception.

By contrast with our concerns for the declining numbers of so many primate species, there are, of course, sources of environmental change in nature which lead to very different kinds of problems. Such cases include major changes in diet, which are administered by humans either inadvertently or intentionally. Responses to change which result from the exploitation by some species of primates of human food, for example, has resulted in a major pest problem in many parts of the world. In order to cope with such problems, at least a two-pronged approach is required. One of these involves an assessment of human attitudes towards, and strategies used for dealing with, pest problems. The other approach requires the delineation and study of the behavioural propensities which facilitate certain primates becoming pests. Ashmore Declue (1988), for example, compared behavioural characteristics among different species of macaques. In the present context, it is of interest that one group of species – namely, rhesus, bonnet and cynomolgus monkeys – readily change their activity patterns and their ranging behaviour to include food from agricultural holdings on a regular basis, despite the risk from attack. In fact, these species readily live in areas which are newly formed, and easily change items of their natural diet for 'novel' ones. Jim Else (Chapter 8) considers both approaches to the problems of primates as pests in his studies of baboons and vervet monkeys in East Africa.

Cases in which supplemented food is intentionally given to primates at regular times in specific locations are referred to as provisioning. The purposes for doing this include attracting animals as tourist resources, and to bring animals into regular contact with their scientific observers. Once again, however, and as with studies of destructive changes in the environment, an appreciation of changes in behaviour and demography are required to understand the influence of the increase in food resources, which may also vary considerably from the natural diet. John Fa (Chapter 7) presents a detailed account of one such case for Barbary macaques (*Macaca sylvanus*) in Gibraltar. This is of special interest because records go back for many years, and comparisons have been made with Barbary macaques living naturally.

By contrast with provisioning, it is the aim of some primatologists to observe their species in conditions which are as undisturbed as possible. The process of habituating animals to become indifferent to the presence of human observers is frequently carried out by fieldworkers, but is rarely considered as such. Caroline Tutin and Michel Fernandez (Chapter 10) present an interesting case in which they do consider habituation as a source of environmental change *per se*, and find, for example, that it reflects the social organizational predispositions of their two species, the western lowland gorilla (*G. g. gorilla*) and the common chimpanzee (*Pan troglodytes*) in Gabon, West Africa.

The contributions in this part of the book, then, all address sources of environmental change which are frequently encountered. They were chosen because they represent some of the most critical with which we have to deal.

6 Forest disturbance and Amazonian primates

ANDREW D. JOHNS

6.1 INTRODUCTION

The economic development of the world's rainforests continues unabated, and is reflected in the increasing importance attached to studies of the ability of primates to survive in disturbed habitats. Most of the geographical range of some primates, such as those of Brazil's Atlantic forests, is already degraded; the same will become true for many more species as we approach the next century. Fortunately, primates are resilient animals; recent summaries of available data suggest that many species may persist in light to moderately degraded forest (Marsh *et al.*, 1987; Johns and Skorupa, in press).

To date, most information concerning the ecology of primates in degraded forest comes from studies in South-East Asia and in Africa. Despite the size of Amazonia and the richness of its fauna, its primates remain poorly studied. Only in recent years have detailed studies of certain of its primates been undertaken (Ayres, 1981; van Roosmalen, 1981; Rylands, 1982; Soini, 1982; Terborgh, 1983; Ayres, 1986). Only three synecological studies have so far been conducted (Mittermeier and van Roosmalen, 1981; Terborgh, 1983; Soini, 1986).

Studies of species-rich primate communities have explained sympatry through a variety of mechanisms, notably feeding specializations and different microhabitat preferences. In West Africa, individual species within mixed cercopithecine groups may diverge only at times of critical food shortages, the emphasis being on separate feeding niches (Gautier-Hion, 1980). Divergent feeding ecology and also divergent use of substrata is evident among South-East Asian species (MacKinnon and MacKinnon, 1980a). In western Amazonia, Terborgh

(1983) considers divergence through microhabitat specialization more evident: six out of eight small-bodied species showed distinct divergence in this way. (Metabolic demands faced by large species are too great to enable specialization in patchy habitats: individual patches would be too small and too much energy would be expended in moving between them.) The importance of habitat variability in maintaining a full community of Amazonian primates might be expected to render the community susceptible to change as a result of habitat degradation through human activities.

The ecology and patterns of occurrence of Neotropical primates in disturbed habitats are poorly documented, although it is evident that some species show preferences for mature or for early successional habitats, and that broad differences may be correlated with various ecological parameters (Johns and Skorupa, in press). This chapter considers features of the ecology and behaviour of species within a single primate community in western Amazonia that account for differences in their observed distribution within a mosaic of undisturbed and disturbed habitats.

6.2 THE STUDY AREA

Data were collected in the north-west corner of the Rio Tefé Forest Reserve in western Brazilian Amazonia (3°32′S 64°58′W) (Figure 6.1). This area has been lightly populated for around 40 years, but the population is limited to the banks of the main rivers and tributaries. One timber-logging operation took place in the vicinity of the settlement of Ponta da Castanha in 1974–5; only 4–5 km^2 forest were logged before the company went bankrupt. The designated role of the area is, however, the sustained production of timber.

The study site was located in the vicinity of Ponta da Castanha and contained unlogged forest and a variety of disturbed habitats (Figure 6.2; see also Figure 6.8).

Primary forest

The unlogged forest on terra firma was of tall, emergent trees reaching 40 m in height, located on undulating relief. Characteristic trees of the upper canopy were *Cedrelinga cateniformis* (Meliaceae), *Parkia nitida* (Leguminosae) and *Eschweilera* spp. (Lecythidaceae); *Oenocarpus bacaba* (Palmae) and *Virola surinamensis* and *Irianthera* spp. (Myristicaceae) were common trees of the middle and lower canopy levels.

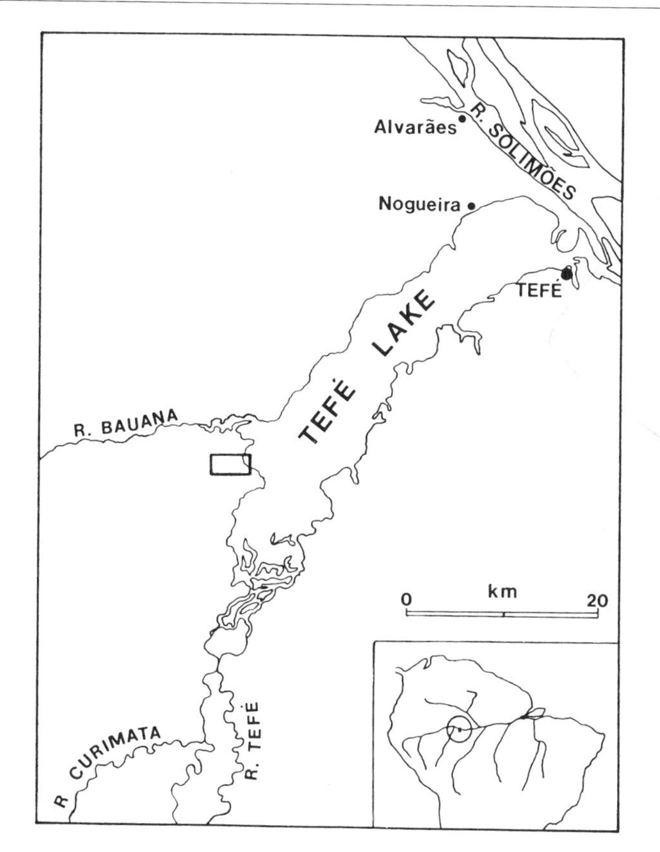

Figure 6.1 Location of the Rio Tefé Forest Reserve study site in western Brazilian Amazonia. The reserve is bordered in the north by the Bauana River and in the east by Tefé Lake and the Tefé River. The boxed area is the study site at Ponta da Castanha, shown in detail in Figure 6.2.

Logged forest

Forest logged in 1974–5, 11 years prior to the study, had been colonized by fast-growing softwoods such as *Inga* spp. (Leguminosae) and *Cecropia* spp. (Cecropiaceae). These had reached a height of 10–15 m, in places forming a dense lower canopy beneath surviving upper canopy and emergent trees.

Because of the bankruptcy of the logging company, the logged forest was divided into two sections: areas where the trees had been cut and dragged out, and smaller areas where trees had been cut but never removed. The former had regenerated to a basal area of 14.9 m^2/ha, the latter had an area of 35 m^2/ha, approximating that of unlogged forest. The logging operation itself (i.e. cutting of three to five trees

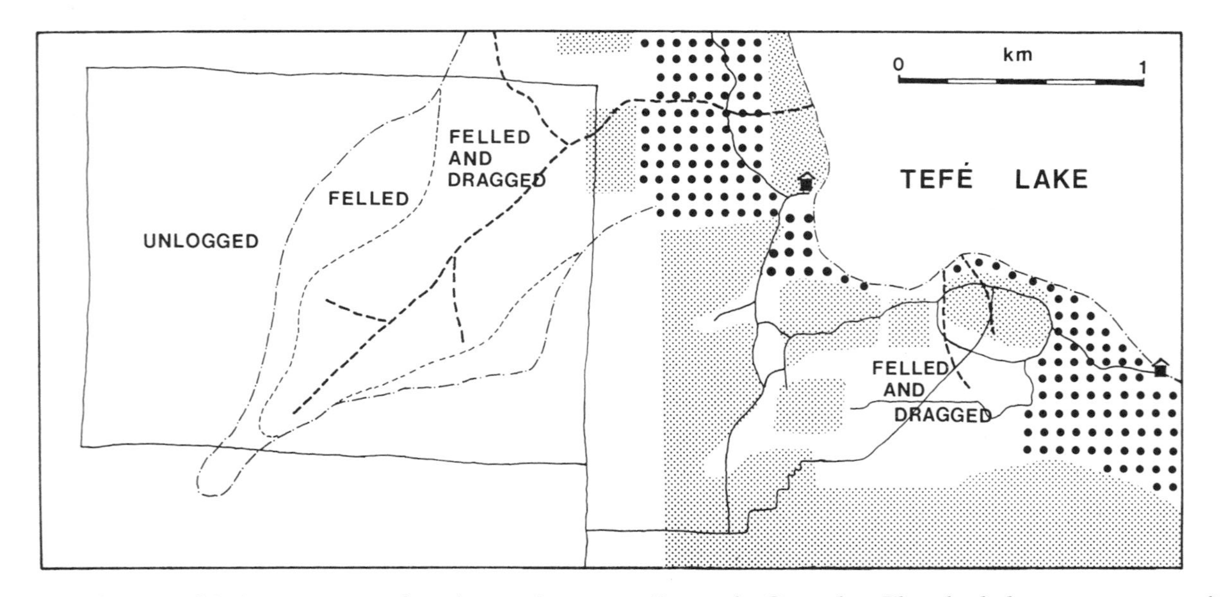

Figure 6.2 Distribution of habitat types within the study area at Ponta da Castanha. The shaded area represents fields and *capoeira*; the stippled area represents plantations of the Brazil-nut tree; the unshaded inland area is forest. The principal trails are shown: the boxed area depicts peripheral trails of a grid 200 × 200 m. The heavy broken lines are principal logging roads.

per hectare) appeared not to exceed the natural rate of loss through tree falls. Damage associated with timber extraction is caused mainly by the building of logging roads and movement of machines through the forest (see Johns, in press). Details of damage caused by logging are given elsewhere (Johns, 1986a).

Cultivated mosaic

Three families, each of around 12 adults and children, cultivated land within the study site. The main crop, manioc (*Manihot* spp.), was grown in a shifting agricultural system. During the year of study, some 7 ha of land were felled and burnt for new cropfields (4 or 5 ha is a more usual annual rate). Regenerating scrub (*capoeira*) growing on old fields is usually re-cut and burnt. Further inroads into logged and especially unlogged forest are rare because of the extra work involved (the rural population is very low in the region and the pressure on the land is slight). Fields grow manioc for 2 or 3 years before being abandoned, and are generally left fallow for more than 10 years. A large proportion of the cultivated area is thus a mixture of regenerating *capoeira* and degraded forest corridors located on swampy ground or along streams.

Brazil-nut trees (*Bertholletia excelsa*) grow wild around the Tefé River, but have also been cultivated for 40 years at Ponta da Castanha. The study site overlaps two plantation areas, within which the natural vegetation is largely reduced to corridors along watercourses and isolated, much-disturbed patches. There is little natural understorey beneath the Brazil-nut trees because of annual cutting and burning (fallen fruits are easier to find if the understorey is removed).

Forest island

Only one large island of natural forest remains in the cultivated zone, covering an area of around 35 ha. Most was logged during the 1974–5 operation.

6.3 STUDY METHODS

Throughout the study period (January–December 1985) field observations were made for around 20 days each month. Time was spent partly in conducting surveys along the 36 km of trail cut through the study area (plus existing trails made by local people), and partly in following primate groups.

Census results were derived from observations of animals made during line transect surveys. Analysis of survey data was primarily by Fourier series estimation (see Burnham *et al.*, 1980). Where insufficient data were available, the simpler King's method was used (see National Research Council, 1981).

Data on primate ecology and behaviour were derived from 10-min scan-sampling of primate associations when the animals were followed. During each scan, data on troop size and composition, activity, vertical level and vegetation type were noted. If animals were observed feeding, the identity of the food plant (if known) and the item eaten were also recorded.

The phenology of fruiting, fl¬wering and new leaf production were investigated in unlogged and logged forests. The trees of > 30 cm girth present in 25 × 25-m quadrats were examined every calender month and the presence of these items recorded. Four quadrats were located in unlogged forest, two in felled forest and two in felled/dragged forest.

It should be noted that the collection of large amounts of information on individual primates was to a large extent compromised by the number of species present, and by the need to conduct extensive surveys at the study site and elsewhere (Johns, 1986a). Information presented in this chapter should be considered as preliminary.

6.4 THE PRIMATE COMMUNITY

Thirteen taxa of primates, comprising 12 species, were sympatric at Ponta da Castanha. These primates are listed, with summarized ecological information, in Table 6.1 (see page 122). Their divergence along basic ecological parameters is illustrated (Figure 6.3).

Several points need to be made briefly about this community. Observations of moustached tamarins (*Saguinus m. mystax)* on the west bank of the Tefé River extend its known geographical range east of the Juruá River (see Hershkovitz, 1977). The presence of Humboldts' woolly monkey (*Lagothrix lagotricha poeppigii*) extends its known range similarly (see Fooden, 1963). The co-occurrence of two forms of *L. lagotricha,* which equate to the 'subspecies' *cana* and *poeppigii,* suggests a need for further investigation to clarify their taxonomic affinities. The distinctive sexually dichromatic squirrel monkey (*Saimiri*) occurring on the west bank of the Tefé River is of unknown provenance at present.

6.5 USE OF DIFFERENT HABITATS

The densities of the various primates in the different habitats were calculated (Tables 6.2 and 6.3). Felled and felled/dragged forests were

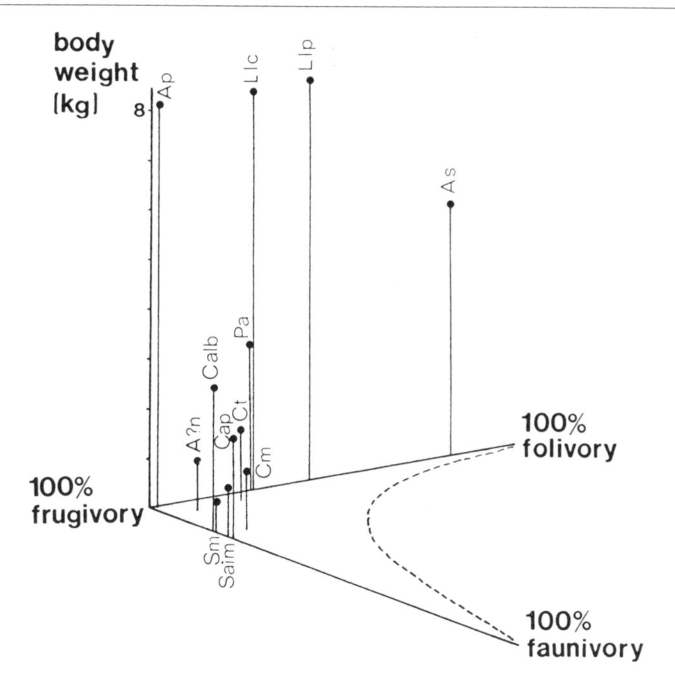

Figure 6.3 Ecological separation of sympatric primates at Ponta da Castanha. Diet is calculated as the number of occasions the primate was observed feeding on a particular food type: this probably underestimates the degree of faunivory (70% of foraging in smaller species is directed towards the capture of animal prey). Abbreviations: Ap, *Ateles paniscus;* A?n, *Aotus ?nigriceps;* As, *Alouatta seniculus;* Calb, *Cebus albifrons;* Cap, *C. apella;* Cm, *Callicebus moloch;* Ct, *C. torquatus;* Llc, *Lagothrix lagotricha cana;* Llp, *L. lagotricha poeppigii;* Pa, *Pithecia albicans;* Saim, *Saimiri;* Sm, *Saguinus mystax.* (No data are available for *Saguinus fuscicollis.*)

combined to give estimates for logged forest; although felled forest is floristically more similar to unlogged forest, abandonment of some felled trees is quite common in typical logging operations. Fields, *capoeira,* Brazil-nut plantations and associated degraded forest strips were combined to give estimates for the cultivated mosaic, although these vegetation types are clearly dissimilar.

Differences in choice of habitat were clear. The large frugivore spider (*Ateles*) and woolly (*Lagothrix*) monkeys avoided cultivated areas and were observed only around the fringes of logged forest (Figure 6.4). The two titi monkeys (*Callicebus moloch* and *C. torquatus*) appeared to occupy different habitat types, with minimal overlap between them (Figure 6.5). To some extent, patterns of distribution were modified by a past history of casual hunting in the region. However, the most

Table 6.1 Primate species occurring at Ponta da Castanha, Brazilian Amazonia, with notes on ecological characteristics

	Saguinus mystax mystax	*Saguinus fuscicollis avilapiresi*	*Aotus ?nigriceps*	*Callicebus moloch cupreus*	*Callicebus torquatus torquatus*	*Saimiri*
Adult body weight (kg)	0.6	0.6	1.0	1.2	1.5	1.0
Feeding strategy						
Trophic group	Frugivore/ insectivore	Frugivore/ insectivore	Frugivore	Frugivore	Frugivore	Frugivore/ insectivore
Speciality	Opportunistic	?	Soft fruits	Soft fruits, secondary vegetation	Soft fruits	Diverse diet
Mean group size	12.9	2.5	3.5	2.4	2.8	53
*n**	80	2	4	35	31	2
Range	1–40	2–3	3–4	1–6	2–10	45–60
Social structure	Fission-fusion	?	Monogamous pairs	Monogamous pairs	Monogamous pairs	Multimale groups
Intergroup relations	Multilayered society	?	Probably territorial	Territorial	Territorial	Mutual avoidance

* Number of occasions on which group counts were considered accurate.

Table 6.1 cont.

	Cebus apella apella	*Cebus albifrons unicolor*	*Pithecia albicans*	*Alouatta seniculus*	*Ateles paniscus chamek*	*Lagothrix lagotricha* subspp.
Adult body weight (kg)	2.0	2.6	3.0	5.0	8.1	8.0
Feeding strategy						
Trophic group	Omnivore	Omnivore	Frugivore	Frugivore/ folivore	Frugivore	Frugivore
Speciality	Diverse diet	Diverse diet	Mesocarps	Opportunistic	Widely dispersed large fruits	Widely dispsersed large fruits
Mean group size	13	25	4.6	3.0	13	26
*n**	41	18	66	17	11	13
Range	1–35	10–35	1–16	1–5	1–30	1–60
Social structure	Multimale groups	Multimale groups	Fission-fusion	Unimale groups	Fission-fusion	Fission-fusion
Intergroup relations	Mutual avoidance	Mutual advoidance	Multilayered society	Mutual avoidance?	Multilayered society	Multilayered society

* Number of occasions on which group counts were considered accurate.

Table 6.2 Estimated density of primates in unlogged and logged forest at Ponta da Castanha (logged forest combines felled and felled/dragged areas)

Species	Primary				Logged			
	Mean group size	*n**	*Encounters/ 10 km walked*	*Density of individuals/ km²*	*Mean group size*	*n**	*Encounters/ 10 km walked*	*Density individuals/ km²*
Saguinus m. mystax	13.7	53	2.2	78	11.0	15	2.5	88
Callicebus moloch cupreus[†]	2.5	2	0.05	0.3[‡]	3.0	2	0.2	1.2[‡]
C. t. torquatus	2.5	15	0.3	2.5	3.1	9	0.5	3.7
Saimiri	60	1	0.7	32	60	1	1.4	81
Cebus a. apella	11.5	25	1.1	11.5	15.3	10	1.9	32
C. albifrons unicolor	24	13	0.4	14	31	4	0.5	31
Pithecia albicans	5.2	39	0.9	9.0	3.9	18	1.2	18
Alouatta seniculus[§]	2.9	10	0.3	0.9	4.0	2	0.05	0.2[‡]
Ateles paniscus chamek[§]	13	8	0.1	1.3	11.3	3	0.2	1.1
Lagothrix lagotricha cana[§]	26	13	0.2	5.0	?	0	0.05	?
L. l. peoppigii[§]			0.1	1.0			0.05	?
Total distance surveyed (km)			554.3				170.1	
Aotus ?nigriceps	3.3	4	0.8	8.5[‡]	3.0	2	2.5	22[‡]
Total distance surveyed (km)			12.8				4.6	

*Number of accurate group counts.
[†]Edge habitat only.
[‡]Insufficient data for Fourier analysis: estimate given is the result of King's method.
[§]Numbers probably affected by hunting in the past.

Table 6.3 Estimated density of primates in cultivated areas and in a 35-ha forest island at Ponta da Castanha

	Cultivated mosaic				Forest island			
Species	*Mean group size*	*n**	*Encounters/ 10 km walked*	*Density of individuals/ km²*	*Mean group size*	*n**	*Encounters/ 10 km walked*	*Density individuals/ km²*
Saguinus m. mystax	11.5	11	0.9	23	14.0	1	0.4	16
S. fuscicollis avilapiresi	2.5	2	0.02	0.08			0	0
Callicebus moloch cupreus	2.5	27	0.6	5.7	1.5	6	0.4	5.6†
C. t. torquatus	3.2	6	0.1	0.3	4.0	1	0.1	1.8†
Saimiri	40	1	0.6	49	40	1	1.4	42
Cebus a. apella‡	16.0	3	0.2	5.6	14.0	3	1.8	61
C. albifrons unicolor	20	2	0.1	5.2†	20	2	0.1	4.8†
Pithecia albicans	2.5	6	0.1	0.9†	3.7	3	0.3	2.4†
Alouatta seniculus‡	4.0	1	0.02	0.2†	2.5	4	0.5	3.8†
Total distance surveyed (km)			560.1				96.9	
Aotus ?nigriceps	3.0	1	0.4	4.6†			0	0
Total distance surveyed (km)			21.6				2.4	

*Number of accurate group counts.
†Insufficient data for Fourier analysis: estimate given is the result of King's method.
‡Numbers affected by occasional hunting.

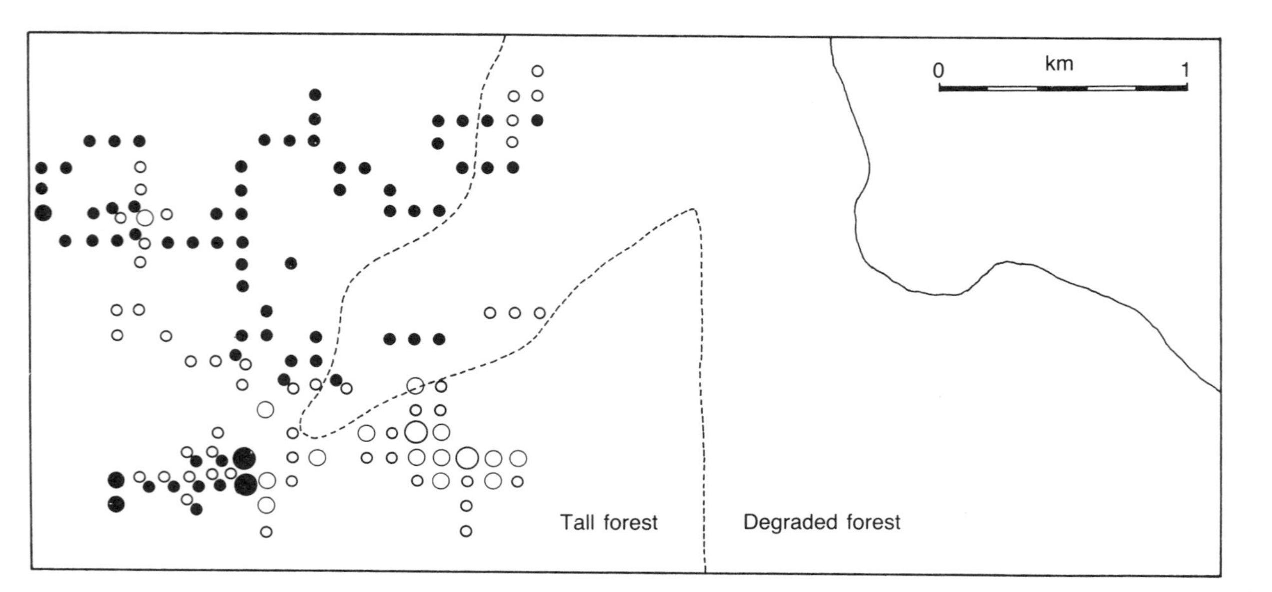

Figure 6.4 Distribution of *Lagothrix* (○) and *Ateles* (●) within the study area at Ponta da Castanha. The numbers of observations in each hectare quadrat are graded: small circle, < 5; medium circle, 5–15; large circle, > 15. The dashed line represents the edge of a tall forest (unlogged and felled but not dragged).

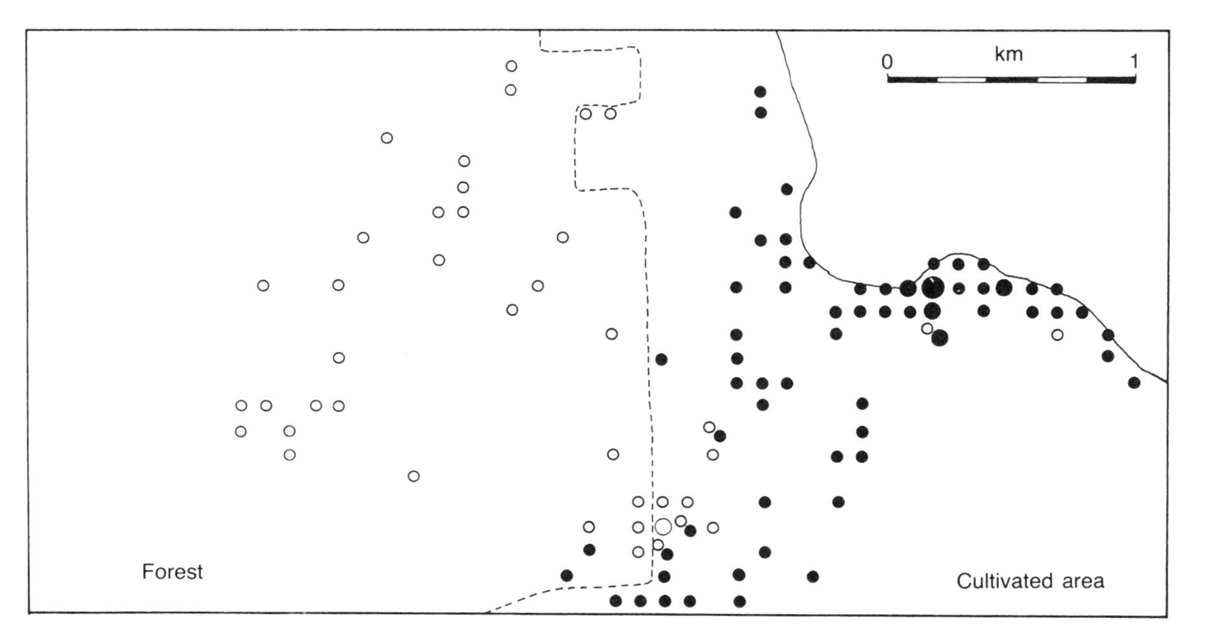

Figure 6.5 Distribution of *Callicebus torquatus* (○) and *C. moloch* (●) within the study area at Ponta da Castanha. The numbers of observations in each hectare quadrat are graded: small circle, < 5; medium circle, 5–15; large circle, > 15. The dashed line represents the edge of *capoeira* or plantations of Brazil-nut trees.

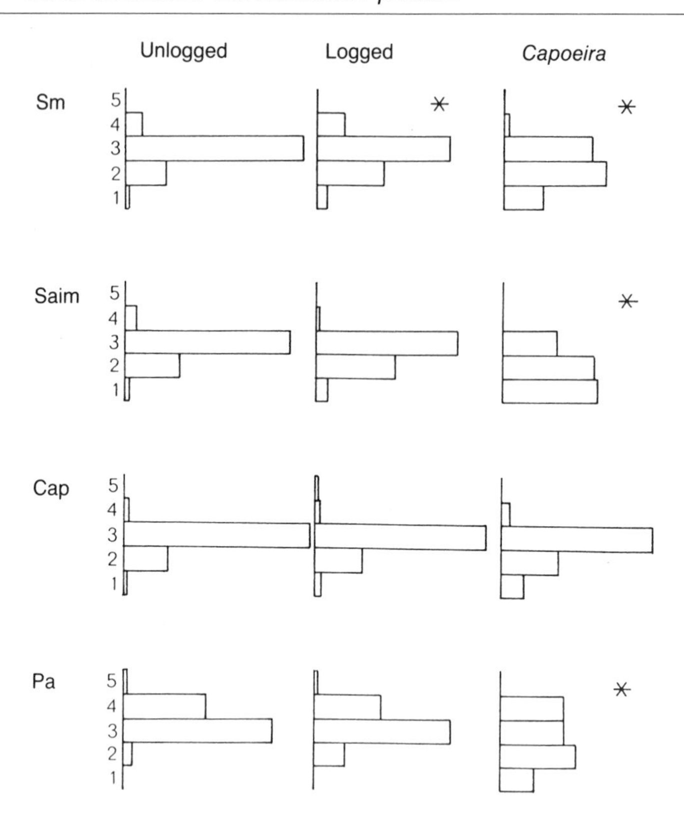

Figure 6.6 Vertical distribution of activity among four common primates in primary and disturbed habitats (abbreviations as in Figure 6.3). Observed distributions in disturbed forest marked with an asterisk are significantly different from those in unlogged forest (Kolmogorov–Smirnov tests, two-tailed, $P < 0.05$). Vertical scale as follows: 1, ground; 2, lower canopy; 3, middle canopy; 4, upper canopy; 5, emergent trees.

commonly shot species were woolly monkeys (*Lagothrix*), howler monkeys (*Alouatta*) and brown capuchins (*Cebus apella*). Woolly monkeys and capuchins are still present, even close to human habitations. During recent years, the abundance of the large rodents *Agouti paca* and *Dasyprocta aguti* in the Brazil-nut plantations (they feed on fallen and rotting fruits) has meant that primate hunting rarely occurs. These rodents are much preferred as a source of meat.

Different habitat profiles require changes in the vertical distribution of activity among primates. For example, the common species – *Saguinus mystax*, *Saimiri*, *Cebus apella* and sakis (*Pithecia*) – all show shifts towards spending more time in lower canopy levels in logged forest and *capoeira* than in unlogged forest (Figure 6.6). This is

Table 6.4 Abundances of selected trees exploited by primates as food sources in unlogged and logged forests at Ponta da Castanha

		*Tree abundance per 0.25 ha**	
Family	*Species*	*Unlogged*	*Logged*
Apocynaceae	*Couma* sp.	1	1
Chrysobalanaceae	*Licania* sp.	1	3
Leguminosae	*Inga* sp.	7	18
Loganaceae	'Cipo fero' (liana)	3	0
Moraceae	*Ficus* spp.	3	1
	Pourouma sp.	1	3
	Brosimum sp.	1	3
Myristicaceae	*Virola surinamensis*	8	5
	Irianthera sp. A	15	4
	Irianthera sp. B	10	3
Palmae	*Oenocarpus bacaba*	19	22
Sapotaceae	*Pouteria* sp.	7	9

*Tree abundance refers to trees/lianas of ⩾ 30 cm girth. All listed species are exploited by primates as sources of whole fruit, mesocarps, arils or seeds.

consistent with the mean tree height, which decreased from 16 m in unlogged forest to 14 m in logged forest and 7 m in *capoeira* (that were calculated from samples of trees of > 20 cm girth along measured transects) and probably with vertical variation in substratum density. Small trees also lack the sturdy substrata required by large primates. This limits them to the upper canopy levels in forest and effectively excludes them from early-stage *capoeira*.

6.6 FOOD AVAILABILITY AND SELECTION

The availability of different foods differs considerably between habitat types, dependent largely upon their botanical composition. A representative sample of trees exploited as sources of fruit or seeds by various resident primates showed considerable differences in abundance between unlogged and logged forests (Table 6.4). Almost all fruit trees were species of closed forest. *Inga* spp. are the only colonizing trees regularly exploited by primates. Of the trees listed, only *Inga* spp. were common in the cultivated mosaic. Some *Oenocarpus bacaba* remained (their fruits are palatable and prized by local people) but the only other common fruit trees in agricultural areas were planted cashews (*Anacardium occidentale*) and passion-fruit (*Passiflora edulis*).

In addition to the lesser abundance of many fruit trees in logged

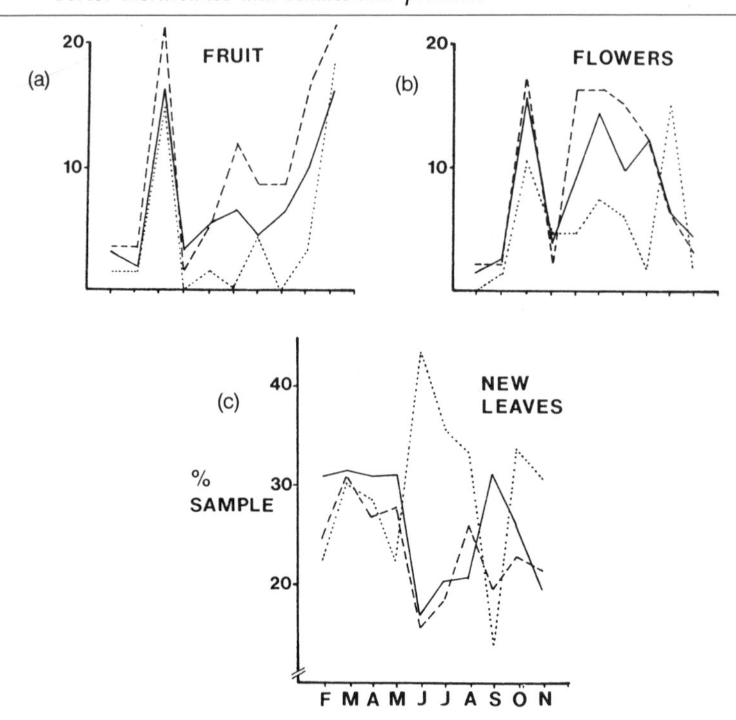

Figure 6.7 Fruiting (a), flowering (b) and new leaf production (c) in unlogged and logged forests: ———, unlogged forest (n = 194 decreasing to 188 as a result of natural treefalls); ------, felled forest (n = 94 decreasing to 93); ......, felled/dragged forest (n = 67 decreasing to 66). Comparing matched samples of 66 trees from each habitat type, there are no significant differences in vegetative activity (Kruskal–Wallis ANOVA: H = 3.52, d.f. = 2, $P > 0.05$) but felled/dragged forest showed significantly less fruiting and flowering than both unlogged and felled forest (Mann–Whitney U-tests: $P < 0.05$ in all cases).

forest, the level of reproductive activity among logged forest trees is less (Figure 6.7). This may indicate that some colonizing trees are not yet mature, although this is not the case for the main colonizers *Inga* spp. and *Cecropia* spp.). It may also indicate that their periodicity of fruiting is longer than is typical of primary forest trees. The greater vegetative activity, compared to reproductive activity that is shown by felled/dragged forest, is probably due to the openness of the canopy (Figure 6.8) and to the higher level of insolation (Johns, in press).

The variation in abundance of different food types is likely to be reflected in the foraging behaviour of primates in the different vegetation types. Insufficient data are available for rigorous comparison. As an example, however, the proportion of foraging time spent at fruit sources by the three common frugivore-insectivores – tamarins

(a)

(b)

Figure 6.8 Logged (a) and unlogged (b) forest in study area at Ponta da Castanha.

(*Saguinus mystax*), squirrel monkeys (*Saimiri*) and capuchins (*Cebus apella*) combined – declined from 20% in unlogged forest to 17% in logged forest and 11% in the cultivated mosaic. The number of observations for each species was 238, 189 and 199, respectively. This probably reflects the reduced availability of fruit (see Figure 6.7), but may be partly due to increases in the availability of certain types of foliage insects.

6.7 DISCUSSION

In the species-rich Amazonian primate communities, different monkeys have resolved potential resource competition in different ways. There is dietary specialization by, for example, pygmy marmosets (*Cebuella pygmaea*), which feed largely on sap (Soini, 1982). There is selection of particular microhabitats (e.g. Goeldi's monkey, *Callimico goeldii*, which occurs primarily in bamboo or scrub forest: Heltne *et al.*, 1981), and differences in foraging behaviour (e.g. sympatric tamarins *Saguinus fuscicollis* and *S. imperator*, which differ in locomotor behaviour and show almost complete spatial separation during dispersed feeding activities: Terborgh, 1983). From a detailed study of a Peruvian community, Terborgh (1983) suggests that each species has a distinct mode of spatial utilization, and that each has particular reactions to conspecific troops which control their ranging patterns. The primate community at Terborgh's site differs in species composition to that at Ponta da Castanha, but also shows major differences in organization. Terborgh (p. 227) gives a compelling argument as to why *Saguinus* should not occur in troops of more than about five animals (see Chapter 11, section 11.4).

Terborgh (1983) considered that as few as 1% of the trees at the Peruvian site sustain a major part of the primate biomass through parts of the year, and that the high seasonal fluctuations in resource abundance was the primary factor limiting density of most primates. At Ponta da Castanha, insufficient botanical information is available to draw substantial conclusions. However, seasonal fluctuations in reproductive activity are also considerable at this site. Common trees such as *Virola surinamensis*, *Irianthera* spp., *Inga* spp. and *Oenocarpus bacaba* have long and overlapping fruiting seasons, but there is a time of year when fruits, particularly fleshy fruits, are in short supply. In heavily disturbed habitats, only small colonizing plants and lianas may be producing fruit or flowers at this time. The primates that are common in the cultivated mosaic are, of necessity, those able to exploit these small and dispersed sources, and also able to subsist largely upon either arthropods or young leaves where necessary.

All the species present in unlogged forest were also observed in logged areas. However, those species making heavy use of the logged forest tended to be those with small body sizes and eclectic dietary habits (notably *Saguinus mystax, Saimiri* and *Cebus apella*). The continued use of logged areas by small-bodied frugivores such as sakis (*Pithecia*) and the night monkey (*Aotus*) may indicate either an ability to utilize fruit of colonizing trees, or the survival of sufficient primary forest trees to induce ranging into the logged section. The low density of howlers (*Alouatta*) throughout the area is puzzling. In Peruvian and Bolivian Amazonia howlers may reach a biomass of 540 kg/km^2 (Freese *et al.*, 1982), compared to 243 kg/km^2 calculated for the whole primate community in unlogged forest at Ponta da Castanha (Johns, 1986a). At an undisturbed forest site 60 km south of the study area, the density of *Alouatta* individuals was also low – 3.3 per square kilometre in comparison to 8.8 *Ateles* individuals and 17.0 *Lagothrix* individuals (Johns, 1986a). Species of *Alouatta* generally survive well in disturbed forest due to an ability to feed opportunistically upon a variety of fruit and leaf material (Johns and Skorupa, in press). They were the only large primates in the cultivated mosaic. Their overall rarity in the Tefé region, however, remains unexplained.

Both unlogged and logged forests at Ponta da Castanha contain very high densities of certain fruit resource trees. Most fruit trees in logged areas were located in the middle and lower canopy levels, however, and many were early successional species. As such, the average size of fruit sources, and perhaps the average fruit size (Foster and Janson, 1985), were smaller. These factors may well limit the use of logged forest by the large sociable frugivores *Ateles* and *Lagothrix*, although these primates did occasionally visit suitable large canopy trees that were not destroyed by the logging process. Since *Lagothrix* is able to feed on leaves, it should be less susceptible to disturbance than *Ateles*. That a higher density of the latter was recorded for logged forest is anomalous. It should be mentioned, however, that *Ateles* persists in certain types of degraded forest in Colombia whereas *Lagothrix* does not (Bernstein *et al.*, 1976). That *Lagothrix* only rarely enters disturbed forest areas at Ponta da Castanha may be partially due to its being a preferred game species, whereas hunting of *Ateles* is largely prevented by local customs.

That all species other than *Lagothrix* and *Ateles* occur in the cultivated mosaic (albeit at low biomass) is unusual. To some extent it may reflect an interchange with the populations in continuous forest. This is certainly the case with the yellow-handed titi monkey *Callicebus torquatus*, which was not thought to occupy territories outside of the unlogged/logged forest, although animals were occasionally observed elsewhere.

The most successful primates in agricultural mosaic areas were, predictably, the same as those that made extensive use of logged forest. *Saguinus mystax, Saimiri, Cebus apella* and *Cebus albifrons* are all common in areas containing a diversity of habitat types. The first two are able to forage through all canopy levels, from the shrub layer of early *capoeira* to the tops of trees in degraded forest corridors, and were notably common. By contrast, sakis were rarely observed outside the corridors of degraded forest and the forest island. Although sakis feed on young leaves and insects to some extent, they rely primarily upon fruits and seeds of forest trees (Johns, 1986b). The night monkey, the other frugivorous species in the cultivated mosaic, has a patchy distribution and was most often seen in the vicinity of planted soft fruit trees (especially *Passiflora edulis*).

The dusky titi (*Callicebus moloch*) presents an interesting case, occurring only in heavily disturbed habitats and never within tall forest, at least at Ponta da Castanha. The species feeds primarily on small fruits, although some studies have recorded significant quantities of young leaves, particularly from lianas (Kinzey and Gentry, 1979; Terborgh, 1983). In heavily disturbed habitats there is a profusion of small fruiting shrubs, often linked with liana mats. Other small primates, such as *Saguinus* spp., may also feed from the fruit sources, but divergence in body size, and thus differential use of substrata, prevents competition with the partially folivorous *Alouatta*. *Callicebus moloch* does not appear to occupy territories overlapping those of its main potential competitor, *C. torquatus*, although there may be some overlap in fringe areas. This is consistent with results from elsewhere (Kinzey, 1981).

A species that occupies a variety of habitats often differs in ecological or behavioural parameters between them (e.g. Kavanagh, 1980). An obvious adaptation is that of group size. The lesser abundance of large food sources in early successional habitats is often reflected in more dispersed foraging among groups (e.g. Johns, 1986c), or in smaller association sizes among species occupying fission-fusion societies. For example, *Saguinus mystax* occurred in associations of as many as 40 individuals in unlogged forest (unusual for a callitrichid, although not unique; see Izawa, 1976) but association size did not exceed 20 in the cultivated zone and was generally much less. Similarly, *Pithecia* formed associations of up to 14 adult animals in tall forest, but never more than 5 in the cultivated mosaic. An ability to adjust social patterns in this way is, of course, indicative of a degree of flexibility to environmental conditions. It may also be a response to hunting pressure (i.e. crypsis: Kavanagh, 1980) but this is unlikely to apply to the aforementioned species since they are not hunted or persecuted in any way.

In addition to flexibility of foraging patterns, species at Ponta da Castanha generally show flexibility of diet (there are actually few primates that do not: Johns and Skorupa, in press). Although several species are specialized to certain microhabitat conditions, specializations are towards exploitation of early successional patches (e.g. *S. mystax*, *S. fuscicollis* and *C. moloch*) and thus enable coexistence with human activities. Only species which obtain most of their food from tall, upper-canopy trees are stressed in disturbed areas, partly because of the lesser abundance of suitable food trees and partly because of the necessity of crossing low regenerating vegetation in order to reach them. These animals are large-bodied and face physical difficulties in microhabitats composed largely of small substrata. As a caveat to this, it should be noted that large animals also tend to be favoured targets of hunters (Johns and Skorupa, in press).

6.8 CONCLUDING REMARKS

Finally, it should be reiterated that the use of heavily disturbed habitats at Ponta da Castanha is facilitated by the proximity of tall forest. The number of primate species that would persist in the area would certainly drop if access to large areas of forest was prevented. At present, the survival of the forest island in the centre of the cultivated zone, the commonness of corridors of degraded forest along watercourses, the predominance of Brazil-nut plantations with patches of natural understorey vegetation and the rarity of primate hunting all contribute to the observed resilience of the primate community. Should any of these factors change, the primate community could rapidly become degraded.

ACKNOWLEDGEMENTS

Work at Ponta da Castanha was conducted as part of World Wildlife Fund US project US–302. It was sponsored in Brazil by the Brazilian National Council for Scientific and Technological Development (CNPq) and by Museu Paraense Emílio Goeldi, Belém. I thank these organizations for their support. I am most grateful for assistance in the field afforded by Victor, João and Joscelinho Azeredo of Ponta da Castanha, and by P. and J. Dunn, J.S. Lourenço, F.C. Novaes, G. Mason, A.P. Nunes, F.B. Pontual and C. Scoggins. I especially thank J.M. Ayres, D.M. de Lima and R.A. Mittermeier. M. Kavanagh and J.P. Skorupa kindly offered comments on drafts of this paper.

7

Provisioning of Barbary macaques on the Rock of Gibraltar

JOHN E. FA

7.1 INTRODUCTION

The energy necessary for all life processes comes directly from foods which are composed of varying quantities of proteins, carbohydrates and fats, all with different energy contents – 4 kcal/g, 4 kcal/g and 9 kcal/g respectively. Moreover, since the caloric needs of a species are constant at any point in time (i.e. there is a standard level required for life processes for each stage of an animal's life), it is the amount of food energy available that determines the condition of an animal in a habitat. In all circumstances, the acquisition of nutrients must satisfy an animal's daily and seasonal metabolic demands, but it must also be regulated by total intake of food relative to energy balances (Robbins, 1983). Changes in the physiology, and consequently in the demography of the animal population, are likely to appear, depending upon whether the amount of food is scarce or plentiful.

Most scientists concur that it is the food supply which primarily controls animal populations (Krebs, 1978), and hypotheses about the mechanisms of population regulation have centred on whether a major role is taken by the animals' social behaviour (first proposed by Wynne-Edwards, 1962), or whether it is the fluctuating resources which affect populations in a strictly density-dependent fashion (Slobodkin, 1961). Distribution of food supply among component animals of a population may be important in relating food resources and population size as alternatives to the more deterministic interpretation of food affecting members of a population proportionately. For example, territoriality and/or social hierarchies can limit population growth as a result of differential survival and/or reproduction caused by food shortage affecting certain individuals more seriously (usually lower-ranking ones) than others. On the other hand, food supply can diminish for all members of a population equally.

Although it is difficult to prove that an equilibrium density of animals exists relative to food supply in the wild, the repercussions of scarcity or abundance of food on a population can be demonstrated. In primates, food shortage has led to sharp declines in vervet monkeys (*Cercopithecus aethiops*) in Amboseli, Kenya (Struhsaker, 1973, 1976; see Chapter 2, section 2.5), and in toque macaques (*Macaca sinica*) in Polonnaruwa, Sri Lanka (Dittus, 1975, 1977, 1979). The identified proximate effects of food scarcity are an increase in juvenile and infant mortality (Dittus, 1975, 1977, 1979; J. Altmann, 1980) and a decrease in female reproductive success with lower conception rates, fewer pregnancy terminations and longer interbirth intervals (Sadleir, 1969a,b; Gaulin and Konner, 1977; J. Altmann *et al.*, 1977). Evidently, any slump in a population's resources has important consequences for the animals' reproductive potential, which is, in turn, expressed in the demography of the group.

There are few reported situations of food abundance in wild primate habitats. Nevertheless, there is evidence that where the monkeys' natural diet has been supplemented by food put out by people this has significant effects on the population. Such artificial feeding or provisioning has been an important tool in detailed studies of the behaviour of various primates (chimpanzees: van Lawick-Goodall, 1968; Kortlandt, 1962; Wrangham, 1974). However, provisioning also has effects on the frequency of behaviours, though not necessarily on the types of behaviours expressed, and provokes long-term changes in population size.

7.2 PROVISIONING

Provisioning is the offer of food by people beyond the natural supply and/or quality of the target animal's environment. The term is used to encompass situations where animals are given items not found in their habitat (or home range) in order to study them at close range, or for other reasons (religious or touristic). Usually, fruits and vegetables are supplied, but cooked foods may also be given.

Food affects consumers through its nutritional components and in the way it is available. The quality of the food supply is judged by the basic ingredients in each food type. However, the dispersion of food, in clumps or spread out, is also important since it determines how far animals need to move to meet their energy requirements. Thus, quality and spatial distribution of food supply actually determine the nature of the food resource. There is no doubt that there are obvious differences in the food supply of provisioned troops and wild ones. In order to understand better the element of separation that might exist between the two types of population, these differences need to be measured. Informed comparisons between the behaviour or ecology of free-ranging and provisioned groups of primates can be made only if the very basis of the assumed difference – the quality and distribution of food – is properly assessed. Unfortunately, when making such comparisons, few studies have considered the significance of ecological parameters (of which food supply is the most important) on the expression of social or other behaviours. Various authors have focused instead on the effect of captivity, which has been very loosely defined as confinement conditions, on behaviour profiles and patterns (Bernstein, 1967, 1970; Drickamer, 1973; Baulu and Redmond, 1980; see also Chapter 19). Only Southwick (1967) and Wrangham (1974) have directly attempted to study the impact of provisioning on free-ranging rhesus monkeys (*Macaca mulatta*) and common chimpanzees (*Pan troglodytes*), respectively.

The food supply of provisioned primates has three common characteristics. The first is that food has a static distribution in space since it no longer shifts within the boundaries of the troop as it moves over its range (Rasmussen and Rasmussen, 1979). Secondly, provisioned food is often found in greater amounts than wild foods. For example, Wrangham (1974) reports that provisioned chimpanzees at Gombe, Tanzania, ate 570 bananas daily: a significant contribution to their natural reserves. A third characteristic is that provisioned foods are nutritionally different from wild ones. This is reflected primarily in the calorific contents of the items and in the higher digestibility of provisioned foods. There is a tendency for provisioned foods to be fruits or seeds (see Malik and Southwick, 1988, for Indian rhesus; and Iwamoto, 1988, for Japanese macaques *Macaca fuscata*.) Nutritional analyses performed by Iwamoto (1988) for the Japanese monkey troop on Koshima Island show that provisioned foods were either richer in protein and lipids (wheat, rice, peanuts and soybean) or more digestible (apple and sweet potato) than wild items (Figure 7.1). Although the nutritional gradient between provisioned and wild foods depends on the food type offered and on the wild foods available, provisioned items are still a more accessible source of energy.

7.3 GIBRALTAR BARBARY MACAQUES: 72 YEARS OF PROVISIONING

Barbary macaques (*Macaca sylvanus*) have probably been in Gibraltar

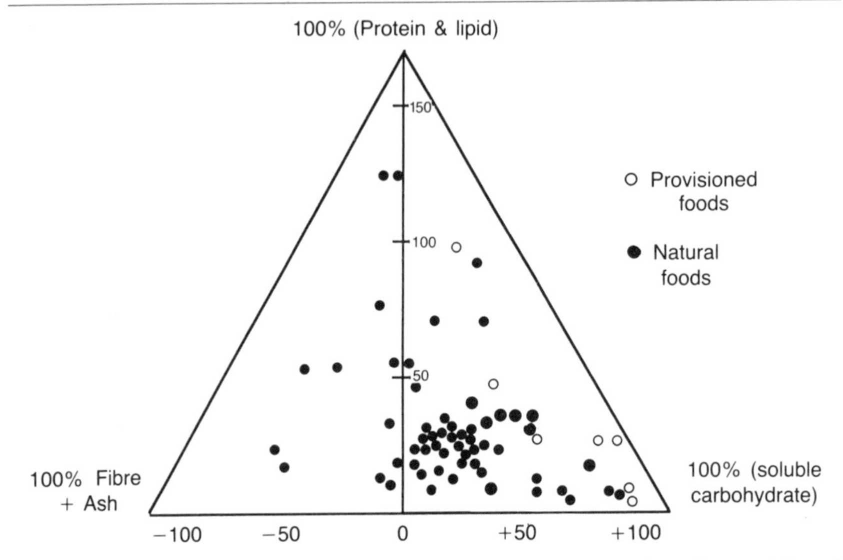

Figure 7.1 Ternary plot of nutritional contents of natural and provisioned foods of Japanese macaques at Koshima Islet, Japan; note that all provisioned foods appear near the top right axis of the graph. (After Iwamoto, 1988.)

since AD 711 although there are no records of their presence until after 1740 (Fa, 1981). Fa (1981) suggests that the Gibraltar macaque population was probably first introduced by the British Garrison between 1740 and 1749 as game sport for resident soldiers. Since 1740, there have been three introductions of monkeys into the population to compensate for excessive culling in 1813, 1860 and 1939–43. With less than 3 km^2 of natural habitat for up to 190 monkeys, which reportedly roamed freely in 1900, Gibraltar was probably too restricted an area with insufficient food and water (Sclater, 1900). The consequent raids and depredations of the monkeys into the town area below led to provisioning in an attempt to confine the animals to the Rock's natural habitats. The upper reaches of the Rock are covered with dense Mediterranean scrub vegetation, secondarily developed from the destruction of a carob (*Ceratonia siliqua*)–wild olive (*Olea europaea*) woodland during the late 1700s. From 1915, the British Army, and later the Gibraltar Regiment, became officially responsible for feeding and culling the macaque population.

The management rationale for the monkeys has changed from merely feeding them at a given spot to restrict the species to the Upper Rock area (the upper reaches of the Rock), to maintaining them as an important tourist attraction (Kenyon, 1938; Fa, 1981) (Figure 7.2). Within these aims it has also been army policy to ensure a minimum

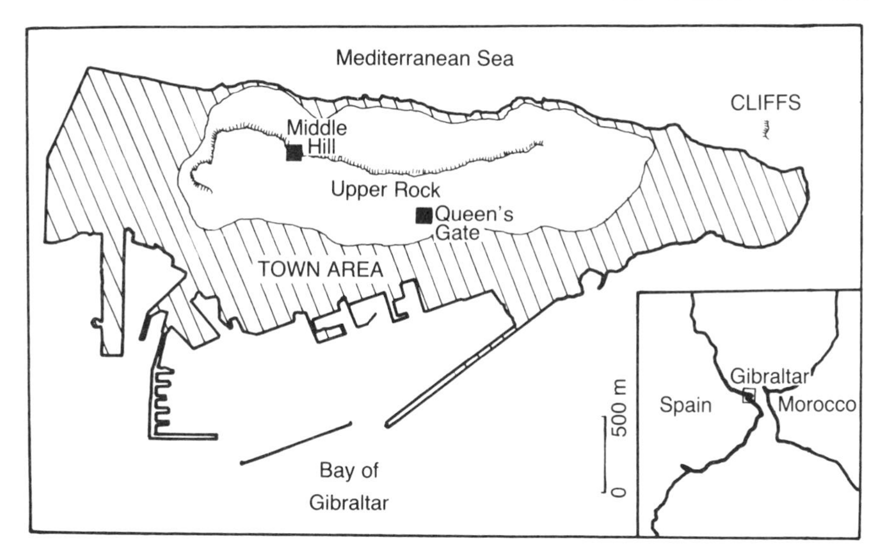

Figure 7.2 Map of Gibraltar showing location of study areas.

number of animals each year. This minimum level, arbitrarily set by the management authorities, has increased from 25 up to 1954 and to 34 thereafter. This is because regular feeding at specific sites has made the animals easier to manage at higher numbers. Since 1946 (Fa, 1984), the Gibraltar monkey population has been composed of two separate groups, one at Queen's Gate and the other at Middle Hill.

7.4 PROVISIONING LEVELS

Levels of provisioning have varied between time periods. The annual expenditure records of the Gibraltar Government on food per monkey show (Figure 7.3), once corrected for inflation (Fa, 1984), that the animals have been fed different quantities during four basic time periods. Food types have changed little since the start of regular feeding after 1936. During a study between 1979 and 1980, Fa (1986) counted 25 different types of fresh fruit and vegetables from the local market given to the macaques at one or more sites within their home range. The natural diet of the animals consisted of 43 plant species, ranging from underground tubers through a variety of leaves to fruits of wild olive (*Olea europaea*), lentisc (*Pistacia lentiscus*) and buckthorn (*Rhamnus alaternus*), found within maquis and garrigue habitats of the Upper Rock. None of the provisioned items are found growing in Gibraltar. The radius within which food was put out did not vary more than 2 or 3 m and provisioning time was usually around 09.00,

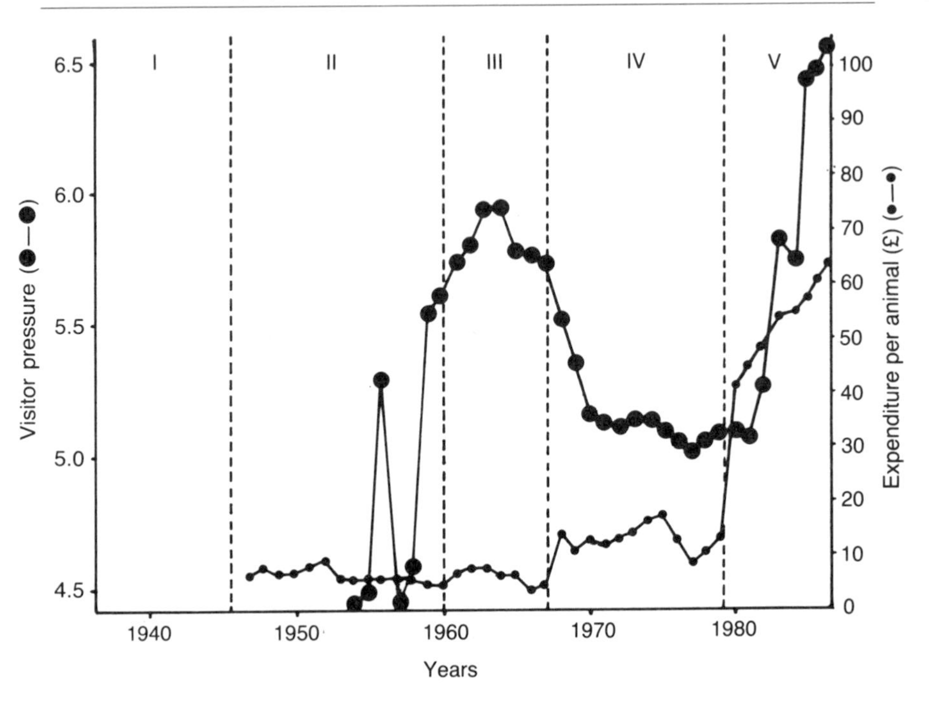

Description of provisioning stages

Stage	*Period*	*Description*
I	1936–46	Provisioned food volume low. Monkeys supplied with two-thirds of food requirements. World War II. No visitors allowed. Garrison cookhouse food included. Monkeys raided town frequently
II	1946–60	Two troops of monkeys formed after 1946. Provisioning at two sites, one for each troop. Scheduling and volume of food similar to stage I. No cookhouse food given
III	1960–7	Increase in provisioning level, due to rise in monkey numbers and food prices. After 1960, Gibraltar monkeys promoted as an important tourist attraction. Both monkey troops visited by large numbers of people. Supplemental feeding by people common
IV	1967–80	Land frontier with Spain closed, leading to substantial increase in food prices but provisioning volumes also raised to prevent monkeys going into town area. Provisioning site of Middle Hill troop moved to higher altitude to stop monkeys entering town. No tourists at Middle Hill after 1970. Queen's Gate still fed by people. Supplemental feeding probably lower due to a fall in tourist numbers
V	1980–7	Ascending levels of supplemental feeding due to increase in tourist numbers since normalization of relations with Spain (re-opening of land frontier) in 1984–1985. Provisioning volumes highest

Figure 7.3 Yearly changes in annual expenditure per Barbary macaque in Gibraltar, as a measure of provisioning level and numbers of tourists visiting Gibraltar each year, taken as an index of supplemental feeding and disturbance by people. (From Fa, 1984.)

irrespective of seasonal changes in sunrise times. Thus, provisioning times fluctuated from 2.5 hours to a few minutes after sunrise, when the animals first became active. Since the macaques located at Queen's Gate were usually fed later, a mean difference of 52 min in provisioning times appeared between the two monkey groups. Throughout the year, the Queen's Gate troop was given a daily mean of 18.96 ± 12.98 kg of food, while the Middle Hill troop received more (25.10 ± 9.50 kg). There was no difference in food volume offered monthly or seasonally. A monthly average of 494 kg for Queen's Gate and 764 kg for Middle Hill, with annual totals of 5929 and 9167 kg, respectively, were put out. Expressed as calories, this represented 12 201 kcal per animal at Queen's Gate (24 more than the calculated energy requirement for a wild Barbary macaque; see Drucker, 1984) and 144 672 kcal at Middle Hill (289 times more).

Apart from food supplied by the management authorities, the monkeys, especially those at Queen's Gate, received supplemental food from people visiting the animals. During 1979–1980, a low tourist year because of a closed frontier with Spain, an estimated total of 42 139 people visited Queen's Gate at a rate of 15.00 ± 6.00 visitors per hour (Fa, 1986). Figures for an open frontier situation in 1987 show that 120.09 ± 82.02 people per hour passed by Queen's Gate and interacted with the monkeys (Fa, 1988).

7.5 THE INFLUENCE OF PROVISIONING

The manner in which a species distributes its time among various activities is important for any characterization of its life style. A study of activity budgets (time spent on each activity per unit period such as a day, month or year), or activity patterns (the distribution of activity proportions across time) can lay the foundations for interrelating ecology and behaviour of a species. Time budgets are shaped by time-limited resources (e.g. food and water) and modified by activities that must occur at particular times of the day (e.g. the need to return to sleeping sites before dark: S. Altmann, 1974). Since diet can influence an animal's use of time by the proportion of structural carbohydrates (e.g. cellulose, hemicellulose and lignin) it contains (see Boyd and Goodyear, 1971), a diet based largely on soluble compounds will tend to release time otherwise used in processing food. All animals through efficient feeding can maximize time as a benefit and minimize it as a cost (see Wolf *et al.*, 1975). Optimal feeding strategies can be promoted by food choice or increased rate of food intake (Hainsworth and Wolf, 1980). Species that must spend a long searching or handling time per

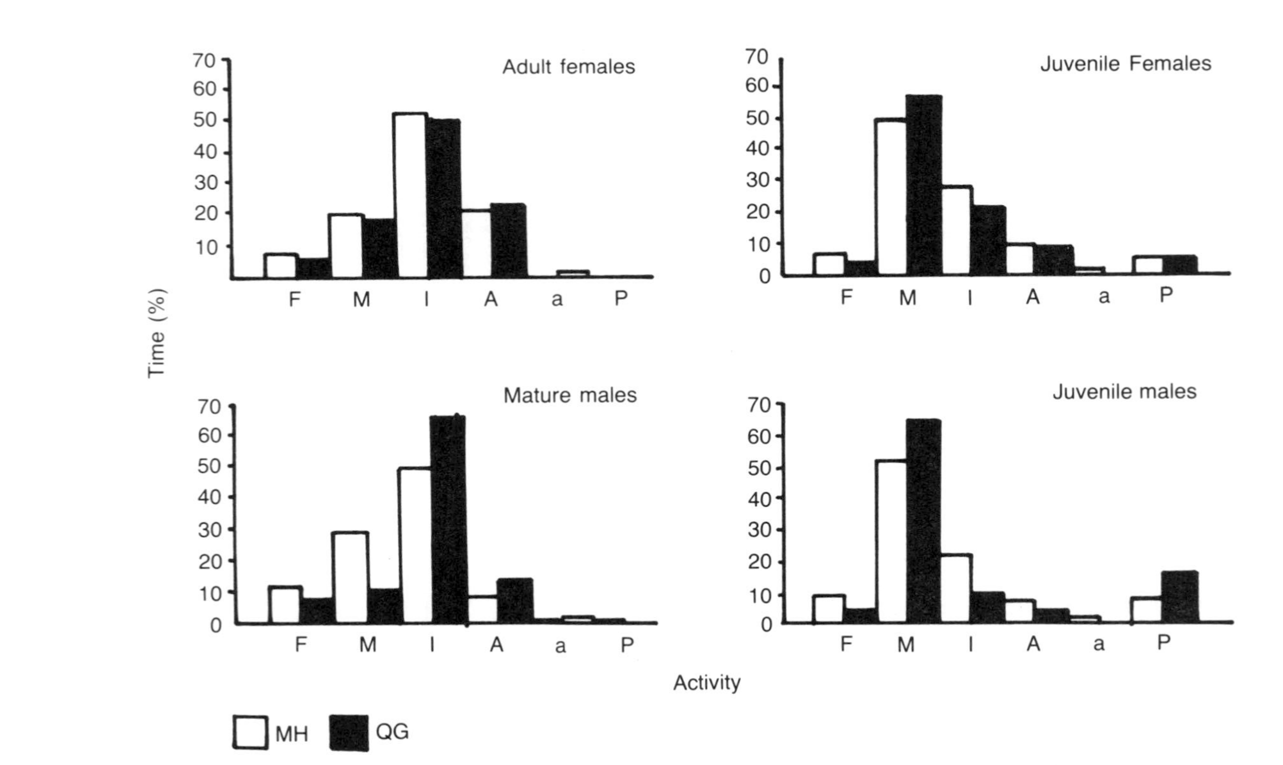

Figure 7.4 Annual activity budgets for different age–sex classes within Queen's Gate (■) and Middle Hill (□) troops of Barbary macaques in Gibraltar ($P < 0.001$). Abbreviations: F, feeding; M, moving; I, inactive; A, allogroom; a, autogroom; P, play. (From Fa, 1988.)

food calorie can gain time when the energy reward per unit time is inflated through provisioning.

A comparison of wild Barbary macaques with the Gibraltar monkeys shows that, while the former spend almost half of their daytime feeding (Deag, 1974 and personal communication), those in Gibraltar spent 7% and 4.8% feeding at Middle Hill and Queen's Gate, respectively (Fa, 1988) (Figure 7.4). Studies of provisioned rhesus macaques in Puerto Rico (Post and Baulu, 1978; Baulu and Redmond, 1980; Marriott, 1988) and of human-fed rhesus in India (Seth and Seth, 1986; Malik and Southwick, 1988), have similarly demonstrated that a decrease in feeding and moving time is associated with the manner and level of feeding. An explanation for this is that the overall caloric gain per unit time and weight is far greater for provisioned foods than for most natural foods. In Gibraltar, provisioned foods contain a mean of 134.08 kcal/g in comparison with 0.04 kcal/g of natural forage (Fa, 1986). In terms of weight per food item, provisioned items weighed an average of 101.92 g as opposed to 0.44 g for natural foods. Hence, it is expected that provisioning will have consequences on the time budgets of artificially fed animals. Provisioned animals are more likely to respond to foods yielding a higher calorific benefit than to wild foods; time budgets of provisioned troops will reflect a shorter 'satiation' time. Correspondingly, because of the static distribution of provisioned food, animals will cut down on foraging time, thus releasing time for other activities such as social interactions. However, activity profiles of the Gibraltar macaques showed no differences in allogrooming proportions between them and wild monkeys, but the provisioned monkeys spent more time resting.

It is also the case that, although the use of time and space by primate groups can be related to the dispersion and predictability of food, the attractiveness of certain foods can sometimes distort the expected relationship between quantity and distribution of the food resource and activity. In Gibraltar, the type and distribution of food can explain the overall activity patterns of the troops. But the offer of very attractive foods, namely sweets, by tourists at Queen's Gate, can explain the lower activity and more restricted use of space in that troop. After examining a number of alternative hypotheses, Fa (1988) concludes that the lack of synchrony in the Queen's Gate activities and the reduced home range (1 ha for Queen's Gate in contrast to 21 ha for Middle Hill), can be explained by the constant presence of people who offer additional food. A further study of this point during a high tourist year (Fa, 1989) confirms that the disruptions to the animals' daily patterns caused by tourists is significant enough to produce alterations of the monkeys' activity schedules. Second, we may consider food as a common cause of aggression.

The acquisition of food and energy by primates is influenced not only by physiological variables but also by social factors. It is known that there is differential access to food resources among wild primates and this has often been used as a measure of social hierarchy in studied groups (Chalmers, 1968; Kawai, 1958). Because food is more dispersed in the wild, intense competition is restricted to situations where food is scarce (see Dittus, 1980). In contrast, provisioned primates concentrate on discrete food patches where competing conspecifics impose the greatest constraints on feeding (Singh, 1969; Southwick *et al.*, 1976). Consequently, social hierarchies within provisioned troops have an important impact on the rates and overall energy intake of the group. In such conditions: (1) feeding bout lengths are influenced by social features of the forager's environment rather than by diet (Lucas, 1983); (2) high-ranking animals can monopolize the resource and dissuade others from the kleptophagy typical of wild primates (Southwick, 1967); and (3) the relative advantage which a high-ranking animal enjoys over a subordinate will appear as a maximization of energy intake.

At Middle Hill, it was found that males spent more time feeding overall than females; a difference largely produced because adult males fed more than any of the other animals (Figure 7.5a). When time spent feeding by the entire troop was correlated with agonistic dominance ranks (results of a cardinal ranking method developed by Boyd and Silk, 1983), only a significant correlation for juveniles appeared (Figure 7.5a).

Estimates of the mean duration of the feeding bout and of ingestion rates (food weight intake in g/min) at Middle Hill showed that adults fed faster than juveniles, but feeding bouts were similar (Figure 7.5b). There was no relationship between social rank and feeding bout duration, or ingestion rates. However, negative correlations showed that feeding bouts and ingestion rates generally decreased with lower ranks (Figure 7.5b). Although aggression increased during feeding, this did not affect feeding bout lengths. Overall, higher-ranking animals received fewer interruptions during feeding than lower-ranking ones, but interruption rates were low for the troop (0–5.67 interruptions per hour per monkey).

The length of time spent by individuals at the feeding point varied between 2 and 14 min (Figure 7.5c). The entire troop spent more than 70% of daytime hours within a radius of 100 m of the feeding point: under 0.1% of the monkeys' home range. Over 80% of nearly 47 000 15-min point samples of positions fell in quadrats containing provisioned food. Duration of stay at the feeding site was related to dominance rank in males but not in females (Figure 7.5c).

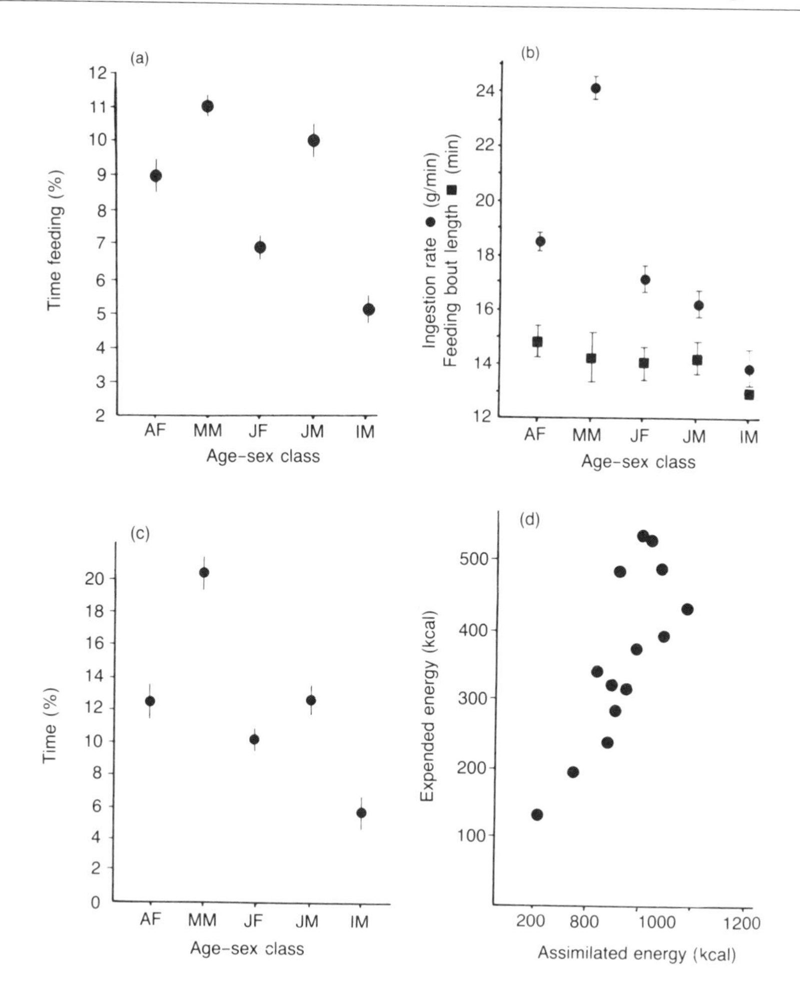

Figure 7.5 Measures of feeding behaviour in the Middle Hill troop of Gibraltar Barbary macaques. Abbreviations: AF, adult females; MM, mature males (includes subadult males only); JF, juvenile females; JM, juvenile males; IM, infant males.

(a) Overall time spent feeding (as mean percentage of an hour) by different age–sex classes. There is a significant negative correlation between dominance rank and time spent feeding by individual monkeys ($r = -0.5$, $P < 0.001$). (b) Lengths of feeding bouts (min) and ingestion rates (g/min) per age–sex class. No significant negative correlation between dominance rank and these variables appeared. There is, however, a clear negative correlation between social rank and interruptions in feeding bouts ($r = 0.92$, $P < 0.001$). (c) Mean percentage of an hour spent feeding at the provisioning site. A significant positive correlation between dominance rank and time spent at the feeding bout appeared ($r = 0.86$, $P < 0.001$). (d) Relationship between mean daily assimilated energy (kcal) and mean daily energy expended by all members of the Middle Hill troop. The general tendency is for animals to consume more energy than they actually expend. An excess ranging from 9 to 38% in consumed over expended energy was recorded by Fa (1986). Higher-ranking monkeys fed more than lower-ranking ones (for all seasons, $r = -0.67$ to -0.80, $P < 0.001$).

Dry weight of both provisioned and natural foods consumed by individual monkeys varied considerably (Figure 7.5d). Older males ate around 1 kg of food per day, adult females 0.5 kg and juveniles between 200 and 700 g. In terms of calories, daily intake ranged from 1000 to 1500 kcal for adult males and females (double the amount reported for wild Barbary macaques), 800 kcal for juveniles and 200 kcal for infants. Dominance rank and mean daily calorific intake were negatively correlated (Figure 7.5d).

Feeding patterns for Middle Hill strongly suggest that priority of access to the feeding site and maximization of energy intake is affected by social rank. Higher-ranking animals stayed longer at the feeding point and generally fed earlier in the day. When lower-ranking animals were permitted to enter the site their bouts were not significantly interrupted by others, but they did not stay as long since most of the food had disappeared by then. Both feeding behaviour and food-energy acquisition at Middle Hill were thus primarily affected by competition with other monkeys, largely as a consequence of the manner in which food was distributed in a discrete patch.

7.6 PROVISIONING, SUPPLEMENTAL FOOD AND DEMOGRAPHY

In populations of rhesus and Japanese macaques, provisioning has stimulated growth rates by elevating natality and/or promoting infant survival. Birth rates for provisioned Japanese macaque groups range from 59% to 73% (Koyama *et al.*, 1975; Sugiyama and Ohsawa, 1982) and have reached 91% in rhesus monkeys (Southwick *et al.*, 1980). Withdrawal of provisioning has depressed birth rates by 35% at Koshima (Mori, 1979) and by 27% at Ryozen (Sugiyama and Ohsawa, 1982).

The two monkey troops in Gibraltar have had different population changes. Modal size for both troops has been 15–20 but mean size for Middle Hill (17 animals) is slightly larger than for Queen's Gate (15 animals). According to data obtained by Fa (1984) from the Gibraltar Regiment records for the period 1936–87, both troops have increased about three times from their starting size (Figure 7.6). The Queen's Gate troop has grown from an initial 5 monkeys to a maximum of 25 in 1953 and fallen to another minimum of 9 in 1969. In contrast, the Middle Hill troop has shown no significant drop in numbers, maintaining itself between 12 and 22 monkeys from 5 founder animals (this troop was formed by a splinter group from Queen's Gate in 1946). Net growth rates for Queen's Gate were positive during 1940–50 and 1970–80 but suffered a deep decline between 1950 and 1970. Middle Hill

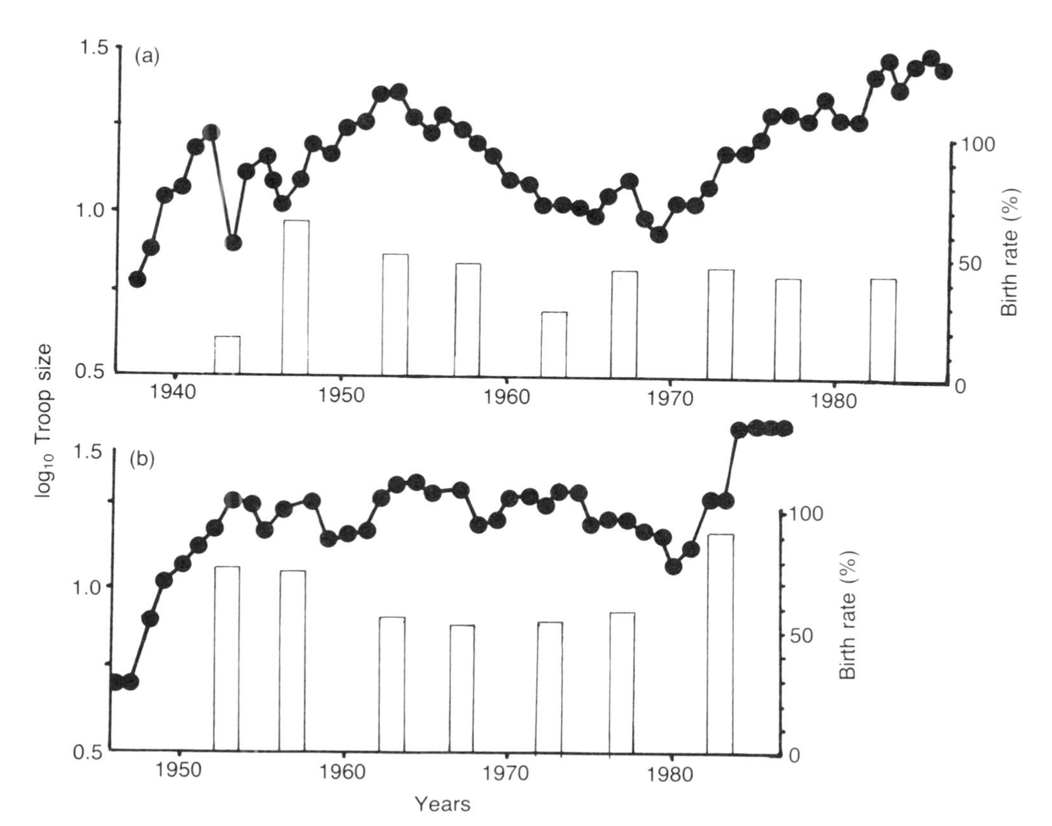

Figure 7.6 Yearly changes in population size and half-decade birth rates for Queen's Gate and Middle Hill troops of Barbary macaques in Gibraltar. (After Fa, 1984.)

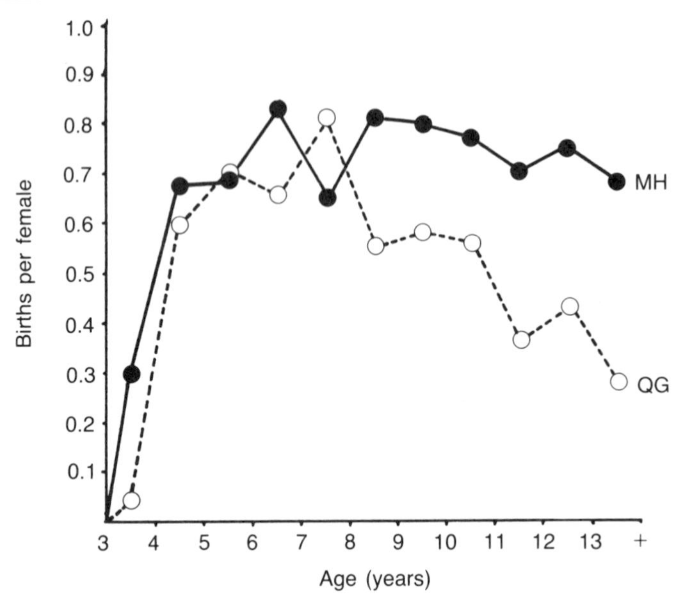

Figure 7.7 Fertility curves for Queen's Gate (QG) and Middle Hill (MH) female Barbary macaques in Gibraltar. (After Fa, 1984.)

grew during 1940–60 but dropped in numbers during 1960–70, after which there was a recovery.

Birth rates in both troops mirrored changes in growth rates; natality averaging 49% for Queen's Gate and 63% for Middle Hill. Statistically significant differences in the female fertility of the troops accounted for the overall differences in birth rate (Figure 7.7). Changes in half-decade birth rates for the troops were synchronous, even reflecting a similar decline for the 1960–70 period. Fluctuations in birth rates resulted from nulliparous females which failed to breed first time, and multiparous ones which remained sterile. On comparing lifetime reproductive success for the females during 1960–70, no difference was observed with other decades. Similarly, annual birth rates of both troops did not correlate with climatic variables (temperature and rainfall), provisioning levels, genetic relatedness or densities. In contrast, Fa (1984) found a significant negative correlation between the number of people visiting the monkeys and demographic changes.

Visitor pressure, which translates into disruptions caused by people feeding monkeys, relates to the observed changes in breeding performance. Restricted access to food provokes aggression in provisioned monkeys. While attacks over common food items are rare, scarce and/or nutritionally attractive foods increases competition (see Wrangham, 1974). Thus, when access to food becomes difficult as a function of an increase in competitors, aggression increases. Because visitors offer the monkeys high-calorie and highly palatable foods (e.g. sweets) throughout the day, the animals have learnt to expect these titbits. Observations by Fa (1988) indicate that this interaction with tourists leads to important changes in the animals' behaviour. Fa (1984) proposed that the manner and level of supplemental feeding of the monkeys are responsible for the fall in breeding. Higher numbers of tourists visiting the monkeys adversely affect breeding performance by increasing tension in the troops. According to Fa (1984), stress in monkeys about to reach puberty could cause retardation of breeding because pubescent-aged females have to compete more for food in order to reach a critical body weight (see Frisch and MacArthur, 1974; Wolfe, 1979). Behavioural modifications instigated by the 'addiction' or conditioned preference for sweets and supplemental foods is thus likely to be responsible for breeding failures. Because the macaques in Gibraltar breed for a period of no more than 4 months, with mating concentrated in 1 month, and because the socionomic gender ratio of the troops (artificially controlled by the Gibraltar Regiment) has usually been 1–2 males for up to 10 females, any distraction from the strictly seasonal mating activities would depress these. In Gibraltar, adult males will abandon oestrous females to feed on items offered by people. Malik and Southwick (1988) have observed a similar phenomenon among rhesus monkeys in India.

7.7 WILD AND GIBRALTAR MACAQUES COMPARED

Although the Gibraltar macaques are managed, they do not differ substantially from wild troops in their troop size. The range of wild troops is between 14 and 48 (Deag, 1974; Fa, 1982; Drucker, 1984; Mehlman, 1984). Hence, the Gibraltar troops fall within the median range of the wild ones. However, in terms of composition, the Gibraltar groups have a median socionomic gender ratio of almost four adult females per male, whilst wild groups have a one-to-one relationship (Fa, 1986). This discrepancy is due to the culling of adult males in Gibraltar to reduce male–male competition, and to lower the incidence of peripheralizations during the mating season.

Behaviourally, the Gibraltar macaques show no difference from wild troops (Fa, 1986) in surface structure (Hinde, 1976). Animals of the same age–gender class remain spatial associates and, as in Deag's (1974) study on wild Barbary macaques, first neighbours are not drawn at random. A clear spatial separation between the genders appears. Time spent grooming is higher for females than for males and, as in the wild, males associate with babies and infants in triadic interactions

Figure 7.8 Adult male Barbary macaque interacting with mother and baby.

(see Deag and Crook, 1971; Deag, 1980) and engage in care-taking behaviours (e.g. Figure 7.8). As for wild monkeys, it is these associations between troop members and infants which primarily modify the group's social organization, although an increase in male aggression in the mating season influences relationships.

The main variable, clearly affected by provisioning, is the time animals need to devote to feeding. A comparison of the time wild (data from Deag, 1974) and Gibraltar Barbary macaques invest in feeding reveals that, while wild monkeys spend most daylight hours looking for food, the provisioned ones tend to level off feeding activities after midday (Figure 7.9). Differences brought about by the static distribution and quality of food do not induce atypical behaviours, but elicit a response relative to the predicability of the resource. This response is usually a reflection of the time released from feeding which the animals can dedicate to social behaviour or resting. In contrast, wild Barbary macaques have to sacrifice socializing time to feed. Interestingly, during Fa's (1986) study, the time Gibraltar monkeys spent socializing did not differ from wild monkeys. However, a negative correlation between feeding and resting appeared, but there was none between feeding and allogrooming. As feeding time decreases in the Gibraltar monkeys, this activity is replaced by resting behaviours, and not by social ones.

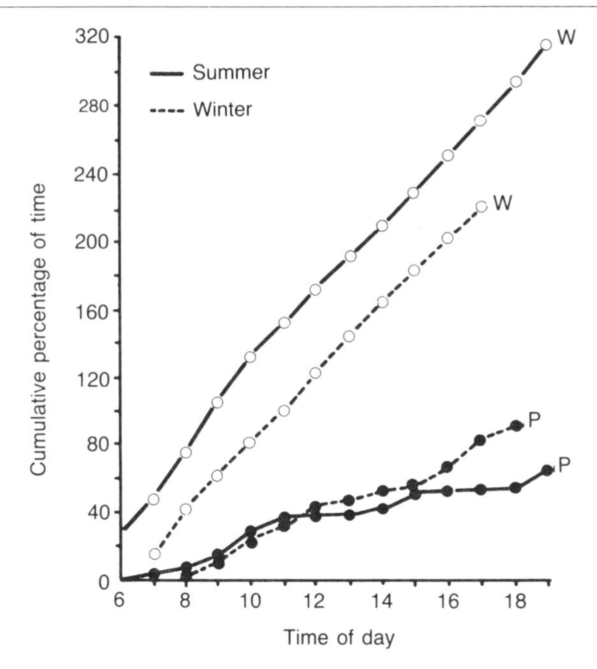

Figure 7.9 Differences in time spent feeding during daylight hours by wild (W) Barbary macaques at Ain Kahla, Morocco and provisioned (P) Gibraltar Barbary macaques. The abscissa shows a cumulative percentage of time dedicated per hour to foraging and ingesting food by each group during summer and winter. (Wild data from Deag, 1974; provisioned data from Fa, 1986.)

Seasonal variability in the Gibraltar macaques' activities shows a marked response to climatic variables because of the time provisioning saves on foraging. Fluctuations in activity cannot be linked to provisioning levels, since these are relatively constant, but they respond to temperatures; an element not seen in wild macaques (Deag, 1974). In fact, both Clutton-Brock (1977) for red colobus (*Colobus badius*) and Post (1978) for yellow baboons (*Papio cynocephalus*), indicate that there is no correlation between activity and climate. Instead, they associate activity profiles with food availability. Faced with the need to forage, wild monkeys will override the effects of endogenous and thermoregulatory processes.

7.8 CONCLUDING REMARKS

Provisioning in Gibraltar has resulted in the adoption of two strategies by the Rock's macaques. First, concentrate on the more abundant, higher calorie and probably more palatable supplementary food;

second, face intense competition for a resource which is highly clumped and to which access is dependent on social rank; or move away from the group and feed on natural foods. Hence, although an animal is able to maximize its energy gain from provisioning to an extent impossible in wild feeding, when energy returns decrease in relation to greater social stress it may no longer be profitable for an animal to stay around the provisioning site. It is, therefore, the dispersion and abundance of food at the provisioning site, the size and composition of the group, including genealogical relationships and even the length of time animals have been provisioned, that will dictate the type and level of response to be expected. Comparisons of aggression and food intake of the Gibraltar monkeys with other studied groups are difficult. Whether the relatively low aggression is attributable to the amount of food available, to group size or to familiarity with feeding on a clumped food resource, is an important point for further study. Knowing an animal's status within the social group provides a useful indicator of its chances of feeding, particularly where, as in Gibraltar, spatial characteristics of the food resource (dispersion) and feeding times (temporal predictability) vary little. Within this framework, low-ranking individuals or members of the younger and weaker age–gender classes may adopt a number of alternatives for minimizing interference from higher ranks. For provisioned troops it may be more important to establish behavioural strategies based on maintaining specific interindividual distances when feeding (Furuichi, 1983) than to choose different feeding sites or food items, as is typical of wild groups (see Post *et al.*, 1980).

Free-ranging monkeys with access to superabundant and high-calorie food and a relative freedom from infectious disease represent important material for studying the effect of resource dimensions on social behaviour, ecology and demography. The further study of primates in natural systems where food is not a limiting factor will, no doubt, provide illuminating comparisons of food-acquisition behaviour in times of shortage and of plenty (Morse, 1980).

8
Nonhuman primates as pests

JAMES G. ELSE

8.1 INTRODUCTION

Wildlife that have become agricultural pests present a widespread problem throughout the world. Traditional control strategies have been, and to a large extent still are, focused on attempts to eradicate the pest species. This has often been very effective in industrial countries, due to technologically sophisticated methods that can result in the actual extermination of a given species within a particular area. Such control is usually more difficult and often unsuccessful in developing countries for a variety of reasons, that include the greater number of potential pest species and the lack of appropriate technology.

There is increasing concern regarding the mass destruction of pest vertebrate species, many of which are in need of conservation. This has led to research into alternative control strategies. This is not a new concept; it is at least as old as the use of scarecrows in Western countries, and the bands of shouting children in many African societies. What has become apparent from such endeavours is that a detailed knowledge of the ecology and behaviour of the target species is of paramount importance to the successful control of animal pests.

Nonhuman primates are often serious pest species in tropical countries. This can be attributed to the intelligence, adaptability and opportunistic tendencies of many primates (Strum, 1986a). These traits also render passive methods of control, such as exclusion from agricultural land by physical barriers, ineffectual. The only techniques that have proved even marginally successful are chasing, shooting, trapping and poisoning the animals, but such measures are wasteful and of limited value; primates are fast learners.

Many serious ecological changes, in countries where primates are indigenous, have been brought about by an increase in the human population and the expansion of agricultural land. In Africa, much of this land was once occupied by wildlife, and those species that have been unable to adapt to the changing environment are being forced into increasingly marginal habitats. Many species are now threatened with extinction outside small, isolated, protected areas. Not all primate species have the ability to adapt to these environmental changes and those that are able to make the adjustment to a different habitat often become pests. Such primates do not restrict their destructiveness to agricultural crops; other activities include raiding homes, gardens and the occasional family chicken-coop in suburban areas (King and Lee, 1987), the killing of small or young livestock, and the raiding of guest rooms and dining halls in tourist lodges (Brennan *et al.*, 1985; Else and Eley, 1985).

The most successful pest species among Old World primates are from the genera *Macaca, Papio* and *Cercopithecus*. In these genera there are three notorious pests: the rhesus monkey (*Macaca mulatta*), the yellow baboon (*Papio cynocephalus*) and the vervet monkey (*Cercopithecus aethiops*). These primates have several features in common, which include a complex social organization; they are highly omnivorous and, while primarily terrestrial, are able to utilize arboreal habitats. The importance of the latter two characteristics for a primate to become a successful pest was pointed out by Kavanagh (1980), who compared the much greater success of vervets to that of the arboreal mona monkey (*Cercopithecus mona*) in adapting to agricultural development in a forest habitat in Cameroon.

The process by which primates become pests also depends upon their behavioural responsiveness in their ability to adapt to a changing ecological environment. This adaptive behaviour provides an excellent opportunity to study and monitor changes in primate behaviour in relation to changes in their physical environment. An understanding of this process provides practical insights into such diverse topics as the ontogenetic plasticity and cognitive capacities of a given species, the dynamics of behavioural change of early humans and, on the management side, new techniques to aid the control of primate pest problems.

8.2 A STUDY OF PRIMATE PEST PROBLEMS

This study is drawn from two projects which included surveys of human attitudes towards, and understanding of, primate pest problems and some measures that are taken to combat such problems. The surveys were conducted by means of questionnaires sent to, and/or interviews held with, people coming into regular contact with primate pests. The first (J. Else and T. Lewis, unpublished work) involved personal interviews with the heads of households (N = 73) in the Gedi Settlement Scheme, a small-plot agricultural area located near the coastal town of Malindi in Kenya. The second survey (Eley and Else, 1984) was conducted initially by mail questionnaires sent to tourist lodges and hotels throughout Kenya (N = 74), followed by personal visits to many of them. To maximize participation and responses, the information requested for each survey was kept very basic. Information requested included the following:

1. Presence of primate pests, species and numbers involved.
2. Type of problems and damage caused.
3. Timing and frequency of visits by primates and resultant raids.
4. Agriculture: details of crops grown.
5. Tourist lodges and hotels: details of eating areas and garbage disposal.
6. Remedial actions taken against primates and effectiveness of such actions.

The aim of both surveys was to determine the extent of primate pest problems, existing attitudes towards primate pests, and the success of measures taken to avert or control such situations. Both projects were supplemented by field observations, and the hotel survey was followed by visits to most of the facilities reporting primate problems, as well as to others known to have pest problems but which had not returned the questionnaire.

The survey of small-plot agricultural holdings in Gedi revealed that most farmers were experiencing serious problems with primates, of which the primary culprit was the yellow baboon (77%), followed by the Sykes monkey (*Cercopithecus mitis*) (20%) and, lastly, the vervet monkey (3%). However, few vervets were present in the study area. No attempt was made to census the number of primates there, or to monitor their home ranges. Group sizes were estimated, on average, as 20 for baboons and vervets, and 10 for Sykes monkeys.

The agricultural products grown by the farmers of Gedi were maize, cassava, beans, sesame seed, cashew nuts, coconuts, cotton, papaya, mangoes and pineapples. All three primate species fed on all of these crops but there were some differences as to the preferred stage of crop development. Raids generally took place daily, especially during July/August when the maize ripens. A total of 172 raids were described in the returned questionnaires, and there were up to three raids per day per farm, which involved at least one primate species. Baboons tended to raid farms from mangrove tidal lands and scrub/

Table 8.1 Methods used by small-hold farmers in Kenya to combat primate pests

Method	*% Using*
Yell/chase/throw objects	100
Dogs	38
Clearance of protective foliage	25
Bows and arrows	1
Snares	Widespread

rock formations, while Sykes and vervets raided from concentrations of tall trees and isolated pockets of forest. There was a relationship between decreasing raid frequency (both the number of days in which raids took place and the number of raids on any given day) and increased distance from refuge areas.

Several respondents reported that all three species raided at approximately the same time each day, but that the actual raids were never carried out by more than one species at the same time. During raids, all species ate rapidly, constantly on the watch for humans or dogs. When challenged and forced to retreat they did so carrying as much food with them as possible. Over 25% of the farmers reporting baboon problems stated that the animals broke into their homes, if they were left unguarded, and ate posho (processed maize meal) and other food items; 20% reported a loss of chickens to baboons; and 10% a loss of young goats. Only one farmer reported problems with Sykes entering his house, and none reported such incidents with vervets.

The methods used by farmers to combat primate pests are listed in Table 8.1. All households practised 'yell/chase/throw objects' to ward off primates but, at the same time, considered them to be highly ineffective. This was especially true when women and children tried to deter baboons, with the former rather than the latter often being the ones to retreat. The use of dogs was regarded as the most effective means of control, and dogs were occasionally shared among neighbours. Clearance of brush and trees was considered to be marginally beneficial. Snares were used as attempts to reduce primate populations in general, rather than to prevent actual raids.

The survey of tourist lodges and hotels showed a high level of primate pest problems of a different nature from those encountered in the agricultural areas (Figures 8.1–8.3). The main problems here were raiding of garbage, entering tourist rooms and central dining areas, and intimidating or attacking tourists. The primate species involved were almost exclusively vervets and baboons. Few lodges experienced serious problems 'at random' during the day from either species.

Figure 8.1 Yellow baboons at a garbage site near a tourist lodge in Kenya. (Photo: James Else.)

Figure 8.2 A baboon eating a bag of flour stolen from a subsistence farmer's hut. (Photo: Deborah Manzolillo.)

Figure 8.3 A vervet monkey raiding food from a table at a tourist lodge in Kenya. (Photo: James Else.)

Table 8.2 Tourist lodges and coastal hotels in Kenya reporting frequency and timing of visits by baboons and vervet monkeys (n = 33)

	Baboons (n = 18)*	*Vervets* (n = 17)*
Frequency		
Daily	6 (33%)	13 (76%)
Irregular	12 (67%)	4 (23%)
Timing		
Morning	5 (28%)	2 (12%)
Afternoon	6 (33%)	4 (23%)
Morning + afternoon	2 (11%)	1 (6%)
All day	2 (11%)	10 (50%)
Unknown	3 (17%)	0 —

*Number of lodges or hotels reporting presence of baboons or vervets.

Rather, problems occurred at specific times, such as during meals. Table 8.2 compares the frequency and timing of visits for these two species, and Table 8.3 lists the type of problem which resulted, together with its frequency.

Proper garbage disposal was a problem at most of the game lodges, because of the remote locations. Garbage was generally thrown into an open, or semi-covered, pit. Although burning was practised by some facilities, the nature of the garbage still prevented adequate disposal.

Table 8.3 Incidence of problems caused by baboon and vervet pests reported in game lodges and coastal hotels in Kenya

	Game lodges (n = 22)		*Coastal hotels (n = 14)*	
	Baboons	*Vervets*	*Baboons*	*Vervets*
Total no. reporting primate pests	19 (86%)	10 (45%)	2 (14%)	13 (93%)
Eating areas*	7 (37%)	6 (60%)	0 —	8 (62%)
Tourist rooms*	6 (32%)	7 (70%)	0 —	8 (62%)
Staff rooms*	7 (37%)	3 (30%)	0 —	6 (46%)
Garbage*	10 (53%)	8 (80%)	0 —	4 (31%)

*Percentages calculated from total numbers.

This was not the case at coastal hotels, where proper disposal techniques were utilized.

Most of the facilities with primate problems made some attempt to control the primates through a variety of methods. The most common were: (1) 'No Feeding' signs, (2) verbally warning tourists of the potential danger, and (3) using employees to chase away the animals. However, the effectiveness of these methods was limited by the fact that tourists like to see primates at close quarters and often encourage them by offering food. Other methods of control showed considerable ingenuity. For example, putting out a leopard skin to ward off vervets when outdoor lunches were being served. This proved to be most effective, although the frantic alarm calls emanating from the vervets in the tree tops did tend to spoil the serenity.

8.3 DISCUSSION

When the survey results were reviewed, different activity patterns emerged between both lodge/hotel and agricultural pests, and between pests at national park lodges and those at coastal hotels. Baboons and vervets in national parks and reserves become lodge pests because of their opportunistic nature. Although they may have an adequate but seasonal supply of natural foods, they are introduced, often initially through poor garbage disposal, to easily obtained, novel (human) foods of high palatability and nutritional quality. As the primates become habituated to humans, food may be given to them by tourists, and the monkeys will raid food supplies put out by lodge staff to attract other wild animals. Once they are on a consistent and enriched diet, primates increase in numbers to the point of possibly exceeding the food-carrying capacity of their natural surroundings

(Brennan *et al.*, 1985; Lee *et al.*, 1986), and thus become permanently dependent on supplemental food. This is exemplified at Amboseli National Park in Kenya, where unprovisioned vervet populations have shown a progressive decline in population density. Between 1964 and 1980 those numbers decreased from 104/km^2 to 68/km^2 (Cheney, 1987), whereas in 1983 provisioned lodge troops had a density of 569/km^2 (Brennan *et al.*, 1985).

In contrast, primates at coastal hotels have rich alternate food sources from the bordering, highly productive, agricultural land that is well within a baboon's wide-ranging travels. In addition, most coastal hotels do not have the garbage disposal problems found at the more isolated tourist lodges. Consequently, baboons around coastal hotels develop completely different behavioural patterns from those inland, and are not a tourist problem (Table 8.3). The coastal baboons rely largely on crop-raiding; they do not generally acquire a taste for hotel food by poor garbage disposal, and their intimidating size makes it unlikely that unhabituated animals will be introduced to food by hotel guests. Moreover, their experiences as agricultural crop-raiders have led them to view humans not as sources of easy food, but rather as serious predators to be avoided where possible. Vervets, on the other hand, with their restricted home range and smaller body size, are a less successful agricultural pest, but are considered a distinct tourist attraction by hotel managers. Tourists, who only visit the Kenyan coast, have very limited opportunities to observe wildlife. The animals are often purposely introduced to novel foods and become habituated by constant exposure to food offered by tourists.

At both inland tourist lodges and coastal hotels, where primates do not view humans as predators, they become increasingly aggressive towards people after habituation. This is especially so when the animals discover that they can easily intimidate tourists into parting with whatever delicacy they may have in their possession. This can lead to ultimate human predation when the animals become intolerable; the monkeys are all then trapped and removed (Else and Eley, 1985). Unfortunately, this is not the end of the story, because new primates move into the empty niche and the cycle repeats itself. In agricultural areas, primates rapidly become wise to trapping techniques and consequently are more difficult to control. However, continued agricultural development ultimately results in a decline in primate populations from the habitat (Southwick *et al.*, 1983).

The results presented from the agricultural survey provide further indications of how a given primate species can adapt differentially, both behaviourally and socially, to accommodate the requirements of its physical environment. For example, S.A. Altmann and J. Altmann

(1970) reported yellow baboon troops averaging from 20 to 90 animals. This is consistent with numerous other reports from various field sites, and Marsh (1985) found two troops of 56 and 82 animals in the lower Tana area, which is close to Gedi. At Gedi, troops of the same subspecies were rarely more than 20 in number, a similar value to that reported by Maples *et al.* (1976) for crop-raiding baboons in other areas along the Kenyan coast. However, since accurate troop censuses were not taken at Gedi, these could have been just foraging parties.

The activity budgets and feeding habits of crop-raiding monkeys also change. Instead of feeding for extended periods over a wide area, the baboons became opportunistic feeders, bolting down as much food as possible in the shortest period of time and then fleeing, carrying food with them. The concentration and high nutritional quality of their new diet allows them to spend less time feeding (Forthman-Quick, 1986; J. Altmann and Muruthi, 1988) and the threat of human predation encourages them to eat fast. Significant reductions have also been reported in the size of the home range, and the number of sleeping sites of provisioned baboon troops (Masua and Strum, 1984; Altmann and Muruthi, 1988).

Numerous scientists have speculated on the effects of ecological, phylogenetic and social factors on primate social organization; an excellent review is provided by Wrangham (1987). Among the general trends that have been documented are the tendencies for the larger terrestrial primates to live in larger groups with more extensive ranges (Crook and Gartlan, 1966; Clutton-Brock and Harvey, 1977), and for the large groups to have more widely scattered food resources (Richard, 1985) and better protection from predators (Clutton-Brock and Harvey, 1977). Such relationships are not tight, with numerous exceptions, and are difficult to interpret (Wrangham, 1987). This is especially true of primates with high ontogenetic plasticity, such as macaques, baboons and vervets. These species can have a range of potential behavioural or social patterns that would not be apparent without specific selective pressures, such as a changing environmental situation.

An obvious way to study the above would be to monitor adaptations to 'unnatural' selection pressures. The pest primate can be considered to exhibit adaptations to unnatural ecological variables. This has resulted in behavioural or activity patterns that would not normally be expected from the species in a more natural setting. For example, troops of crop-raiding baboons tend to be significantly smaller in number. This is not what one would expect with an abundant food source, but is a necessary adaptation for their new raiding/foraging strategy, which includes coping with humans as predators. Even if these were actually foraging parties, temporarily split off from the main troop, one would still expect their size to be directly related to the size of the food resource, as has been observed with chimpanzees (T. Struhsaker, personal communication). Forthman-Quick (1986) reported that the tendency of certain baboons to seek human foods appeared to be a major factor in the fission of the parent troop being studied at Gilgil, Kenya. Previous reports, based on more detailed behavioural observations of crop-raiding primates, have described several other behavioural adaptations, such as vigilant behaviour, use of sentinel animals before and during a raid, and decreased vocalizations to accommodate their new predator, people (Maples, 1969; Maples *et al.*, 1976; Kavanagh, 1980).

Unusual ecological pressures can also be used to confirm the existence of suspected behavioural patterns which are normally difficult to discern. Richard (1985) discouraged the use of the concept of a predictable food source, which would imply that primates have a preconception of food distribution in space and time, giving them a foreknowledge of resource availability. In a natural situation, this would be extremely difficult to document or to differentiate from an animal which merely monitors potential resources at frequent intervals, or even wanders randomly. However, this has proved not to be the case when a concentrated food resource is available for only short but consistent times (Forthman-Quick, 1986). In the setting of a tourist lodge, food availability can be very predictable, and primates can home in on this with remarkable accuracy. For example, we previously reported (Else and Eley, 1985) baboons arriving at a famous Kenyan forest lodge precisely at 4 p.m. every day for afternoon tea, which is served on the open roof. Numerous similar incidences were recorded with both baboons and vervets. Without question, both species are very adept at monitoring the precise timing and location of a predictable, concentrated food source.

8.4 CONCLUDING REMARKS

Primates which successfully adapt and become pests tend to be social, terrestrial species that are generalized, omnivorous feeders and opportunists. Those ecological pressures forcing primates to become pests elicit changes in social and behavioural patterns, many of which do not occur in a more 'normal' environment. The study of such changes can provide insight into relationships between ecology and social organization, and the variety of such changes indicates that the diversity of social organization of these species may be much greater than has been previously documented.

ACKNOWLEDGEMENTS

I am grateful to Trish Lewis, who conducted the personal interviews at Gedi; to Diann Eley, who spent a considerable amount of time collating the hotel returns; and to J. Altmann, M. Else, D. Manzolillo, T. Struhsaker and L. Leyland for reviewing the manuscript. Both surveys received partial funding from the Animal Resources Branch of the National Institutes of Health and the Yerkes Regional Primate Research Center.

9

Rehabilitation of captive chimpanzees

ALISON C. HANNAH and WILLIAM C. McGREW

9.1 INTRODUCTION

As wild populations of primates continue to decline, we have become more aware of the importance of proper management of captive populations (Fritz and Nash, 1983; Seal, 1986; Foose *et al.*, 1987). In some cases this awareness has led to attempts to release captive primates into their natural environment. However, critics (e.g. MacKinnon, 1977; Soave, 1982) have questioned the justification of such projects in the light of presumed, more-pressing needs to channel time, effort and money into preserving remaining habitats and populations of wild primates. We submit that these two approaches need not be thought of as being in competition, and that if primates can successfully adapt to a new environment, this has important implications for primate conservation (McGrew, 1983).

As now used, the term **rehabilitation** covers a variety of procedures: **Release** means to set free captives, often with little or no follow-up of their fate. Sometimes this is sufficient, as with some island sites mentioned later. **Repatriation** refers to cases in which animals are returned to the country of origin, usually from inappropriate temperate climes to more hospitable tropical ones. This may mean an improvement in the standard of living, even if full freedom is not achieved. For example, some of the individuals in Brewer's (1978) report (mentioned later) were ex-pets from Europe. The term **translocation** usually involves a shift from one wild site to another, with minimal time spent in between, in captivity. By definition, these are wild-born individuals, unlikely to acquire behavioural abnormalities by being in short-term human contact. The term **reintroduction** may appear to be a misnomer, in the sense of implying 'to introduce again', but the term is in fact used frequently in the general context of 'restoration'. Finally, there is **rehabilitation** in the strict sense of training behaviourally inadequate individuals in skills which allow them to survive with greater independence. Minimally, this may mean resocializing them for group-living outdoors, as with Pfeiffer and Koebner's (1978) project mentioned below. Maximally, it involves coaching individuals to find and process food and water, to detect and avoid predators and other hazards, to seek and make shelters, to practise healthy personal habits, and to produce and rear offspring; in short, to lead normal independent lives. This ideal has yet to be realized, although Brewer's attempts (1978) come close to it.

9.2 PREVIOUS REHABILITATION PROJECTS

Rehabilitation studies have involved three of the four species of great apes: the mountain gorilla (*Gorilla gorilla beringei*), the orang-utan (*Pongo pygmaeus*) and the common chimpanzee (*Pan troglodytes*). However, it seems that the only attempt to restore gorillas to the wild involved an infant mountain gorilla which had been confiscated from poachers and cared for for 3 months (Fossey, 1981, 1983). An unsuccessful attempt to introduce it to a wild group was followed by a second successful introduction into another wild group. But, a year later the infant died of pneumonia after a long period of heavy rains.

There have been five rehabilitation projects involving orang-utans. All concerned wild-born animals that had been either confiscated from hunters when very young or had been kept as pets for several years and then donated to the project. However, in the published descriptions, there is little distinction between the animals which entered the project at an early stage, and those which had been kept as pets for a long time.

Details of specific rehabilitation techniques are not available for every project, but they all involved a similar general procedure of setting up a feeding station in the forest, taking young orang-utans for walks, encouraging them to climb in trees and to become more independent. Individuals which were felt to be capable of surviving on their own, but did not leave the rehabilitation centre by choice, were sometimes transferred to another area.

The first orang-utan rehabilitation project began in Sarawak in 1961 (Harrisson, 1963) with three infants but it ended, for political reasons, in 1964. Two individuals which survived were transferred to a newly established rehabilitation station in the Sepilok Forest Reserve in Sabah (De Silva, 1971). From 1964 to 1969, 41 orang-utans were involved; 10 died, 7 left the area and 24 remained either at the station, or visited it occasionally from the forest. A more recent report by Payne (1987) states that nearly 200 orang-utans have been involved in this project, of which 'only a small proportion' have fully adapted to life in the wild.

Two further projects were set up in northern Sumatra by Borner (1979) and by Rijksen and Rijksen-Graatsma (1975); Rijksen, (1978). Of 31 individuals involved in the latter project between 1971 and 1974, 8 were killed by a clouded leopard, 4 died, 6 disappeared and 4 were transferred to another area of forest. Only two individuals left the rehabilitation centre of their own accord, and were known to have successfully integrated into the wild population. The other seven remained near the centre. Borner (1979) commented that in 4 years, at least 100 orang-utans were successfully rehabilitated between the two projects, whereas Rijksen (1978) reported that in 5 years, fewer than 100 individuals were rehabilitated. Perhaps Rijksen's criteria for successful rehabilitation were stricter, since he included positive social contact with wild orang-utans, as well as regular independence from provisioned food.

A fifth project is that of Galdikas (1975, 1980), who has worked with at least 14 infant orang-utans in the Tanjung Putung Reserve in Kalimantan (Borneo). We have been unable to find much information on the success or failure of her work, although Galdikas does say briefly that animals that had been kept as pets for several years could take a little longer to become used to the forest. Interestingly, even a female that had been kept for 6 years in a small cage and could not walk when she arrived at the rehabilitation centre, was successfully rehabilitated, consorted with a wild male, and gave birth. Hence, although past experiences have led to differing lengths of stay at rehabilitation centres, at least some orang-utans from almost every background have been successfully rehabilitated.

There have been six projects involving the rehabilitation of common chimpanzees (see Table 9.1). They vary widely in factors such as the background of the individuals, the size of the release areas (from an island of 0.13 ha to free-range in a national park), and the amount of preparation and follow-up carried out. All have shown some successes and some failures. In most instances, release sites have been islands, so that the problem of predators, as was noted in the case for some of the orang-utans, could be reduced or eliminated. However, one must still consider the background of the individuals in relation to the specific environment into which they are released.

The first of the chimpanzee projects began in 1966, when 17 apes were released into Rubondo, a forested island of 2400 ha in Lake

Table 9.1 Projects involving the rehabilitation of chimpanzees

Release site	*Date released*	*No. released*	*Ages (years)*	*Background*	*Pre-release preparation*	*Provisioned*	*Follow-up*	*Adaptive behaviour**	*Outcome*
Rubondo I., Tanzania (2400 ha)	1966–9	17	4–12	Wild-born from zoos	X	X	X	1,2(?)	Reproducing population on island
Ipassa I., Gabon (65 ha)	1968–72	8	4–8	Wild-born from lab.	X(?)	+	+	1,2,4,6	Two escaped; others removed
Niokolo-Koba National Park, Senegal	1973–5	8	1–6	Wild born (6) Captive-born (2)	+	+	+	1,2,4,5,6	Moved to island in R. Gambia
Baboon I., Gambia (490 ha)	1979	9	1–13	Wild-born (7) Captive-born (2)	+	+	+	1,2,4,6	Added to above group
0.13 ha I., Florida	1975	8	4–11	Wild-born (6) Captive-born (2) from lab.	+	+	+	2 (crude)	?
3 islands, Liberia (6, 27, 28 ha)	1978 1983 1985	18 24 22	5–20+	Wild-born from lab.; some pets	+	+	+	1,2,3,5	Reproducing populations on islands

*Adaptive responses: 1, eating foods; 2, nest-building; 3, ant-eating (without tool-use); 4, ant-dipping/termite-fishing; 5, stone tool-use (for nut-cracking); 6, predatory behaviour.
X: no
+: yes

Victoria, Tanzania (Grzimek, 1970; Borner, 1985). All the chimpanzees were wild-born but had spent between 3.5 months and 9 years in captivity under conditions which varied from solitary confinement in small cages to group housing. They were released in four groups from 1966 to 1969, with no preparation before their release and no close monitoring after their release. They were from European zoos, and most were not wanted because they had become aggressive towards people. Not surprisingly, the main problem following release was that they attacked people. One, and possibly two, adult males were shot because they continually sought out and attacked the game warden who lived on the island. The exact survival rate of those released is not known, but there is now a second and possibly a third generation on the island. Those chimpanzees who were born there avoid people, and aggression towards them is no longer a hazard. Interestingly, of the original 17 animals, at least 2 females were still alive in 1985, and there was then a free-ranging population of at least 20 chimpanzees (Borner, 1985).

The next rehabilitation project involved the release of 8 chimpanzees from a medical research laboratory onto a 65-ha island (Ipassa) in Gabon in 1968 (Hladik, 1973, 1974). All were wild-born and had been in captivity for periods which varied from a few months to several years. Although they were given supplemental food, particularly when there was low fruit production, they found many natural foods for themselves. Apart from fruit, for example, they ate leaves and stalks, ants, small mammals, birds and birds' eggs. They also built nests in trees each evening in which to sleep. These chimpanzees remained on the island until 1978 when they discovered that they could wade across the river to its banks at low tide. Most of the group was captured and taken to a medical research laboratory, but at least two or three individuals escaped. One female which escaped was later observed with an infant (A. Gautier, personal communication).

Brewer's (1978, 1982) rehabilitation project represented a different approach. This was the first attempt to release chimpanzees into a wholly natural site. She worked intensively with a small group of mostly young chimpanzees. She taught them the skills needed for survival by demonstration; showed them which foods to eat and, when necessary, the techniques needed to get these foods. These included stones to crack open hard-shelled fruits and pods, and twigs for termite-fishing. She also taught them how to build nests in trees, and actively discouraged them from sleeping on the ground. Their training began in the Abuko Nature Reserve in The Gambia, and continued at Mt Assirik in the Niokolo-Koba National Park in Senegal. At Mt Assirik, the apes became increasingly self-sufficient, but they began to encounter wild chimpanzees in the area who reacted aggressively towards them. When the wild chimpanzees began to come to the camp area and attack the newcomers, Brewer decided to remove them for their own safety. In 1979, they were taken to islands in the Gambia River, where they remain today in a group of over 30 chimpanzees, which includes those of another worker, J. Carter, and more recent additions. Several of the females have bred and reared offspring.

Carter's (1981) project is linked to that of Brewer. She worked with a group of nine chimpanzees, of which some were captive-born and some wild-born. She began with an 18-month training period at Abuko Nature Reserve. There, the captive-born chimpanzees in particular had to be taught to accept wild foods. They were then moved to a 490-ha island in the Gambia River, where Carter lived with them and taught them many of the things that Brewer had taught her animals. Moreover, one of her chimpanzees had been raised in a human family with no contact with other chimpanzees (Temerlin, 1975), and had to learn appropriate social behaviour. She needed more intensive training than the others but eventually progressed.

All of the rehabilitation projects for chimpanzees mentioned so far have produced at least some individuals who were able to survive in a natural environment without provisioning. They show that individuals of very varied experiences may be 'rehabilitated'. On the other hand, it is perhaps hardly surprising that in extreme cases of environmental and social deprivation, success in rehabilitation is severely limited.

The eight chimpanzees of Pfeiffer and Koebner's (1978) group in Florida, for example, had all been kept in single cages. Six of them had at least some visual and auditory contact with other chimpanzees, but the other two (a male and a female) had been confined to small, dark, sheet-metal cages for several years; both showed motor deficits and various stereotypies. However, they did learn to climb when they were transferred to a larger cage with other apes but, although the male eventually performed less-stereotyped behaviour, the female remained extremely withdrawn.

After 4 months of living together in various combinations, the eight chimpanzees were kept together for 5 months before being transferred to a 0.13-ha island in a safari park in Florida. The female from the most deprived environment remained in an enclosure there for 4 weeks until she was coaxed out with food and the door was blocked behind her. Six weeks later she was found dead. Apparently, she failed to avoid being bitten by a hippopotamus which lived in the moat surrounding the island. Another two individuals, who showed stereotyped rocking in captivity, reduced this behaviour after release. Although the social life of the male from the most deprived environment remained limited, he did show some improvement in locomotion and in social behaviour. Moreover, the island for this group was small and bare, and cannot be compared with the environmental enrichment of other projects. Even so, most of the chimpanzees showed reduced abnormal behaviour in the relatively better conditions.

9.3 A REHABILITATION PROJECT IN LIBERIA

In 1974, the Laboratory of Virology of the New York Blood Center established a research laboratory (Vilab) using chimpanzees in Liberia on the west coast of Africa, an area within the range of the western subspecies of the common chimpanzee. In 1978, groups of chimpanzees which had completed safety tests for hepatitis vaccine (each chimpanzee can be used only for one vaccine trial) were released onto natural offshore islands. The plan was that they should remain there for a period of adaptation, and eventually be released into protected areas, such as national parks, in West Africa. At present, Vilab maintains three such island groups, with a total of 58 chimpanzees. The most recently formed group, released in June 1985, has been more closely studied than previous ones (Hannah, 1986), and is discussed below as a specific case.

This group was made up of 22 apes (*Pan troglodytes verus*), 10 males and 12 females, ranging from 5 to over 20 years of age. Table 9.2 lists the animals and gives personal details. They were all born in the wild; some had been kept as pets for up to 5 years, one had been kept in a zoo for several years, and the others had spent up to 8 years at Vilab.

In some ape rehabilitation projects there is little mention of the apes' behaviour before release. There are no baseline data with which to compare their behaviour after release, and to give a measure of adaptation to a new environment. It is an important aspect of the project in Liberia that there was an opportunity to make such comparisons. Hence, before release, the chimpanzees were studied in their groups at the Vilab laboratory. They were housed in three separate outdoor enclosures, each with an area of 160 m^2, and most of the individuals scheduled to be released on the same island were first introduced at the laboratory.

Data were collected by focal animal sampling at 1-min intervals during observation sessions of 20 min. Point samples and one-zero samples were used on a checksheet with 11 behavioural categories, which included affiliative behaviour, aggressive behaviour and stereotypies. Scan samples of interpersonal spacing were also recorded at 5-min intervals. At other times, such as when a new group member was added, data were collected ad lib. Decisions about which individuals should be kept together, and on the order in which they should be released, were based on these data. This period of data collection enabled the observer (A.C.H.) and the chimpanzees to become familiar with one another. For some of the younger or more insecure individuals, attachment to the observer was an important bond soon after release.

The group of 22 chimpanzees was split into four subgroups for release (see Table 9.2). This reduced the possibility of individuals becoming lost. Once members of the first subgroup knew the island, individuals which were released subsequently often travelled with those that already knew their way around. The first subgroup contained the younger and lower-ranking individuals. The adults were added in subgroups at 3-week intervals, with the highest ranking adult males being released last. If individuals became ill at any stage they were returned to the laboratory.

The release site was a 9.7-ha island (island A) surrounded by mangrove swamp, which allowed access to another island (island B) of 17.4 ha (Figure 9.1). The area of dry land available to the chimpanzees was therefore 27.1 ha; the surrounding area of estuarine mangrove swamp added another 57.9 ha, which gave the animals a total amount of about 85 ha in which to live. The apes travelled through the mangroves in order to reach island B. They also used the mangroves to make nests. Both islands were completely forested with dense vegetation in many areas. Paths were cut to allow human access, and the chimpanzees used these paths, as well as other smaller ones that they made for themselves.

The chimpanzees were always fed at the same site on island A. This was an area at the edge of the river covering 18 m^2 at high tide and 30 m^2 at low tide. There was also a water tap linked to a water drum, which was filled in the dry season when the river water was too salty for drinking. In addition, there was a cage at the feeding-site in which individuals were kept before being released. It was 3 m long x 2 m wide × 1 m high and was made of wire mesh, with a wooden floor and roof (Figure 9.2).

The chimpanzees were transported anaesthetized to island A by boat, and shut in the cage at the feeding-site to recover. They remained there overnight and were let out early in the morning and observed throughout the day. This allowed time to find individuals who had wandered off alone and become lost. The whole group was also gathered together before the observer left the island in the

Table 9.2 Information on 22 chimpanzees released on islands in Liberia, June–August 1985

Name	*Sex*	*Age (years) at capture*	*Age (years) at release*	*Years as pet*	*Sub-group released in*	*Radio-collared**
Group 1						
(a) Cruella	f	4.0	7.0	0	1	X
Pim	m	2.0	5.0	0	1	X
Reagan	m	3.0	7.5	0	1	X
(b) DmW	f	1.5	8.5	4.0	2	X+
Franco	m	1.0	9.5	4.0	1	+
(c) Daniel	m	0.5	10.5	5.0	4	X
Knut	m	2.0	6.5	>1.0	1	+
Maki	f	0.5	9.5	0	3	X
Trokon	m	0.5	8.5	3.0	1	+
(d) Hermaphrodite	m	0.75	5.5	0.25	1	+
Total (7m, 3f)		Mean = 1.58	Mean = 7.8	Mean 1.73		
Group 2						
Blamah	f	1.5	7.5	0.5	1	X
Brutus	m	1.0	8.5	3.0	4	X
Carolla	f	2.5	9.5	1.0	2	X+
Goldilocks	f	2.5	7.0	0.5	2	X
Grace	f	?	>20.0	0	3	+
Helen	f	1.0	5.5	0.5	1	X
Houdina	f	3.0	9.5	4.0	2	+
Maria	f	1.5	6.5	0.5	1	+
Meryn	m	0.5	5.0	0	1	+
Popeye	f	?	6.5	<1.0	1	X
Samantha	f	1.0	9.5	0	3	+
Sokomodo	m	2.0	9.5	>1.0	4	X
Total (3m, 9f)		Mean = 1.65	Mean = 8.7	Mean 1.0		

Group classification: 1a, disappeared on first day of release; 1b, returned to laboratory; 1c, became ill and died; 1d, crossed canal and was killed by chimpanzees on adjacent island; 2, still survive on island (5 years later).
*X+, not wearing a radio-collar when first released but had one fitted at a later date (six collars were available and these were transferred from one chimpanzee to another when necessary).
X: not fitted with radio collar
+: fitted with radio collar

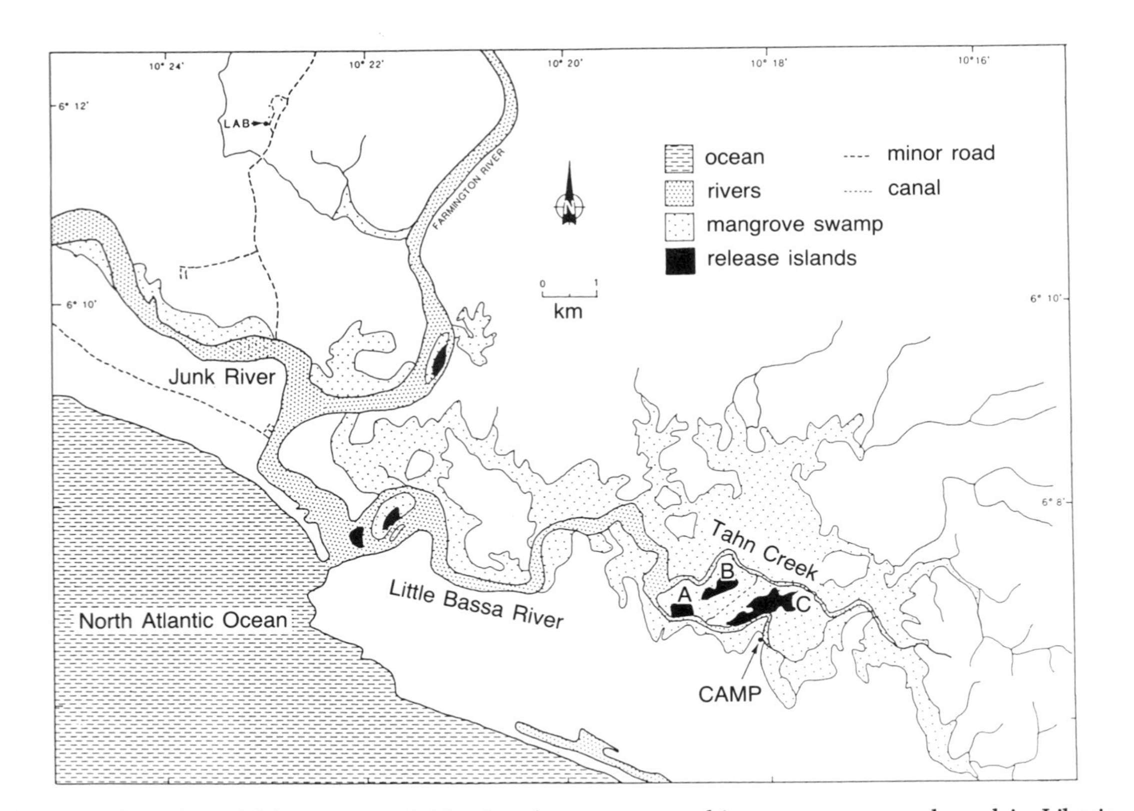

Figure 9.1 Location of laboratory and islands where common chimpanzees were released in Liberia.

Figure 9.2 The cage at the feeding site of the rehabilitation area in which chimpanzees were kept before being released. (Photo: Alison Hannah.)

evening so that, if they chose, they could remain together until the following morning. (Chimpanzees will often attack others recovering from tranquillization, even when they know them.) The cage allowed for safe recovery if there were already apes there. It also made it easier for individuals who had not lived together at the laboratory to become acquainted. The three adult males, which were the last individuals to be released, had not been introduced to some of the younger group members before release, in order to avoid aggression in the relatively confined spaces of captivity. Moreover, these males were kept in the cage for a week. During this period, the younger males of the group especially, came to the cage and acted submissively towards them. By the time the three adult males were released, a dominance hierarchy had been established and there was no fighting.

Finally, anxious individuals used the cage as a sleeping site when they were first released. Two females, Grace and DmW, slept inside the cage (with the former taking in branches to make a nest), and Blamah (female) and Meryn (male) slept on top of it. Before long, the three of these chimpanzees who remained on the island slept in trees, but the familiarity of a cage may have helped them in their initial adjustment from captivity.

In addition to the protection afforded by the cage, some individuals were fitted with radio-collars. This allowed them to be tracked if they did not come to the feeding-site on feeding days; which was every day

Figure 9.3 A field assistant walking around the rehabilitation site with newly released chimpanzees. (Photo: Alison Hannah.)

when first released. The collars were especially useful at the beginning when the chimpanzees were finding their way around and learning to come to the feeding-site when they heard a boat arrive. Animals were found by radiotracking many times in the early weeks. The provision of collars added to the expense of the project, but considerably improved the chances of survival. For example, comparing three releases, the percentage survival was 50% when no radio-collars were used, 60% when 30% of the group was collared, and 95% when the whole group was collared. There was a total of six radio collars available, but optimal use was made of them by transferring them from members of the first subgroup released to the next, as soon as appropriate. Hence, although only some animals were wearing collars, when they split into small groups to travel, at least one member of each party was usually wearing one. Carolla probably would not have survived had she not been wearing a collar. When first released, she always travelled alone and had to be tracked every day before she finally learned to come to the feeding-site. One day she sat in a tree calling continually until the observer found her; she then followed the observer to the feeding-site. Five years later she is still on the island (but now without a collar) and travels with the others and comes to the feeding-site on feeding days.

The observer (A.C.H.) also came to supervise the releases more closely than had previously been done. Hence, the site was visited every day after release, and whole days were spent there. The presence of a familiar person probably made a difference to the younger or more insecure group members. They followed the observer until they knew their way around and felt confident about travelling on their own. It also meant that any illnesses which developed could be spotted quickly and treated.

Walking around with the chimpanzees also gave the opportunity to collect data on their adaptation to their new environment (Figure 9.3).

9.4 NATURAL ADAPTIVE BEHAVIOUR OF THE REHABILITATED CHIMPANZEES

Various categories of adaptive behaviour were observed. They were recorded opportunistically under excellent conditions; the chimpanzees could be approached to within 1 m and easily followed.

The first adaptation to the release site was to taste leaves and small fruits. When first released, the chimpanzees were given bread and fruits daily, and did not have to find foods for themselves. However, five of the first nine individuals released very soon began to eat leaves. Over the next few days the remaining apes also tasted the leaves that the others ate. When more animals were released, they closely watched the others eating and tasted the same leaves and fruits. Newcomers did not eat much on their first day, but gradually increased their intake. Learning which natural foods they could eat seemed to be a process of trial and error as far as the leaves were concerned, but all fruits were eaten. Groups released later ate the foods which were already being eaten by the first group.

When all of the chimpanzees were on the islands and were foraging for leaves, seeds, nuts and fruits, the number of days during which provisioned food was provided was gradually reduced from seven to three per week. However, because the natural food could not provide a balanced diet for the chimpanzees, the number of provisioned days was not reduced below three per week.

The next behaviour to develop was nest-building. At the laboratory, all chimpanzees had been given freshly cut branches to encourage nest-building, and to provide baseline data on their capabilities to do so. Sometimes, lower-ranking individuals were limited in their attempts by the small number of branches that they could obtain. The biggest nests were made by the more dominant animals. However, all the chimpanzees had at least attempted to build nests in captivity. These varied from thick piles of interwoven leafy branches to crude circles of small twigs, with the individual sitting in the centre carefully arranging them. Chimpanzees also sought to sleep above the ground by lying on tyres balanced on the bars of a climbing frame, but in large groups many had to sleep on the ground as there were insufficient tyres for all of them.

Observations on nest-building on the islands were sometimes difficult to achieve. It was hard to follow the animals when it began to get dark, particularly if they went into the mangrove swamp. In fact, the tall mangrove trees close to the river were often used for sleeping. However, five animals were soon seen to sleep in nests, and all the chimpanzees were eventually known to sleep in trees after an early period of sleeping at the cage or on the ground.

Two weeks after release an adolescent female was first seen picking at the nests of weaver ants in order to eat the ants and ant larvae inside. The rest of the group observed her and a number of others began to eat ants.

Tool-use to crack open palm nuts was also observed (Hannah and McGrew, 1987). This was initiated by an adult female on her first day after release. Figure 9.4 shows the female using a stone to crack a palm nut which she had placed on the stand for the water tap. An adult female and an adolescent male were watching her. Two females also started to crack nuts on the same day, and over the following weeks most of the group members showed the technique. No nut-cracking had occurred on the islands until the innovative female was released, after which it gradually spread through the group until all but two group members cracked nuts.

Nut-cracking also spread to different locations on the islands. The chimpanzees carried stones to suitable sites. Again, females were the first to show such innovative behaviour, and they also picked up the behaviours more readily. A group of chimpanzees that became ill and died were those which, in general, did not learn some of the behavioural patterns that required more skill, such as nut-cracking and ant-eating. This is not to say that they had to be able to crack nuts or

Figure 9.4 An adult female chimpanzee using a stone to break open a palm-nut fruit and being watched by another adult female (foreground) and an adolescent male. (Photo: Alison Hannah.)

eat ants in order to survive, but it may generally show that they were not adapting as well as the other animals.

9.5 OTHER CHANGES IN BEHAVIOUR IN THE NEW ENVIRONMENT

As well as learning to adapt to their new environment, other changes in behaviour were seen in the apes. For example, attacks were much less frequent, perhaps because the chimpanzees had so much more space in which to live. In the more restricted space at the laboratory, and when the dominant male of a group became excited, as just before feeding time, he displayed by pant-hooting, walking bipedally, and slapping the walls. At the climax of the display he charged around the enclosure chasing others, slapping any within reach. If an individual became cornered he usually jumped on its back and slapped it. On the islands there was much less aggression in a similar situation. If the dominant male was already at the feeding-site on island A when the observer arrived with food, there were no such displays. He only displayed occasionally if he arrived at the feeding-site later than the others. On these occasions, he charged into the feeding area and any others not already in the trees would climb into them in order to avoid him. The way in which food was given out also helped to reduce quarrelling between individuals. At the laboratory, food was thrown into the outdoor enclosure. Some attempt was made to throw food to particular individuals, but this could not always be done accurately and squabbling over food often occurred. On the island, the observer handed food to each individual, generally in order of descending rank, and aggression rarely occurred.

Two of the original group of 22 animals showed stereotyped behaviour before they were released. A male, for example, showed the following:

- **Body-rock**: sat with knees bent, legs in front of body, arms extended to sides with hands on ground and rocked torso from side to side.
- **Squat-walk**: sat in the above position but changed location by moving feet forward, raising body slightly and shifting torso forward to meet legs, thus moving around while seated. Body-rock and squat-walk often occurred together in series.
- **Poke-eye**: prods left eye with right thumb.
- **Pinch-palm**: pinch left palm between right thumb and forefinger.

These patterns were recorded systematically. He showed at least one of these stereotypies on 26% of samples of behaviour, the most frequent being body-rock, which occurred on 14% of samples.

The other chimpanzee to show stereotyped behaviour in captivity was a female who sat in a corner with her legs flexed and each arm crossed over and resting on the opposite knee. She also rocked her head backwards and forwards with jerking movements. This behaviour was observed on 25% of the observations. The male often performed stereotypies when not involved in other activities, whereas the female's stereotyped behaviour seemed to be more related to stress, since it usually occurred following an aggressive display or an attack by the alpha male of her group.

After release, both individuals soon stopped these behaviours. The male found his new environment very stimulating. When the others were walking around he enthusiastically ran ahead and back again. Moreover, when he had been on the island for only 3 weeks he fell ill and was returned to the laboratory, where he resumed his previous stereotyped behaviour. For the female, the source of stress had been left behind. She resumed rocking during a temporary return to the laboratory (when she was not a target of aggression as she had been previously), but not to the same extent.

9.6 INITIAL LOSSES FROM THE REHABILITATED GROUP

Individuals who did not survive on the islands, or who were removed because they seemed incapable of subsisting there, could be divided into the following categories:

1. *Disappearance*. Three chimpanzees (Cruella, Pim and Reagan) rushed from the release site and were never seen again. Nine others freed at the same time also left the release site and went into the bush, but over the next 3 hours rejoined the observer when she was walking around. Cruella and Pim tended to avoid people, and this may explain why they were never found in any searches over the following weeks. They may have died of illness or starvation following the change in diet from the laboratory to the bush. They did not come to get provisioned food.
2. *Illness*. Three chimpanzees (Trokon, Maki and Knut) were returned to the laboratory at different times, but showed similar symptoms on the island. They all had diarrhoea of bacterial origin and loss of appetite. They were treated for the diarrhoea but never regained their appetites, and would eat and drink very little even when offered a wide range of favoured foods and drinks. They were given occasional intravenous infusions to prevent dehydration but, after up to 4 weeks of continual weight loss, all three eventually died. The fourth individual who died (Daniel) had shown a slight loss of appetite but otherwise seemed healthy, before he was unexpectedly found dead.
3. *Removal*. One chimpanzee (Franco) was returned to the laboratory only 5 days after release because he seemed physically incapable of surviving. He was presumed to have suffered some neurological damage in early life, and his movements had always been slightly uncoordinated. This became more apparent on the islands where he often fell over on uneven surfaces and could not keep up with the others. He was returned to Vilab, where he had always walked without problems.

   DmW was also returned 4 days after release, because she was found weak and lying down with a respiratory infection. (Other group members with respiratory infections, but otherwise healthy, were given daily medication in milk on the island.) On arrival at the laboratory DmW acted normally and soon recovered, but remained there for 3 weeks until the next group was taken to the island. She then spent 20 days on island A, but was again found weak and lying down. She was returned to the laboratory, where nothing could be found wrong with her. After 4 days she was released for the third time, but after 10 days was found weak and lying along a tree branch. On this occasion she had some cuts, suggesting that she had been fighting, and for several days she could not move her head properly. On full examination she was found to be pregnant. Since this was her first offspring she remained at the laboratory, to be kept together with a mother and infant, a procedure which was found to be useful for nulliparous females. She successfully raised her infant at the laboratory.

4. *Intergroup aggression*. In the dry season at the beginning of 1986, the water level dropped in a man-made canal separating the group of chimpanzees on islands A and B from another group on an adjacent island (C). The apes soon discovered they could wade across this canal at low tide. Two adult females and an adolescent male from the group under study on islands A and B crossed to island C, which contained 15 chimpanzees aged between 6 and 11 years. The next day the two females were returned to their group, but the male was still missing. After a search, he was found dead on island C close to the canal. As he had many wounds, he had presumably been trapped on this island at high tide and was attacked and killed by the chimpanzees from the other group. After this incident, the study group was returned to the laboratory for the remainder of the dry season.

9.7 A COMPARISON OF 'FAILURES' AND 'SUCCESSES'

When the failures and successes of Table 9.2, in terms of the chimpanzees who did not survive or were returned to captivity and those that remained on the islands, are compared with reference to such characteristics as gender, age at capture, age at release, and the number of years as a pet, the only statistically significant difference is that between males and females in each group. Of the original group of 22, a higher proportion of females successfully adapted than males. Moreover, although in some cases the reasons for this are not known, there are data to indicate that females in general are more adaptable than the males. The females initiated behaviour such as nut-cracking and ant-eating; they also learned some techniques of foraging more readily than males. At a later stage, when another 18 chimpanzees were added to the study group, so that data were available for 35 animals, only 2 of the 20 females did not learn to crack nuts, whereas 8 of the 15 males did not do so.

Success and failure of rehabilitation may also be compared by reference to when individuals were released. It happened that 7 out of 10 'failures' were in the first subgroup. As previously mentioned, this can be explained in part by the fact that additional group members can travel together with chimpanzees who already know their way around, and are less likely to get lost. Of the three individuals who disappeared and were never found, all were members of the first subgroup released. One clear way in which to eliminate this kind of loss would be release only radio-collared individuals in the first group. There is another factor to be considered. Individuals who are added at a later date are not only able to make use of the others' knowledge of the geography of the island but they can also observe which natural foods the others are eating and how they obtain these foods. In other words, they have models on which to base their behaviour.

In the end, however, the likelihood of success or failure is probably related to a wide variety of factors for each individual, including age (at capture and at release), gender and housing conditions while in captivity.

9.8 CONCLUDING REMARKS

What can we now say about the prospect of rehabilitating chimpanzees into new, more natural, environments? At present, there are no systematic data to answer the question, but answers may be given in general terms. Hence, the ideal candidate for rehabilitation would be born in the wild, captured late, never have left the home country, and would have lived in captivity for a short time with fellow chimpanzees in good conditions. If such an animal is selectively rehabilitated (in the sense of specific training) into a habitat which is suitable for, but devoid of, established conspecifics, the chances of success are probably high. On the other hand, the worst candidate would be born in captivity but nursery-reared after neonatal separation, and kept in isolation from its conspecifics until adulthood. If she or he is released, without pretraining or follow-up, into an unsuitable habitat or a habitat already occupied by hostile conspecifics, then early death is almost certain. Moreover, and to be realistic, we will find that some proportion of captive apes are beyond reclamation and cannot be rehabilitated; they must likely remain in lifelong custody.

Most prospects lie between the extremes, and other factors must be taken into account. Apes, 'humanized' by captivity, cannot be expected to act 'sensibly' from the outset of free-ranging. This means siting rehabilitation projects away from local people, preferably on islands, for the safety of both parties. If release islands are large enough for the chimpanzees to become totally self-sufficient, and not require any human contact, then the next generation should have a natural fear of humans. This is the case on Rubondo Island in

Tanzania, which was mentioned previously. Perhaps only then should release into a free-ranging site be considered.

Again, there are issues which relate to health. Apes in captivity may have acquired contagious pathogens, inadvertently or otherwise, which could be fatal to their wild counterparts. One solution is to release infected apes only into areas where they will meet no conspecifics. Another is, of course, to cure them of all ills before their release.

Our rehabilitation studies showed that the chimpanzees adapted well to living on the island sites in Liberia, and that we could make improvements to the release and follow-up techniques to aid their survival.

Many of the animals showed that they have the potential to meet and to master the demands of new environments. It is interesting to consider the extent to which they may be advantaged by their persistent flexibility of responses compared to many other primates. Data are generally preliminary but projects which have involved releases or translocations of wild-born macaque species (Kawai, 1960; Carpenter, 1972; Southwick *et al.*, 1984), squirrel monkeys (Jerkins, 1972; Bailey *et al.*, 1974), vervet monkeys (V. Wilson, 1980) and baboons (Strum and Southwick, 1986) have had varying degrees of success.

Among species of the great apes it is perhaps worthy of special mention that, from a general review (see section 9.1), rehabilitation releases for orang-utans seem to be more successful than for chimpanzees. Differences in social organization and the relative lack of aggression, both to conspecifics and towards humans, may be important factors here. Orang-utans can be more easily released into an area where there is already a wild population without worrying about them being attacked, because these apes are widely spaced and, for the most part, solitary. Also they do not show aggression towards humans, which has occurred in some chimpanzee release projects.

Thus, rehabilitation of great apes can be done, particularly for wild-born apes, after being kept in appropriate conditions in captivity. Such release projects can help in the conservation of a species. Conservation of remaining wild populations and habitats should be the first priority, but rehabilitation projects should not be thought of as competing with the preservation of wild populations. Rehabilitation projects can be used as educational tools, and there is always the possibility that, in the future, reproducing populations of apes maintained on islands could be released into suitable free-ranging sites. In the last analysis, whether or not the principles of rehabilitation can be converted into practice depends on the resources and willpower of us, their captors.

10 Responses of wild chimpanzees and gorillas to the arrival of primatologists: behaviour observed during habituation

CAROLINE E.G. TUTIN and MICHEL FERNANDEZ

10.1 INTRODUCTION

Wild primates show a variety of responses when primatologists arrive to study them. Some are very shy and flee rapidly, while others lack fear and are easy to approach and observe. Habituation is the term used to describe the acceptance by wild animals of a human observer as a neutral element in their environment. The process is rarely described, as it is commonly regarded as a means to an end; namely, the progression to a state that allows the natural behaviour of a species to be observed and documented.

It is clear from the long-term studies of common chimpanzees (*Pan troglodytes*) at Gombe (Goodall, 1986) and Mahale (Nishida, 1979), in Tanzania, and that of mountain gorillas (*Gorilla gorilla beringei*) at Karisoke, in Rwanda (Fossey, 1983), that both species of African ape can become totally habituated to the presence of observers. However, the chimpanzees at both Gombe and Mahale were habituated with the help of artificial provisioning. They were given choice food items such as bananas or sugarcane as positive incentives to accept the presence of humans. Details from other studies, where provisioning was not used, indicate that chimpanzees can be very difficult to habituate. For example, partial habituation has been achieved at Kibale (Uganda) through years of exposure to primatologists studying other species (Ghiglieri, 1984), and also at Taï Forest (Ivory Coast), where the nut-cracking sites that attract chimpanzees regularly to a few specific locations allow predictable and frequent encounters with observers, and at Bossou (Guinea), where peaceful coexistence with farmers has included some unintentional provisioning (Albrecht and Dunnet, 1971; Boesch, 1978). Other populations of chimpanzees in the savannah-dominated habitat at Mt Assirik (Senegal) and in dense forest at Okorobiko (Equatorial Guinea) however, have remained very shy after several years of study (Jones and Sabater Pi, 1971; Tutin *et al.*, 1983).

Mountain gorillas represent a different case. In addition to the Karisoke study groups, other groups have been habituated, in a relatively short time, with no provisioning, elsewhere in the Virunga Volcanoes (Rwanda and Zaire) and at Kahuzi-Biega (Zaire) (Goodall, 1979). On the other hand, western lowland gorillas (*Gorilla gorilla gorilla*), living in tropical forest, have never been successfully habituated, despite studies lasting at least 12 months at three sites (Mt Alen in Equatorial Guinea (Jones and Sabater Pi, 1971), Campo in Cameroon (J. Calvert, personal communication) and Lopé in Gabon (this study).

The ease of habituation of a particular population of primates appears to depend largely on the following factors. First, the nature of any previous experience with humans; second, the structure of the habitat; and, third, the behaviour of the species when faced with an unfamiliar intruder into their environment.

It is rarely possible to know the precise history of a particular population of primates: to discover how much, if any, previous experience they have had with humans, and to be sure of its nature. It is likely to be more difficult to overcome a specific fear resulting from a negative experience with humans, such as hunting, than a general fear of the unfamiliar. In our study area in the Lopé Reserve, in central Gabon, western lowland gorillas and chimpanzees occur sympatrically at similar population densities. Lopé has been a game reserve since 1946 and it is unlikely that the apes have ever been hunted by poachers, but they probably had had some experience of humans in that certain areas of the forest were selectively logged in the late 1960s. However, we reasoned that because the two species share a common habitat and history, any difference in their responses to our arrival was likely to be a result of species-specific behaviour.

The social organization of gorillas and chimpanzees at Lopé resembles that of populations elsewhere. The stable, cohesive group of gorillas with several adult females, their immature offspring and usually a single, fully mature male (the silverback), contrasted with the larger community of chimpanzees, which have approximately equal numbers of adult males and females. All of the members of a chimpanzee community rarely assemble; they spend their time alone or in subgroups of varying size and composition, which may change membership several times in a single day. The only exception to this is the mother–offspring grouping, in which almost constant association endures until the offspring reaches adolescence.

The first aim of our study at Lopé, which began in 1984, was to habituate gorillas and chimpanzees to our presence. We ruled out the use of provisioning and hoped that repeated, peaceful encounters would convince the apes to tolerate our presence at progressively closer distances. However, we experienced great difficulty in locating our study groups on a regular basis, due to their low population densities (about one per square kilometre), and their lack of visibility in the dense forest. Both species are also highly frugivorous, which means that they range widely and are impossible to track most of the time. Hence, progress toward habituation has been very slow.

10.2 STUDY METHODS

Observers searched for gorillas and chimpanzees by walking along an extensive network of animal paths, listening for calls and looking for the apes or their fresh trails (feeding remains, faeces, nests or tracks). When they were located, the apes were approached and an effort was made to observe them, and to be seen by them, in nonthreatening circumstances – namely, to be clearly visible at an acceptable distance. However, despite careful precautions, the low visibility in much of the study area meant that apes were sometimes encountered unexpectedly and, when this happened at close range, it was alarming to both parties.

Data were collected freely. Whenever possible, they included the age–gender class, and a description of all observed individuals; the distance between the apes and the observer; the heights above the ground, if in trees; and all vocalizations and observed behaviour. It was usually clear whether or not an individual animal had detected the observer. Data on gorilla responses to observers were taken from at least four social groups, whereas that of chimpanzees certainly came from one community of at least 50 individuals. Definitions of behaviour patterns shown in response to observers are given in Table 10.1.

Adults of both species could usually be identified by gender, although adult female and blackback male gorillas were sometimes confused. We rarely knew the gender of immature gorillas and, while that of juvenile chimpanzees could often be identified, no separation by gender was made for the data on immatures in the following analyses. We included only cases in which the age–class (and gender for all but immatures) of the individuals that detected the observer

Table 10.1 Definitions of behaviour patterns shown in response to observers in the Lopé Reserve, Gabon

Gorillas

Loud vocalizations: roars, threat barks and screams

Charge: rapid noisy running approach, either direct or oblique, towards the observer

Monitor: surveillance of the observer by repeated, quiet, approaches to stare, interspersed with movement back to the group

Curiosity: includes two, or more, of the following elements: staring, head swaying, moving to obtain a clearer view of the observer, chest beating, slapping tree trunk and clapping hands, or a hand and a foot

Avoid: changing direction, or descending a tree in order to move either rapidly, or at normal walking speed, away from the observer

Ignore: no discernible response shown; after glancing or staring at the observer, the individual continues with previous activity

Hide: either moving behind vegetation (sometimes hiding the whole body, but often only the face), or pulling vegetation in front of face or body to form a screen.

Chimpanzees

Charge, Curiosity, Ignore and *Hide:* same as for gorillas

Loud vocalizations: 'eagle' wraaghs, waas or screams

Soft vocalizations: hoo or whimper

Flight: rapid jumping or sliding out of a tree or running at speed along the ground causing much noise

Stealthy retreat: slow, cautious and almost silent descent from tree or avoidance on the ground

Approach/Wait for another: after a glance or stare at observer, chimpanzee moves directly to another chimpanzee and makes physical contact, or turns towards another chimpanzee and waits for it to approach before both move away from the observer

were identified. More than one response was recorded from the same encounter if an animal's detection of the observer was judged to be independent of the others, and not a result of a previous response by a conspecific. In the analysis, only the initial responses of the detector were considered.

10.3 RESULTS

Table 10.2 lists the responses made by different age–gender classes of gorillas to an observer. 'Curiosity' was the most frequently observed response, occurring in 29% of cases. This was followed by avoiding behaviour (Avoid), which occurred in 20% of cases, and then 'Loud vocalizations' by 18% of the cases. 'Hide' and 'Charge' (with no vocalizations) were both uncommon responses. They were shown in only 2% and 1.7% of the cases, respectively. When age–gender classes

Table 10.2 Responses of members of gorilla groups to observers at Lopé, Gabon

	Age–class					
Response	*SB*	*BB*	*AF*	*SA/J*	*INF*	*Total*
Loud vocalization	28	4	7	2		41
Loud vocalization and Charge	28		1			29
Charge	4					4
Monitor	17	2				19
Curiosity		2	21	44		67
Avoid	1	2	26	15	2	46
Ignore	2		11	3	1	17
Hide			1	4		5
Total	80	10	67	68	3	228

Abbreviations: SB, silverback males; BB, blackback males; AF, adult females; SA/J, subadults and juveniles; INF, infants.

are compared, a striking polarization emerged, in which silverback males showed different responses to those of adult females and immatures. The few data on blackback males indicated an intermediate position. Adult female and juvenile gorillas either avoided observers or showed curiosity towards them, while silverbacks reacted most often by giving loud vocalizations and frequently charging or monitoring the observer, placing himself between his group and the intruder. The few cases when gorillas other than adult males responded to the observer with loud vocalizations were stressful encounters when the gorillas were met unexpectedly in dense vegetation at distances of between 2 and 10 m.

Table 10.3 lists the responses to an observer shown by the different age–gender classes of chimpanzees who were members of subgroups of at least two individuals when they were encountered. Overall, 'Flight' (shown in 34% of cases) was the most frequently observed response, followed by 'Approach/Wait for another' (22%), 'Stealthy Retreat' (9%), and 'Loud' and 'Soft vocalizations' (both for 7%). 'Hide', 'Ignore', and especially 'Charge' and 'Curiosity', were shown very rarely. Differences between age and gender classes were less striking than in the gorillas, but adult females did 'Approach/Wait for another' much more often than did adult males (43% and 9% of cases, respectively). Loud vocalizations were only given by adults and adolescents in close-range encounters of less than 15 m. Juveniles screamed on detecting the observer at greater distances (20–25 m) when they were in trees in which no adult female was present.

The distance between the apes of both species and the observer,

Table 10.3 Responses of members of chimpanzee subgroups to observers at Lope

	Age–class						
Response	AM	AF	AdlM	AdlF	JUV	INF	Total
Loud vocalization	6		2		4		12
Quiet vocalization		2		8	2		12
Charge	1	1					2
Curiosity	1				1		2
Flight	27	23	2	2	4	1	59
Stealthy retreat	12	3	1				16
Ignore	1	2	1				4
Hide	2	3		2			7
Approach/Wait for another	5	26			3	5	39
Total	55	60	6	12	14	6	153

Abbreviations: AM, adult males; AF, adult females; AdlM, adolescent males; AdlF, adolescent females JUV, juveniles; INF, infants.

when the latter's presence was detected, had consistent effects at the extremes. All encounters at 5 m or less caused fear and alarm, which was expressed by loud vocalizations and rapid flight. Defensive behaviour, as in charges at the observer by silverback gorillas and, very occasionally, by adult chimpanzees of both genders as well as adult female gorillas, also occurred in close-range encounters. On three occasions, chimpanzee subgroups 'mobbed' observers discovered at very close range. Conversely, in encounters where mutual observation was established at more than 100 m, the apes either ignored the observer or showed curiosity, but such cases were extremely rare. Most encounters were at intermediate distances, and no clear relationship emerged between the actual distance from the observer and the response shown.

Other factors were found to be important in determining the responses. These included whether the ape was on the ground or in a tree. For example, both species were more alarmed when in trees. Other factors were their position with respect to other group members and, for immatures particularly, *vis-á-vis* the mother. Isolated individuals showed more fear than those close to conspecifics. The possibility for evasion was also important in that less fear was evident in denser vegetation, whereas the greatest fear was shown in open areas such as marshes and savannahs. Again, in special cases, the activity of the apes was influential. For instance, chimpanzees eating meat showed exceptional tolerance of observers, as did adult male chimpanzees in apparent consort relationships. Factors that we were

Table 10.4 Comparison of communication with other group members following detection of observers

Impact	Adult males		Adult females		Immatures	
	Gorilla (n = 62)	*Chimp* (n = 52)	*Gorilla* (n = 58)	*Chimp* (n = 59)	*Gorilla* (n = 65)	*Chimp* (n = 20)
None	3	65	71	31	72	25
Complete	97	35	12	25	3	5
Selective	0	0	15	44	25	70

unable to document probably also played a role, such as the apes' position with respect to the boundaries of their home range.

In many encounters, observation and/or audition of other group members immediately after the detector's response allowed us to assess the communicatory impact of the response in terms of alerting conspecifics to the observer's presence. Three types of communicatory impact of a detector's response were recognized: *None,* where all others in the group remained unaware of the observer's presence; *Complete,* where all others were alerted to the observer's presence; and *Selective,* where some were warned and others in the group remained unaware of the observer's presence.

Table 10.4 shows the communicatory impact of the responses of the different age–gender classes of gorillas and chimpanzees. In 193 of the 228 observations of gorillas' responses, the communicatory impact could be determined. In 49% of these there was none; 38% were complete and 13% were selective. However, differences between the age–gender classes were striking, and the responses of silverback males alerted all members of the group in 97% of cases.

The 18 cases where the communicatory impact of the silverbacks' response could not be determined were those when the silverbacks silently monitored the observer. It is likely, however, that communication did occur, as in all cases the group moved away from the observer. The communicatory impact of the responses of adult female and immature gorillas was similar. Hence, in 72% of cases no other gorillas were alerted and the whole group was alerted infrequently. The responses which communicated selectively did so through the auditory components of curiosity (chest-beating, slapping and clapping), or from running during avoidance, which drew the attention of nearby gorillas to the observer.

The communicatory impact of 149 responses for the chimpanzees could be determined. In 45% of cases there was none; 27% were complete and 28% were selective. The responses of adult male chimpanzees alerted all members of the detector's subgroup slightly more often than did those of adult females (with 35% and 25% of the cases, respectively), but adult male responses warned none of the other members of their subgroup, twice as often as those of adult females did (65% versus 31% of cases). All cases of selective communication by chimpanzees involved known or presumed mother–offspring pairs. When an adult female chimpanzee detected an observer, in 44% of cases she approached an infant, took it in a ventral or dorsal carrying position, or waited for a juvenile to approach her before leaving. The reverse situation, of an immature approaching an adult female immediately on detecting an observer was shown in 40% of cases. In a further 30% of cases, loud or soft vocalizations by an immature attracted the attention of an adult female, which was presumed to be the mother, who approached the immature and became aware of the observer's presence. Other members of the subgroup ignored the distress vocalizations of immatures in all but one case (Tutin *et al.*, 1981).

10.4 DISCUSSION

First, it is of interest to consider how gorillas and chimpanzees perceive intruding primatologists. The likely alternatives seem to be that observers are considered as: potential predators, potential competitors for food, or as an unfamiliar species with neither positive nor negative associations.

In the tropical forests of Gabon, the only potential predators of the apes, apart from humans, are leopards (*Panthera pardus*), which occur at low density and are largely nocturnal. At Lopé, we found evidence of predation by leopards on apes only once. The victim was a juvenile gorilla debilitated by illness. Predation is possibly a rare event, but a leopard might be expected to take advantage of an easy opportunity to kill, which arises from any lack of vigilance by the apes. We have observed only one interaction between a leopard and apes at Lopé, when a large subgroup of chimpanzees gave loud alarm calls and climbed high in trees on discovering a leopard in the process of eating a large duiker (*Cephalophus sylvicultor*).

The few descriptions of responses to potential predators by chimpanzees and gorillas given elsewhere show that chimpanzees, when severely threatened, actively defend themselves and use loud vocalizations and noisy charges to intimidate the predator (Gandini and Baldwin, 1978; Tutin *et al.*, 1981) and, in an extreme case, chimpanzees were observed to attack and kill a leopard cub (Hiraiwa-Hasegawa *et al.*, 1986). The intimidation display (charge and loud vocalizations) of silverback gorillas is a well-documented response to human hunters, and adult gorillas, when threatened or wounded, have been known to attack humans (Schaller, 1963; Sabater Pi, 1966; Cousins, 1983).

Of the potential competitors for food among the sympatric fauna, forest elephants (*Loxodonta africana cyclotis*) are the most significant in terms of biomass. They occur at a mean density of 1.5 per square kilometre in the study area. The elephants are relatively frugivorous, and dietary overlap is particularly extensive between them and gorillas, as it includes the common herbaceous plants as well as many species of fruit (Williamson, 1988). Our observations indicate general mutual tolerance between apes and elephants, but on five occasions gorilla silverbacks responded to the approach of elephants with intimidation displays. In all cases, the elephants departed rapidly.

Relations between the two species of ape are characterized by mutual avoidance but, again, silverback gorillas occasionally direct intimidation displays at approaching chimpanzees (Tutin and Fernandez, 1987). In sum, gorillas sometimes react with intimidation displays to species that are likely to be competitors for food.

If the apes in our study area had had no previous exposure to humans, the reactions that we observed reflect their responses to a novel experience. For adult apes, new experiences of this type must be extremely rare as all of the sympatric species of mammals must be very familiar. Fear of the unknown is probably an adaptive response but the costs of avoiding the area occupied by the intruder will vary, depending on the resources that brought the apes there. It is likely that apes will be less willing to leave a rare resource, such as a large fruiting tree, than a common one, such as a thicket of herbaceous plants.

In close-range, unexpected encounters with us, the behaviour shown by gorillas and chimpanzees was similar, with both giving intimidation displays of great effectiveness. Chimpanzees showed intimidatory displays only in close-range encounters, and it seems likely that in these circumstances observers were perceived as a danger equivalent to that posed by a potential predator. Apart from this, fortunately small, subset of close-range encounters, the responses of gorillas and chimpanzees to an observer's presence differed, as did those of the different age–gender classes of each species. While we cannot be sure of how we were perceived by the apes, it seems likely that we were initially unfamiliar. As time passed, some encounters gave negative associations to our presence, some were neutral, but none were positive. Given this, the cost of interrupting ongoing activities at any particular encounter may have been very important in determining the response of the apes. Support for this comes from the exceptional tolerance shown by chimpanzees encountered while eating a black colobus monkey (*Colobus satanus*) which they had just killed; and that shown by adult chimpanzees on five occasions when accompanied by a single adult female. In all these apparently consort relationships, the adult male detected the observer but did not move away until the adult female (who had not detected the observer) finished feeding and approached him. The relative costs of avoidance could rarely be assessed, but they are unlikely to have exerted consistently different influences on the two species.

The striking differences between the two species in their responses to observers seems to be directly related to their different forms of social organization. Whenever a silverback gorilla detected an observer he showed behaviour that protected his group: alerting them to the observer's presence and often approaching to place himself closest to the perceived threat. When gorillas, other than the silverback, detected the observer their responses of either avoidance or curiosity rarely alerted other group members. The frequency with which adult female, and especially immature, gorillas showed curiosity (which often included approaching or climbing in order to obtain a clear view of the observer) indicates that the consistently protective role of the silverback reduces the fear shown by other individuals. If, as seems likely, the social organization of the gorillas at Lopé resembles that of the mountain gorillas, the silverback is the father of all infants and young juveniles in his group.

The responses of adult male chimpanzees to an observer rarely acted to protect conspecifics. On the other hand, adult female chimpanzees very often responded to the observer's presence with behaviour that protected infants or juveniles (apparently their own offspring). This manifestation of maternal behaviour was seldom shown by gorilla females, and was never observed as an initial response after detecting an observer. It seems that the paternal behaviour of the silverback gorilla alleviates the need for adult females directly to protect their immature offspring.

Nicolson (1977) compared some aspects of mother–infant behaviour in chimpanzees in three conditions: a small cage with a single (subordinate) companion; a large enclosure with a group of six; and in the wild at Gombe. The amounts of time that mothers spent in contact with their young infants was least in the caged environment and most in the wild, with the enclosure mother being intermediate. This suggested that some aspects of maternal behaviour are influenced by the environment, and that in a small cage with minimal potential danger the mother was less solicitous of her infant. A similar explanation may apply to the differences that we observed between gorilla and chimpanzee mothers at Lopé. The gorilla mother is in a stable social environment, and the silverback actively protects all of his group from any external threats or disturbance. In multimale chimpanzee communities, with constantly changing subgroups, adult males show no consistent defensive or specific paternal behaviour, and the mother has sole responsibility for the safety of her offspring.

This difference may have other effects, and we observed that when a warning was given by any member of a group, young gorillas looked at the silverback, while young chimpanzees looked at their mothers.

10.5 CONCLUDING REMARKS

Such observations are pertinent to the care of gorillas and chimpanzees in captivity and should strongly discourage any separation of mothers from their normal groups. It is almost certain that maternal behaviour will be less efficient in a totally 'safe' and unchallenging environment, and most cases of maternal neglect, or even abuse, occur when mother–infant pairs are kept in social isolation.

ACKNOWLEDGEMENTS

We thank the L.S.B. Leakey Foundation, World Wildlife Fund–US and World Wide Fund for Nature International, and especially the Centre Internationale de Recherches Médicales de Franceville, for supporting our research; Mr R. Dipouma and Alphonse Mackanga of the Direction de la Faune, Ministère des Eaux et Forêts for permission to work in the Lopé Reserve. We are very grateful to the following colleagues, who contributed observations while working at Lopé: Catherine Bouchain, Jean-Yves Collet, Anna Feistner, Stephanie Hall, Ann Pierce, Liz Rogers, Ben Voysey and Liz Williamson. We also thank Bill McGrew and Liz Williamson for constructive comments on the manuscript.

11

Primate conservation and wildlife management

SIMON K. BEARDER

11.1 INTRODUCTION

Despite the intelligence, adaptability and opportunism for which primates are renowned, many species are increasingly unable to cope with the pace of change induced by the rapidly expanding human population. They are undergoing a catastrophic decline in numbers and diversity, along with most other nonhuman inhabitants of the tropical and subtropical regions of the world. This chapter summarizes the seriousness of the plight that faces wildlife in the tropics, with particular reference to simians. It examines ways in which they can be helped to survive into the twenty-first century in the light of their ability to withstand some change; and it outlines the advantages of peaceful coexistence between humans and our closest living relatives.

More is probably known about nonhuman primates than any other single order of mammals. Their importance is reflected by their extensive use in research by anthropologists, psychologists and biomedical scientists. Their shared ancestry with humans means that they are used as models for interpreting the adaptations of early hominids, and for reconstructing earlier phases of human evolution, as well as for understanding human physiology and disease. The simian specialization for a diurnal lifestyle (extreme visual acuity, colour vision, reduction in reliance on the sense of smell), which is unusual among mammals, including the prosimians, has made them of special interest in comparative studies of sensory and perceptual abilities, and the brain. Their diverse expressions, complex interactions and subtle long-term relationships make them a target of research on the principles that govern social organization (Smuts *et al.*, 1987). The practical and academic value of primates has, in turn, led to concern from conservation biologists. If it proves impossible to protect nonhuman primates, which are relatively well understood, what hope is there for other species of larger mammals, and, ultimately, for humans?

11.2 THREATS TO PRIMATE DIVERSITY

Two major problems beset estimation of the conservation status of the primates. One is the obvious difficulty of collecting reliable data on numbers and distribution for arboreal species in dense habitats. The other is the lack of agreement on the naming of species, which complicates any comparison of results from different authorities (e.g. the *IUCN Red Data Books*, see Lee *et al.*, 1988; and the *Convention on International Trade in Endangered Species of Wild Fauna and Flora*, CITES, 1977).

The most comprehensive survey of primate distribution, abundance and conservation is that of Wolfheim (1983), who considers the interaction of factors contributing to the decline of all nonhuman primates, which she divides into 149 species, including 116 simians and 33 prosimians. Wolfheim makes a useful distinction between proximate and ultimate influences on the population status of each species. The extent to which present-day influences affect primate populations depends on species-specific characteristics. The most important ultimate component of a species' status is the size of its original geographic range, followed by its degree of habitat specialization, its body size, its population density and behavioural traits. Thus, in general, a large primate with specialized habits, and habitat requirements occupying a small geographic range at low density, will be most vulnerable – for example, the mountain gorilla (*Gorilla gorilla beringei*), the bonabo (*Pan paniscus*) or the orang-utan (*Pongo pygmaeus*). Proximate factors responsible for changes in primate populations may be naturally occurring events, such as modifications in climate or topography (the formation of new watercourses or mountain ranges), which generally take place very slowly. These changes, however, are usually overshadowed by human activities, which can alter the prevailing conditions very rapidly. These include deforestation, due to clearing for agriculture and timber, selective logging, the use of herbicides, the spread of exotic vegetation, warfare, water control projects and damage by livestock, and human predation such as hunting, pest control and the collection of live animals for trade (Wolfheim, 1976, 1983).

The advantage of Wolfheim's approach is that it attempts to assess the contribution of a range of interacting variables which either tip a

Table 11.1 Species within each simian family that fall in three major status-rating categories (from Wolfheim, 1983)

Family	*'Threatened'**	*'Vulnerable'*	*'Safe'*
Callitrichidae (16 species)	6 (38%)	5 (31%)	5 (31%)
Cebidae (25 species)	11 (44%)	6 (24%)	8 (32%)
Cercopithecidae (65 species)	39 (60%)	14 (22%)	12 (18%)
Hylobatidae (6 species)	5 (83%)	1 (17%)	0
Pongidae (4 species)	4 (100%)	0	0
Totals (116 species)	65 (56%)	26 (22%)	25 (22%)

*Wolfheim subdivides the 'threatened' rating into three categories of severity, which are omitted here.

species towards extinction or increase its likelihood of survival. This results in a species by species estimate of the degree of threat on a five-point scale. The derived status ratings of simian species are summarized in Table 11.1. The interaction between constellations of factors affecting any given species is evidently complex and difficult to assess, but the attempt to do so provides the most detailed appraisal of conservation status that is available for any group of mammals. According to this analysis, 56% of simians are 'threatened' (i.e. they are likely to face extinction if current trends continue) and 22% are 'vulnerable' (i.e. they are likely to become threatened if conditions worsen). This leaves only 22% that are considered 'safe'.

The 25 simian species that are not considered to be in danger are: Goeldi's monkey *Callimico goeldii*, and the marmosets *Callithrix argentata* and *Cebuella pygmaea*; two tamarins, *Saguinus fuscicollis* and *S. midas*; two howler monkeys, *Alouatta caryaya* and *A. seniculus*; the night monkey (treated as one species, *Aotus trivirgatus*); two capuchins, *Cebus albifrons* and *C. nigrivittatus*; the squirrel monkey (treated as one species) *Saimiri sciureus*; two saki monkeys, *Pithecia monachus* and *P. pithecia*; five guenons, *Cercopithecus aethiops*, *C. ascanus*, *C. mona*, *Erythrocebus patas* and *Miopithecus talapoin*; three baboons, *Papio anubis*, *P. cynocephalus* and *P. ursinus*; and four langurs, the *Presbytis aygula–melalophus* group, *P. cristata*, *P. entellus* and *P. phayrei*. Although there is room for disagreement on the species divisions and the precise extent of threats to each one (see CITES, 1977; Lee *et al.*, 1988), this list essentially includes those forest species which are relatively small and/or wide-ranging (callitrichids, cebids and talapoins), and those species which can survive, to some extent, away from forests (e.g. vervet monkeys, patas, savannah baboons and langurs). For most species, however, the effects of habitat alteration and human predation have tipped the balance towards decline, even when they appear well equipped to survive, as in the case of macaques which can often live in close proximity to human settlements, but in so doing are vulnerable to exploitation.

The overriding importance of the proximate factors which have brought primates into this perilous situation have been discussed in detail (Harper, 1983; Wolfheim, 1983; Oates, 1985; Mittermeier *et al.*, 1986; Mittermeier, 1986a; Stevenson *et al.*, 1986; Eudey, 1987; Mittermeier and Cheney, 1987). Suffice it to say that more than 80% of primates spend most of their time in trees and nearly all primates depend on trees for sleeping and defence from predators (Napier, 1971). The major habitats occupied by primates, the tropical forests of the world, cover only about 7% of the land surface, but they contain approximately half of the total number of plant and animal species. They are presently being destroyed at such a rate (20–40 ha/min) that, at a conservative estimate, there will be no more than isolated pockets remaining by the middle of the next century (see Figure 11.1) (Sommer, 1976; Myers, 1980, 1986; Iltis, 1983; Western and Pearl, 1989). Even in areas where the forest canopy is still intact, such as in central and western Amazonia which has only recently been invaded by modern technology, long-sustained hunting pressure by the human inhabitants of these regions has resulted in the collapse of populations of many larger birds and mammals (Terborgh, 1986b). Consequently, primate conservation is not secure, even in areas which appear from a distance to be intact and pristine.

11.3 ACTION NEEDED TO HELP ENDANGERED SPECIES

If the future diversity of primates is to be secured, it is the proximate causes of their destruction that need to be addressed (see Oates, 1985; Mallinson, 1986a; Mittermeier, 1986a; Eudey, 1987; Mittermeier and Cheney, 1987; Oates *et al.*, 1987). Wolfheim (1983, p. 751), for example, concludes:

> Habitat reserves offer the only reasonable hope for the survival of most primates. Our concern, effort, and money should be directed toward establishing more reserves, obtaining a better understanding of primate populations in nature, and educating more people about the beauty, value, irreplaceability, and need for conservation of tropical wildlife.

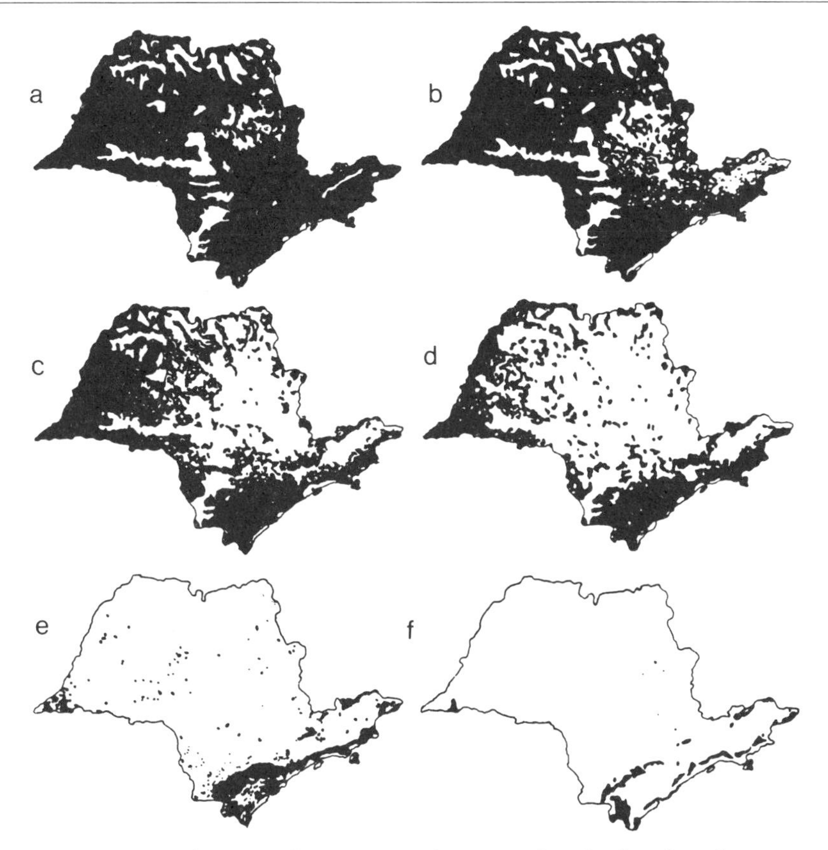

Figure 11.1 Map of São Paulo State, southeastern Brazil, showing the proportion of forest cover (dark areas) (a) in the primitive situation, (b) in 1886, (c) in 1920, (d) in 1952, (e) in 1973 and (f) the prospect for 2000. (From Victor, 1975.)

Primates should not be considered in isolation from other species; the modern threat is to entire ecosystems and the pressing need is for properly financed international cooperation to avert global declines of catastrophic proportions (Bishop, 1980; Fischer, 1982; Soulé, 1986; Myers, 1988). The importance of maintaining intact ecosystems lies in the fact that they represent complex, interdependent webs of life in which the removal of one species has potential repercussions on many others (May, 1979; Frankel and Soulé, 1981; E.O. Wilson, 1988). This interdependence is most clearly seen in the relationships between plants and insects (Janzen, 1978), but primates play a role in pruning vegetation, as pollinators and in the dispersal of seeds (Richard, 1985).

There is, therefore, concern that a focus on saving individual species detracts from the more important goal of maintaining ecological integrity. For example, the creation of new parks to save a given species may simply divert resources from existing ones which have greater prospects for long-term viability (Western, 1986). Furthermore, an overemphasis on completely protected nature reserves, which generally tend to be small or else difficult to police effectively, may cause other possibilities to be overlooked. Forest destruction is frequently contrary to economic benefit, since it eliminates the sustainable use of forest products and the protective cover which they provide (water catchment, soil retention, etc.) (Struhsaker, 1981; Weber, 1987).

Most wildlife occurs outside reserves and national parks. There is considerable scope for species survival in the land adjoining reserves and in other rural areas, either where the wildlife, including primates, already complements human welfare, or where it can be developed in a way that maintains diversity while allowing sustainable exploitation; for example, in game-ranching areas and forests in which fruits, nuts, meat, timber and medicinal products are selectively harvested (Strum, 1986; Western, 1986).

In practice, as certain species of special aesthetic, scientific or medical value become increasingly rare, they act as catalysts for concerted conservation efforts at many levels. This is particularly the case with primates, which are effective in attracting public interest (see Oates, 1985; Eudey, 1987). Essentially, five categories of action are recommended in the case of endangered primates (Mittermeier, 1986a/b):

1. Locating and determining the status of target species.
2. Improving prospects for their survival in the wild through research, protection, the elimination of illegal trade and, if necessary, genetic management.
3. Increasing awareness at all levels of the importance of the species and its habitat.
4. Establishing conservation-oriented training for citizens of countries where endangered species occur.
5. Developing integrated captive breeding programmes, if viable and politically feasible.

Indeed, it can be argued that, in the face of declining wild populations, captive breeding is also the only justifiable option for nearly all primates that are to be used for biomedical purposes, or held in captivity for research, education or recreation.

11.4 PRIMATE STUDIES AND CONSERVATION

This volume is concerned with the way in which simians respond to environmental change. Knowledge of the limits of their behavioural adaptability is of obvious importance in the rescue of selected species, but it also plays a role at the level of the ecological community. How can the study of primate responses to environmental change contribute to appropriate conservation action in the broader realm? Such knowledge: (1) gives an early indication that all is not well; (2) provides evidence of how best to plan forest management when that is deemed necessary; (3) serves as a focus of attention for conservation education, and (4) provides a basis for developing a significant tourist attraction. Each of these dimensions will be considered in turn.

Primates as indicators of environmental health

It has been argued (e.g. see Chapter 1, section 1.1) that primates are generally unspecialized and adaptable. However, it is also the case that, as a group, the primates are an interesting paradox. On the one hand, they are obviously well equipped to modify their behaviour in the face of a wide variety of changing circumstances (see Chapters 1, 2, 6, 7, 8, and 9; Richard, 1985; Marsh *et al.*, 1987). On the other hand, they are poorly suited to recovery from population declines. The characteristics of a relatively large brain, a long period of infant development and a long lifespan, which provide the potential for behavioural adaptability, are frequently associated with a low reproductive turnover (Martin, 1975), and the need for each individual to learn a great deal during its lifetime (see Chapter 3).

Ironically, therefore, species which have evolved considerable behavioural flexibility also have very low fecundity. Fecundity is lowest in the larger bodied species, and these are generally the most seriously threatened; they are poorly placed for rapid colonization, or to replace their numbers when depleted (note that the callitrichids are an interesting exception to this rule). Furthermore, the high proportion of learning that occurs in a social context, including the transmission of traditions across generations, can result in primates being particularly vulnerable if their social systems become disrupted. Although low fecundity and the importance of social traditions are not confined to primates, but can also be found in other animal groups (e.g. carnivores, ungulates, cetaceans, some birds, etc.), it is generally true that when primate densities begin to decline there is cause for serious concern. This can lead to conservation initiatives that can benefit not only primates but many thousands of other animals and plants. This is illustrated by a wide range of primate-centred projects initiated by conservation bodies, such as the World Wide Fund for Nature and the Fauna and Flora Preservation Society.

Primate ecology and forest management

Contrary to initial impressions, it emerges that the relatively constant climatic conditions of tropical forests are typically associated with unpredictable production of resources by their vegetation (Terborgh 1986a/b). Periods of flowering and fruiting, which are frequently synchronized in a way that ensures cross-pollination and enhances seed dispersal, may be separated by gaps of up to a year or more when there is little or no production of fruit and flowers. Different plant species appear to be responding to the same environmental signals, which do not necessarily follow a seasonal pattern but may be due to unpredictable fluctuations, such as abrupt dry spells or periods with an unusual drop in night-time temperature. The result is that animals that concentrate on these food sources (primates, birds and bats) are faced with irregular fluctuations in their availability and abundance.

Detailed long-term studies of primates and other frugivorous species at the Cocha Cashu forest reserve in Peru (Terborgh, 1983) have been used to illustrate the particular importance of a relatively small number of plant species that are vital to survival ('keystone plant resources': Terborgh, 1986a/b). Five primate species that eat the same fruits for most of the year (two capuchins, *Cebus apella* and *C. albifrons*; a squirrel monkey, *Saimiri sciureus* and two marmosets, *Saguinus imperator* and *S. fuscicollis*) shift their diets at times of food scarcity; in this case a dry season from May to July. They subsist on a small number of keystone plant resources (palm-nuts, figs and nectar). At this time, each species tends to exploit a different plant source and there is also divergence in their use of animal prey. Other predominantly frugivorous species (many other mammals and birds) appear to rely on the same few resources during the dry season which, therefore, take on a special significance. It is concluded that a mere 12 plant species, less than 1% of some 2000 at Cocha Cashu, sustain almost the entire frugivore community for 3 months of the year. Indications that a small number of keystone species are, similarly, vital during periods of scarcity in other tropical forests, together with the fact that frugivores represent a high proportion of the larger vertebrate biomass in these forests (Emmons *et al.*, 1983; Terborgh, 1983), mean that keystone species have important implications for the management potential of forests. For example, the keystone species occupy only a very small part of the space in a forest and, if they are allowed to remain intact, it may be possible to maintain something of the diversity of tropical wildlife in multiple-use areas outside the scattered reserves (Leighton and Leighton, 1983; Terborgh, 1986b).

Primate adaptability and conservation education

Attempts to release primates into the wild provide good examples of a powerful link with conservation education. Irrespective of the controversy that exists over the relative merits, or rates of success, of different kinds of project (Konstant and Mittermeier, 1982; Caldecott and Kavanagh, 1983; Harcourt, 1987b), there is no doubt that they emphasize the importance of conservation of the natural habitat. They may also draw attention to what can be achieved if attention is given to details of the animals' requirements (see Chapter 9).

The coordinated programme of captive breeding, preparation and release of golden lion tamarins (*Leontopithecus rosalia*) in the remnant Atlantic coastal forest of Brazil provides a good example of conservation education, which has had an effect on people in many countries. The international management plan established by the Golden Lion Tamarin Cooperative Management Committee includes efforts to protect and restore the last refuges for this species. In addition to long-term field studies, which include rehabilitated individuals from the captive population, there is a programme of medical, behavioural and genetic management of the species in captivity, both in Brazil and internationally. The work is underpinned by a well-managed education programme centred in Brazil, but which has ramifications worldwide (Kleiman *et al.*, 1986). As a result, one of the most seriously threatened animals, which numbered around 100 individuals in nature, has been saved from extinction. Captive populations grew from 70 animals in 12 zoos in 1972 to nearly 400 individuals in 50 zoos by 1984, with an annual increase of 20–25%. The wild population now has a chance of recovery. Collaboration and public enlightenment have been the keys to this remarkable success, in addition to the devoted work of particular individuals and their supporting institutions. It is an unfortunate fact that many other primates are fast approaching the need for similar rescue plans, for which this project will provide a blueprint (Mallinson, 1986).

Field stations in key conservation areas, whether built for release programmes or for a combination of research, environmental education and recreation, have proved to be a relatively inexpensive and extremely effective means of protecting habitats, or ensuring that they are used in a sustainable way (Butynski, 1985; Mittermeier, 1986b;

Figure 11.2 The endangered woolly spider monkey or muriqui (*Brachyteles arachnoides*) of the Atlantic forest of Brazil. (Photo: Simon Bearder.)

Aveling and Aveling, 1987). Here, primates are valuable as 'flagship' species in establishing such centres, and in attracting the necessary funds to promote environmental education. Good examples among many others are the muriqui or woolly spider monkey (*Brachyteles arachnoides*) (Figure 11.2) in Brazil and the yellow-tailed woolly monkey (*Lagothrix flavicauda*) in Peru (Figure 11.3) (see references in IUCN/SSC Primate Specialist Group Newsletters). There is considerable potential for the development of cooperative links between zoos or wildlife parks and field centres in the countries from which captive animals originate. Such links can provide feedback between those with the interest and resources to support conservation initiatives, and the sites of interest. The establishment of further collaboration is a matter of urgency.

Figure 11.3 The yellow-tailed woolly monkey (*Lagothrix flavicauda*) of the Peruvian Andes. (Photo: Andrew Young.)

Primate responses to the proximity of people

A final area in which primates contribute to habitat protection concerns their ability to tolerate the presence of people who benefit from watching undisturbed wild animals in their constantly changing environment of a complex society. The tourist potential of primates obviously varies with species. Some are more attractive than others, or are more easily habituated (see Chapter 10). It is also true that uninhibited animals can become a menace (see Chapter 8), and the situation may be reversed so that tourists become detrimental to the primates through inappropriate feeding, excessive disturbance or the transmission of disease. Nevertheless, the potential for successful primate-related tourism is well illustrated in the case of mountain gorillas (Harcourt, 1986; Aveling and Aveling, 1987).

Mountain gorillas lend themselves to close approach by humans as their habitat is sufficiently open, and they do not usually travel far in the course of a day. This allows them to become habituated through regular, cautious approach. In Rwanda and, more recently Zaire and

Uganda, international conservation efforts have halted the decline of relict populations of this subspecies (around 500 individuals). They may even be on the increase, mainly as a result of the development of tourism. This example serves to illustrate that even in Africa, where the human population increase is 3% per annum, where thousands of square kilometres are converted to agriculture each year, and where timber extraction is on the increase, it is possible to conserve wilderness areas if there are tangible benefits to substantial numbers of local people and there is no significant drain on the treasury. Coupled with the protection of gorillas is an increasing public awareness of the importance of forest reserves for human welfare.

Other primates are less easy to follow or habituate. Chimpanzees are perhaps at the opposite extreme to gorillas; they can take months to approach in the first instance, and their wide-ranging movements mean that, without artificial provisioning, they can be hard to find even when habituated (Goodall, 1986). Forests can seem uninviting to the uninitiated, but the rewards of watching wild primates are considerable. The future development of the touristic and educational potential of wild primates will make a vital contribution to their continuing survival.

11.5 ADVANTAGES OF PEACEFUL COEXISTENCE WITH WILD PRIMATES

One species of primate, *Homo sapiens*, appears to be unique in the extent of its adaptability. The human body, however, is not obviously better equipped than other primates to withstand dramatic changes to its environment. Rather, survival in a diversity of habitats is made possible by the extent to which humans can change their environment to suit their needs. But this ability to control nature is not infinite. Change to the environment, and particularly the accelerating pace of change resulting from modern technology and the exploitation of new sources of energy (e.g. hydroelectric projects, nuclear and fossil fuels), is having a dramatic effect upon other species, and pushing their adaptability to the limits. It is becoming increasingly obvious that human disturbance of the ecosphere not only compromises the survival of thousands of other species, but it represents a most serious global crisis that threatens to engulf the perpetrators of the change.

The long-term survival of both human and nonhuman primates depends, among other things, on the conservation of forests. The once-rich mantle of tropical forest protects an inherently fragile region of the globe that easily changes into desert. Tropical forests provide an enormously complex and irreplaceable reservoir of genetic potential which can be exploited for food, timber and medicine; and their function in absorbing and regulating the flow of tropical rainfall has far-reaching implications (Myers, 1983, 1984; Sutton *et al.*, 1983). The myth of boundless fertility of the soils supporting tropical forests has been exploded; their destruction is often irreversible (but see Jordan, 1986). The dramatic decline in sustainable productivity in areas which have been cleared of forest has brought serious consequences for local people and for global temperatures and rainfall patterns (Myers, 1984; Caulfield, 1985). Only as people appreciate the reality of this situation will they act to do something about it. The principles of what to do are well established (IUCN, 1980), but it remains to be seen whether the action advocated by conservation biologists can be decisively implemented (Soulé, 1986; Myers, 1980, 1988; Western and Pearl, 1989). The plight of nonhuman primates is intimately linked with our own, and the extent to which we are able to ensure their survival provides an important barometer to the success of conservation policies.

11.6 CONCLUDING REMARKS

The threat of extinction looms menacingly over many primate species; their ability to adapt to changing circumstances is frequently being grossly exceeded by the extent of human impact in the form of habitat destruction and hunting. Their future undoubtedly depends on effective conservation of wild populations, with knowledge of the limits of their adaptability, but it is the preservation of intact forest and woodland ecosystems that will benefit all primates, including humans, in the long term. The task of reversing forest destruction may seem daunting, but it has to be done without delay if the productivity and stabilizing function of forests are to be maintained for the future. The longer it takes, the greater will be the loss of species diversity. The world will not quickly recover from such impoverishment, and the consequences upon coming generations will be severe.

Part Three

Environmental Change in Captivity

Golden lion tamarins (*Leontopithecus rosalia*) in an outside cage in the marmoset complex at the Jersey Wildlife Preservation Trust (see page 220). (Photo: Phillip Coffey.)

As with the contributions in Part Two, those in this part were chosen because they are of topical interest, and represent growth points in theoretical and/or practical issues.

Within that general perspective, Leah Scott (Chapter 14) describes some inexpensive and easily managed ways in which the otherwise unstimulating environments of common marmosets may be enriched. Her's is a commonsense approach to a problem in laboratory research where primates are used in a variety of biomedical experiments, as in pharmacology, and monkeys are housed individually. Common marmosets are a robust species in many kinds of captive environment, some of which give minimal opportunities for social contact and physical exploration. Leah Scott addresses questions of improving welfare on these terms. It is an approach which, given the situation, is laudable and realistic, and it is to be hoped that many research workers will follow her example.

From a different standpoint, Margaret Redshaw and Jeremy Mallinson (Chapter 12) describe the progressive attempts made within the resources of the Jersey Wildlife Preservation Trust to stimulate natural patterns of behaviour in two species of endangered primate, the lowland gorilla and the golden lion tamarin. Their concern is with changes in behaviour which result from changes in the patterns of physical and social resources available to the animals, given that the size, cognitive abilities and natural lifestyles of the species are strikingly different. The work on golden lion tamarins is of additional interest in the light of the release programmes for the species in a natural habitat in Brazil (see Kleiman *et al.*, 1986). It is important from many points of view that representatives of primate species, especially endangered ones, should be maintained in ways which preserve their cognitive, social and physical skills.

It is also the case that we may learn a good deal about the natural behaviour of different species by presenting them with specific and well-defined challenges in captivity. Dorothy Fragaszy and Lee Adams-Curtis (Chapter 13) consider questions of ecological validity and novelty in experimental tasks, and show how benign challenges can be used to study sensorimotor skills and the influences of the social context in which they develop, as well as to enrich the environments of capuchin monkeys. These animals are ideal candidates for such investigations; they are highly manipulative, socially attentive and 'bright'. It is of interest, however, to find that once again, good clear empirical evidence for the acquisition of a skill by socially mediated learning is as elusive as ever.

Other contributions in this part of the book interface behavioural and physiological responsiveness. Much concern about environmental change, for example, particularly that which is felt, justifiably or not, to be deleterious to the life of an animal, involves the concept of 'stress'. David Warburton (Chapter 18) discusses the mechanisms involved and distinguishes the 'stress response' – namely, 'a complex pattern of cognitive and physiological change that prepares an animal for action' – from the concept of distress, which, by definition, refers to detrimental effects of prolonged environmental stressors. His discussion encompasses some aspects of species differences in these regards and recognizes the importance of these responses for health.

Species differences in responsiveness are further discussed in exemplary detail by Sally Mendoza in her review (Chapter 17) of the long-term work on titi monkeys and squirrel monkeys carried out at Davis in California. These species differ significantly and consistently in their social organizational responsiveness and in their interactions with the physical environment, which clearly influence the ways in which each of them responds to environmental change. Moreover, behavioural responsiveness is reflected in differences in the pituitary-adrenal and autonomic responses, which, within ameliorating influences of the social context within which they occur, 'creates a dynamic loop between the social system, the participating individual, and its internal regulatory processes'.

In addition, there are two contributions on the relationship between environmental change and reproductive status in marmosets and tamarins. The mating systems of this family have many unusual features for simian primates, and there is much current interest in them. Following on from an earlier paradigm (French and Snowdon, 1981), Jeffrey French and Betty Inglett (Chapter 15) describe responsiveness to unfamiliar conspecifics in two species of tamarins. For example, in cotton tops unfamiliar females are treated with indifference by the breeding females of their social units, compared with the aggressive behaviour shown by female golden lion tamarins under similar circumstances. These differences are interpreted in terms of the ways in which reproductive exclusivity is maintained by breeding females in each species. Hence, female cotton tops are able to inhibit ovulation in other females, whereas female lion tamarins do not do so and, therefore, may use aggressive behaviour to defend their breeding status actively.

Given the interest in mechanisms of reproductive control in callitrichids, David Abbott and Lynn George (Chapter 16) present a specific set of cases which clearly demonstrate the absolute relationship between social status and reproductive condition in female common marmosets. As with no other female primate known so far, the suppression of ovulation, or its commencement, is rapidly switched in either direction by a change in social status. A more detailed and comparative discussion of these fascinating issues is given in David Abbott's (Chapter 4) other contribution in the first part of this book.

12

Stimulation of natural patterns of behaviour: studies with golden lion tamarins and gorillas

MARGARET E. REDSHAW and JEREMY J.C. MALLINSON

12.1 INTRODUCTION

Captive environments for primates have improved dramatically over the last 20 years. It remains the case, however, that all too many can only be described as remarkably dull. They lack stimulation, and apparently take little account of the needs of the caged inhabitants. Concrete or tiled boxes, with a few fixed metal structures, relate poorly to what is known of the behaviour, social relations and complex knowledge of the environment that these animals use in the wild. With such caging, the result is often a poor exhibition, of limited educational value, in which the animals look and often behave inappropriately. Earlier this century there might have been the excuse that little or no information on the natural behaviour was available; but the last 30 years have seen a vast increase in the number and quality of field studies on an ever-widening range of species.

The kind of 'improvements' made to captive environments depend, for example, on whether the aim is to imitate nature and attempt to reproduce a natural environment, or to use technology to entertain and occupy the animals. The use of token systems, television sets and computer keyboards have been in vogue, although the long-term benefits of these devices are unclear and opinions vary as to their usefulness. Certainly, the public can be entertained in this way, particularly with interactive devices, but if we are really interested in education – specifically, conservation education – then these appear to have little to offer. Interactive exhibits which are of marginal educational value, at best aid anthropomorphism and, at worst, can encourage an unwarranted familiarity with species that may be dangerous. More importantly, they may do little by way of furthering respect for an animal and passing on knowledge of the special adaptations that fit it for the particular ecological niche it occupies in the wild. Many animals, including primates, are capable of learning tricks, but in the current climate of conservation awareness and of primate needs the emphasis should be on encouraging as much natural behaviour as possible (see Chapter 19). Although individuals are caged, they are not circus animals. In line with this perspective, the aim at the Jersey Wildlife Preservation Trust is to maximize the acquisition and practice of naturally occurring skills, and to help the animals capitalize on all the environmental and social opportunities available. The exhibits and management methods described provide a complex and appropriate social grouping, in an environment that is sufficiently large and interesting, with a routine and a diet that encourage considerable amounts of foraging as well as feeding activity.

It must be emphasized that such captive environments are not static. Change comes from the outside, as with any improvements in the art of animal management, and from the inside, where there are dynamic forces in the form of maturation, changing social relationships, and seasonal variation.

In considering the effects of environmental change and improvement, the golden lion tamarin (*Leontopithecus rosalia*) and the western lowland gorilla (*Gorilla g. gorilla*) were selected as case studies for widely differing reasons. The first species is considered to be on the brink of extinction in the wild state, a situation in which captive breeding has already been able to play a valuable role (Kleiman *et al.*, 1986; see also Chapter 11, section 11.7), while the second species, because of its intelligence and relatively close phylogenetic relationship to humans (Redshaw, 1978), is among the primates most likely to suffer the consequences of poor conditions in captivity.

12.2 MANAGEMENT AND HUSBANDRY OF GOLDEN LION TAMARINS AT THE JERSEY ZOO

Background to the current project

A pair of golden lion tamarins was exhibited at the Jersey Wildlife Preservation Trust as long ago as 1961–2, but no breeding was recorded at this time (Mallinson, 1964). However, the experience gained with this species and the wide range of other species of callitrichids kept and bred at the Trust during the 1960s and 1970s has been invaluable in developing guidelines for the psychological and physical well-being of these small-bodied South American primates.

Numerous aspects of the management and husbandry of the collection of marmosets and tamarins have been changed or modified since the early 1970s. In the absence of any established guidelines, marmosets and tamarins were initially housed as pairs in one inside unit. This consisted of a string of mesh cages each measuring approximately 240 × 140 × 210 cm high, and separated by double wire. Branch material, for example, was at first provided in the form of 'a tangled mass of branches' at the top of the cage, in which the marmosets and tamarins found it difficult to manoeuvre. Subsequently, this was changed and a series of straighter, sloping branches was provided that would allow the fast quadrupedal running and jumping form of locomotion that is typical of this family. A substrate of sawdust was provided and servicing was via a long corridor which also functioned as a public viewing area. The room temperatures ranged from 17 to 21°C, though this was raised in specific areas to as much as 29°C with the use of infrared lamps. Details of the management and diet are reported in Mallinson (1971, 1975a), King (1975), Carroll (1982) and Allchurch (1986).

In summary, the animals were kept largely in pairs, in relatively close proximity to other species and conspecifics, with little possibility of privacy, and without natural vegetation or sunlight. At this time the management of marmosets was concerned with the very real practical difficulties associated with keeping these tropical animals in considerably cooler climates. Skeletal problems associated with the unusually high requirement for vitamin D in callitrichids (King, 1975, 1976) tended to take priority over their psychological needs. At the Trust the methods of ensuring an adequate supply of this vitamin have gradually changed. From 1964, injections were given at intervals of 10–11 months, but since 1971 supplementation by oral administration has been used for animals kept inside. It has been reduced and phased out for those kept with access to outside areas. New World Primate Pellet, which is an integral part of the diet, also contains vitamin D_3.

Successful breeding can, and indeed has, occurred in widely differing circumstances, but such success is not necessarily a good indicator of suitable conditions. A captive environment has often been an impoverished and overcrowded one in which the range and frequency of normal behaviour are reduced. Awareness of the deficits and limitations of the laboratory-like conditions led to experiments with outside accommodation (Mallinson, 1971, 1977). During the summer, and subsequently for a winter, some animals were put out on a small waterfowl lake and others were put into a large, planted outdoor cage with unheated, though waterproofed, boxes for sheltered accommodation. In general, they appeared to thrive, with thicker coats and higher levels of activity. Behaviours not previously seen included appropriate freezing and hiding responses to airborne predators, swimming, and catching and eating small birds.

The present situation

In designing the new facility for marmosets and tamarins, the first part of which was in operation in 1975, a systematic attempt was made to provide optimal captive conditions. The aim was to ensure successful long-term, multigeneration breeding and to facilitate the expression and practice of as wide a range of natural behaviour as possible in these increasingly endangered monkeys. A detailed description of the facility is given in Mallinson (1975b). An immediately obvious consequence of the change in accommodation was a marked increase in the rate of parent-rearing amongst all the species of breeding marmosets and tamarins that were kept at that time. This did not include golden lion tamarins.

However, golden lion tamarins have been kept again at the Trust since June 1978, when a pair was received on breeding loan from the Smithsonian Institution's National Zoological Park, Washington DC. Since that time, Jersey Zoo has cooperated in the international plan for management of the captive population (Kleiman *et al.*, 1986; Mallinson, 1986a). Further pairs have been received, and more than 40 infants have been successfully reared to date (January 1990). Each pair is housed separately, with up to five offspring in a group. Further details of the management of this species are given in Carroll (1982).

The housing for each group of golden lion tamarins comprises a heated indoor cage and an outside area. The outside cages have a floor area of between 8 and 9 m^2, and are between 2 and 2.5 m high, with a natural earth base that is planted with grass and growing shrubs, creepers and plants (see the first page of Part Three). Natural branches are also provided, making a network of perches. The cage is constructed from 2.5-cm weld-mesh and opaque screens to prevent physical and visual contact with other groups. The inside is a secure, heated area, with a central service corridor not accessible to the public. Individual cages are provided with perches, a nest box and an infrared spotlight. The outside areas are used by the animals all year round unless, as occurs only rarely, there is ice or snow. Even in near freezing temperatures they go outside and after chilling seem to warm up rapidly by basking in the warmth of the hot spots inside.

The changes in the accommodation described above, with its considerably increased space and variety, as well as associated acclimatization to a temperate zone, aimed to provide many of the social and environmental conditions that would assist a captive breeding population to be selfsustaining. As Konstant and Mittermeier (1982) have emphasized, if reintroduction is to be a viable possibility, the maximum of behavioural variability must be maintained in captive colonies.

Strategy for reintroduction

The importance of a captive species retaining as many of its natural characteristics as possible was illustrated when some captive-bred lion tamarins that had been kept in small cages escaped from captivity in Brazil into the surrounding forest (Coimbra-Filho and Mittermeier, 1977). One animal stayed in the undergrowth and did not climb into the trees, while others were very clumsy and, as a consequence, were easily recaptured. The behaviour contrasted with that of those which had been kept in a large cage and appeared to adapt better to wild conditions.

The current project to reintroduce golden lion tamarins in Brazil is now successfully using family groups that have experienced the captive conditions described above (Beck, 1987; Mallinson, 1987). However, now that reintroduction of this species has become a reality, further efforts are being made in Jersey to provide a yet more natural habitat as a training ground for other groups that could be future candidates for reintroduction.

Examination of the information gained from the earlier stages of the reintroduction project suggested areas where the experience of captive animals could be improved, as well as indicating which kind of social unit increases the chances of survival (Beck, 1987). For example, younger animals have been found to survive longer, and members of age-graded family groups appear to fare better than paired adults. As with many primate species, new skills are more easily introduced to, and acquired by, juveniles in a group. Perhaps not surprisingly, older animals were found to be slower to learn training tasks where the goal was to forage for natural foods and to locomote on natural vegetation. Captive-bred tamarins reared and housed in environments with milled wood furniture in fixed configurations seemed reluctant to use natural vegetation and appeared deficient in orienting through a natural forest. Though captive-bred animals appear to have an innate response of avoidance to overflying raptors and mob when the raptors are perched, methods of dealing with other dangerous animals have to be learnt by trial and error.

The groups are released into a nest box mounted 3–5 m up in a mature tree and provided with an adjacent embedded food source in the form of a 'feeder'. This device, which can be roped into position, is constructed from lengths of plastic tubing, which are filled with bananas and marmoset diet, and the ends are then plugged. Smaller holes in the sides of the tubes allow for foraging and access to the food. For the first day or so feeding takes place close to the nest box, but, as each group explores further afield, so the travel distance to the food source is gradually increased. This supplementary feeding is continued on a daily basis for the first 10 months of reintroduction, and then marmoset diet alone is provided on alternate days. Invertebrates, such as meal worms, are also made available from early on by locating these in nearby bromeliads.

With this kind of management strategy, naturally occurring food and nest holes are encountered and foraging becomes a habit. Unfortunately, some of the released animals from Jersey are inclined to follow dropped food onto the ground, a behaviour that has no adverse repercussions in enclosed cages, but which in the forest could put them at risk from ground-dwelling predators, such as snakes.

Captive behaviour in more 'naturalistic' situations

The successes and the changing strategies employed in the rehabilitation project, which is now using more cost-effective 'on the job' experience for the animals rather than the original, relatively time-consuming prerelease training, have confirmed the validity of some aspects of the captive management practices and indicated where changes might be made. With this kind of emphasis in mind, a small island was constructed at Jersey in the middle of a waterfowl lake, on either side of which were wooded areas containing groups of lemurs. The island itself has a surface area of approximately 90 m^2. Heated indoor accommodation ($1.8 \times 1.0 \times 1.5$ m high) raised above the ground is provided, and the surrounding turfed area has young trees (swamp cypresses and weeping willow), logs and ropes (Figure 12.1).

The aim was to create a more naturalistic 'open' situation, with increased behavioural scope. After a group of golden lion tamarins was established there, a key focus of interest was their response to, and use of the various environmental features of the island. Comparisons could then be made with the behaviour of similar groups of lion tamarins, as well as other species of marmosets and tamarins which are housed in the wired-in enclosures of the marmoset complex buildings.

Time-scan data were collected at 5-min intervals. Each sample included information on individual activity, location and type of

Figure 12.1 The golden lion tamarin island at Jersey Wildlife Preservation Trust. (Photo: Phillip Coffey.)

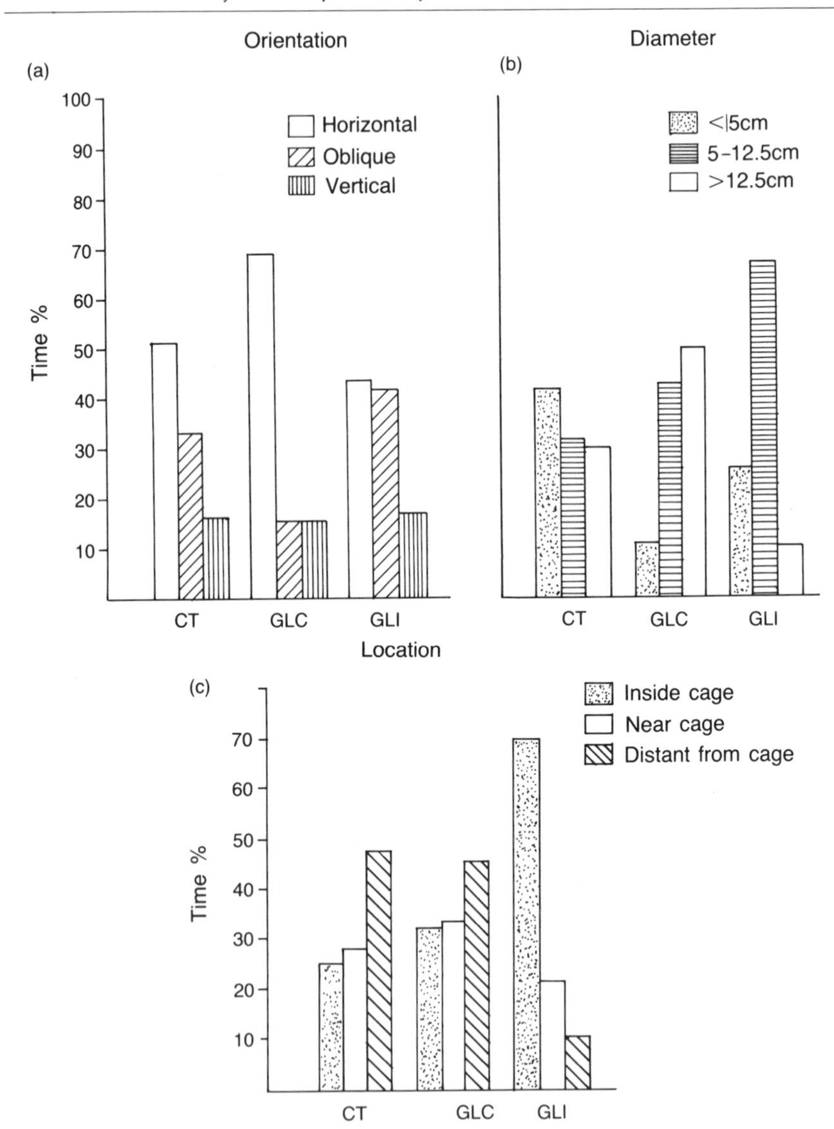

Figure 12.2 The type of support preferred (a,b) and use of cage area (c) at Jersey by golden lion (GLC) and cotton-top tamarins (CT) in the complex, and by golden lions on the island (GLI).

substrate used. A group of lion tamarins was first relocated on the island in May, but observations on all the groups were made between 08.30 and 17.00 h during July and August.

The behaviour of the tamarins with regard to use of the caging, and the dimensions and orientation of preferred support, is illustrated in

Figure 12.2. In addition to showing how they behave on the island and in the complex, data are also presented on cotton-topped tamarins (*Saguinus o. oedipus*) housed in the same complex. From Figure 12.2 it can be seen that the tamarins on the island spent most of the time for which they were observed in the inside accommodation. This may be a consequence of the unfamiliarity and openness of the island environment, but as this family group had already been in occupation for 3 months they might have been expected to have habituated to the new circumstances. However, in comparison with the cotton-tops, even in the safe environment of the wired cages, the golden lions spent more time inside. That the island situation may be experienced as a more dangerous one by the lion tamarins is perhaps shown by the data on the time spent on the ground. In the complex, this ranged between 6 and 11% of the time, and contrasted with the cotton-tops and similar data on silvery marmosets (*Callithrix a. argentata*), for which the figures were 3% and 2%, respectively. On the island, the tamarins were very infrequently seen on the ground and the comparable figure is less than 1%.

The data for the use of the different types of support, which are also presented in Figure 12.2, show, not unexpectedly, that the larger and heavier golden lion tamarins (600–700 g weight), both in the complex and on the island, used the thinnest supports least of all, and less frequently than the smaller and lighter cotton-topped tamarins (450–500 g weight). In the cage, the lion tamarins clearly preferred the thickest branches, though on the island the situation is complicated by the fact that there were very few supports available of the largest category. Hence, the use of the intermediate sized perches is likely to have been exaggerated. A similar situation has arisen with the orientation of the supports used. In the cage, the lion tamarins showed a marked preference for horizontal branches, but on the island the availability of these was reduced, apparently forcing greater use of obliquely sloping branches, many of which were positioned at quite a steep angle, as Figure 12.1 shows. They were at times seen to use the forks of the young trees while resting individually, though there was little room for the kind of grooming sessions observed on broader, horizontal perches. The tamarins may also have been reluctant to use open branches, as in such a situation they could have been vulnerable to predation by gulls.

That the lion tamarins on the island were not seen to use the loosely fixed ropes may be a reflection of their lack of experience with this material, since the usual zoo cages were not fitted with these. Alternatively, in view of the morphological adaptations of the hands and feet of the callitrichids and the mode of locomotion usually adopted, it seems that such furnishings, unless sufficiently tightly strung to resemble natural branching, may be unsuitable.

In the wild, it seems that although the *Leontopithecus* species can exploit less well-established secondary growth forest, they rely on tall primary or mature secondary forest for sleeping holes and as a source of animal prey. At present, the island accommodation is more like sparse secondary forest in the process of becoming established. However, with the provision of a wider range of supports and an increase in cover above ground level using flowering creepers and shrubs as well as the continued growth of the young trees already planted, a more suitable and 'natural' environment will result.

In the marmoset complex cages, activity on the ground usually consists of foraging amongst the vegetation. This also takes place at other levels within the cage, where there is bark to be peeled off, and where there are creepers and flowering shrubs which attract insects. Most of the species in captivity seem to spend some time searching for live food, though the amount of effort expended in doing so varies. Observations indicate that outside the cage, silvery marmosets spend around 12% of their time foraging; for lion tamarins, the figure is 15–20% and for cotton-topped tamarins it is 25–30%. In addition to active searching, the lion tamarins were also found to spend 8% of the time intently watching and scanning the grassy cage floor from a perch, apparently on the lookout for live food. The comparative figure for the other species was less than this, at around 4%. Foraging bouts seemed in general to be quite short, averaging around 10 seconds. Amongst young animals these were particularly likely to be interspersed with bouts of play chasing and other social activities.

Golden lion tamarins in the natural situation are said to take larger animal prey than other callitrichids, and with their relatively long hands and fingers they are described as more manipulative foragers, searching in holes, breaking open rotten wood and prising off bark (Rylands, cited in Macdonald, 1984). It is thus perhaps not surprising that of all the callitrichid species kept in the collection at Jersey, the lion tamarins appear to most favour the live food offered to them (J.B. Carroll and C.C. West, personal communication). In contrast to the other species they show little or no hesitation in taking and feeding on all items, including pink mice, meal worms, crickets and locusts. Data in the zoo records indicate that supplementary items, notably sparrows, tits, earthworms and field mice have been taken in the outside areas. So far on the island, there has been little opportunity for such skills to be put into practice. Ducks from the surrounding pond, which visit the island to feed on dropped items of food, have elicited little in the way of response other than avoidance, though a noisy aggressive mobbing action by the group occurred in response to a moorhen. Perhaps, with an increase in vegetation cover and more experience in the open, the taking of young birds will be observed here too.

Evaluation of management practices

The success of the management practices currently in use can be judged in a variety of ways. The animals seem less stressed; they appear healthier and more active; stereotyped behaviour is rare; they reproduce well; and parent-rearing is the norm. As far as the environment is concerned, the use of planted and perched outside areas have provided behavioural opportunities of which all the species, including the golden lion tamarin, take good advantage. A wider range of flowering plants and shrubs, in conjunction with a policy of less vigorous pruning, and the addition of more dead and rotting branch material would provide greater cover and stimulate yet more activity in the outside areas. It might be harder work for the visiting public to find and watch the animals but, like the island exhibit, it is likely to function as a more effective tool for conservation education.

In the future, there will undoubtedly be further changes in the environmental facilities provided for marmosets and tamarins at the Trust taking more account of the species differences in biology and behaviour. As has occurred recently, the most useful leads for these may come from field studies, rehabilitation and reintroduction projects. Nevertheless, regular and improved monitoring of animals in the captive situation is vital if the effects, beneficial or otherwise, of 'improvements' and changes are to be assessed adequately. Only in this way will it be possible for captive breeding to contribute to conservation directly, by providing the best prepared animals for projects such as the golden lion tamarin reintroduction programme.

12.3 MANAGEMENT AND HUSBANDRY OF LOWLAND GORILLAS

Background to the present project

With small-bodied animals like the marmosets and tamarins the physical wear and tear on the captive environment is minimal in comparison with the havoc that a large-bodied primate like the lowland gorilla *Gorilla g. gorilla* can produce. Though the design requirements are thus very different, the underlying principles of management policy are similar. Hence, as with the marmosets and tamarins, changes in the captive environment for apes have been made with the aim of maximizing the behavioural opportunities, and increasing the chances of the animals retaining and practising as many of their natural manipulatory and social skills as possible.

In the wild, gorillas inhabit a forest environment that is a complex three-dimensional one in which they may spend nearly half of their day feeding on over 50 different plant species and some invertebrates, and range over an area of approximately 5 km^2 (Harcourt, 1987c). This contrasts dramatically with the situation of most gorillas in captivity, where movement is constrained and diet is limited.

Like many other zoos, Jersey has kept apes in relatively small cages, largely made of concrete, metal and glass. However, the present situation for lowland gorillas is rather different from this and, as with the callitrichids, many factors have combined to contribute to the changes that have taken place.

The gorilla complex at Jersey, built in 1971, covered a relatively small area and contained three small inside cages (each measuring 3 × 3.6 × 4 m), which also functioned as sleeping dens. Only a small group of animals could be kept and, with the public able to look into the dens and outside areas from two or three sides, little privacy was available. Heated shelves gave some degree of comfort and night bedding was provided. In the outside area (measuring 15 × 10 m) there was some concrete and metal climbing apparatus, but relatively little room in which to charge about or to retreat. The effective size of the outside area was reduced as a consequence of it sloping quite steeply, though there was a pool at the lowest level in the area.

The design of this building, which was built to house an adult male gorilla and two females, followed quite closely the conventions of ape accommodation provided by other reputable zoos. Viewed from the present, however, its inadequacies seem obvious. Despite the deficiencies, the animals mated and from the resulting pregnancies nine live births took place there. Inexperience, isolation and other factors resulted in the first six infants being hand-reared, though, subsequently, infants were cared for by their own mother. Details of the management and diet used at this time are recorded by Mallinson *et al.* (1973). The limitations of the building, and individual differences in temperament, resulted in the adults being separated for long periods, particularly the young silverback male (Johnstone Scott, 1984).

Requirements of gorillas in captivity

The size and strength of great apes like the gorilla can preoccupy the minds of those designing enclosures. However, from an animal welfare point of view the physical well-being and psychological needs of apes should be paramount in considering what features ought to be incorporated in a new facility. Safety and finance, though important limiting factors, are none the less secondary. The aim must not simply be, as has occurred in many cases, to design just another showcase for an exotic animals' shop window.

Changing attitudes within zoos generally, and a wealth of field data on the natural behaviour and ecology of mountain gorillas, as well as the successful births at the Trust, led to increasing awareness and pressure for changes in the gorilla housing and management. Though it might have been possible to increase the level of entertainment and interest in the old complex by the use of various devices (Maple, 1979; Markowitz, 1979), this was regarded as neither sufficient nor appropriate with these intelligent and sensitive great apes.

The situation in Jersey was not unusual, for many zoos have a history of problems in gorilla management, specifically in the areas of infertility, rearing, individual incompatibility and the habits associated with boredom, stress and a lack of stimulation. If one asks the question: 'What do captive gorillas need in order to be physically and mentally healthy, and to reproduce and rear offspring?', there do seem to be some basic requirements that have emerged from work with captive and wild animals. These needs include being able to do the following:

1. Forage extensively and feed frequently.
2. Climb, nest-build and handle a wide range of objects and materials.
3. Rest in sunny, shaded or sheltered places as prevailing weather conditions and inclination permit.
4. Be private, to have places in which to retreat and sufficient space to avoid others.
5. Socialize and be part of a group containing the full range of age–sex classes of individuals.

In an attempt to take account of as many of these needs as possible, the aim has thus become one of building up and maintaining a large social group, consisting of adults and their offspring, and possibly additional females, with as natural a lifestyle as possible (Johnstone Scott, 1984).

Changing fashions in ape accommodation have resulted in a move away from rocks and concrete grottos in favour of larger, more interesting inside and outside areas. Regrettably, there is still a tendency to put the vegetation behind, rather than in with the animals, providing the visitors with a realistic back-drop, but giving no fun or stimulation to the apes.

The needs just listed can be responded to in different ways. In some facilities, social groups of gorillas are kept in large, high, weld-mesh cages containing considerable quantities of apparatus, with a base of deep straw litter. Others have larger, wooded areas (where many of the trees are protected) surrounded by a moat, often with the additional use of electric fencing.

The present situation

In Jersey, an answer was developed which included a large landscaped outside area of approximately 1950 m^2 surrounded by a vertical wall (Figures 12.3 and 12.4), together with two inside exhibition areas of 7.3 × 4.8 m, three smaller off-public dens, a small crush cage and an adjacent kitchen and food preparation area. The inside areas have power-operated doors, underfloor heating and are furnished with metal platforms and numerous sisal ropes. (A detailed description of the facility is given in Mallinson, 1980.)

The environmental changes implicit in the design of the new gorilla complex have inevitably been linked with changes in management routine, social grouping and nutrition that have taken place at the same time. A more natural, yet varied diet, with an increased proportion of vegetables, whole plants, and more cut forage than previously, has gradually been put into use (King and Rivers, 1976; Allchurch, 1986). The Trust now has an organic farm adjacent to the zoo which grows many items suitable for the gorillas, such as sunflowers, French beans, chard and bamboo. Cut forage includes oak, horse-chestnut, hawthorn and willow (Hicks, 1976).

Transfer to the new facility took place in stages. Initially a group of three subadult animals was moved to the new building in December 1980 (Bowen, 1980). The silverback male, and an adult female and infant were moved in May 1981, followed by another adult female with her newborn infant and older juvenile offspring in July of the same year. Details of births, as well as other transfers and moves occurring since this time, are given in Johnstone Scott (1984).

Initial responses to the new accommodation

Data on the effects of the move on the subadults (a male and two females aged between 5 and 6 years) were collected when the animals only had access to the inside areas, as the landscaping and planting of the outside enclosure had yet to be completed. Nevertheless, the information gained is of some interest as this part of the new accommodation

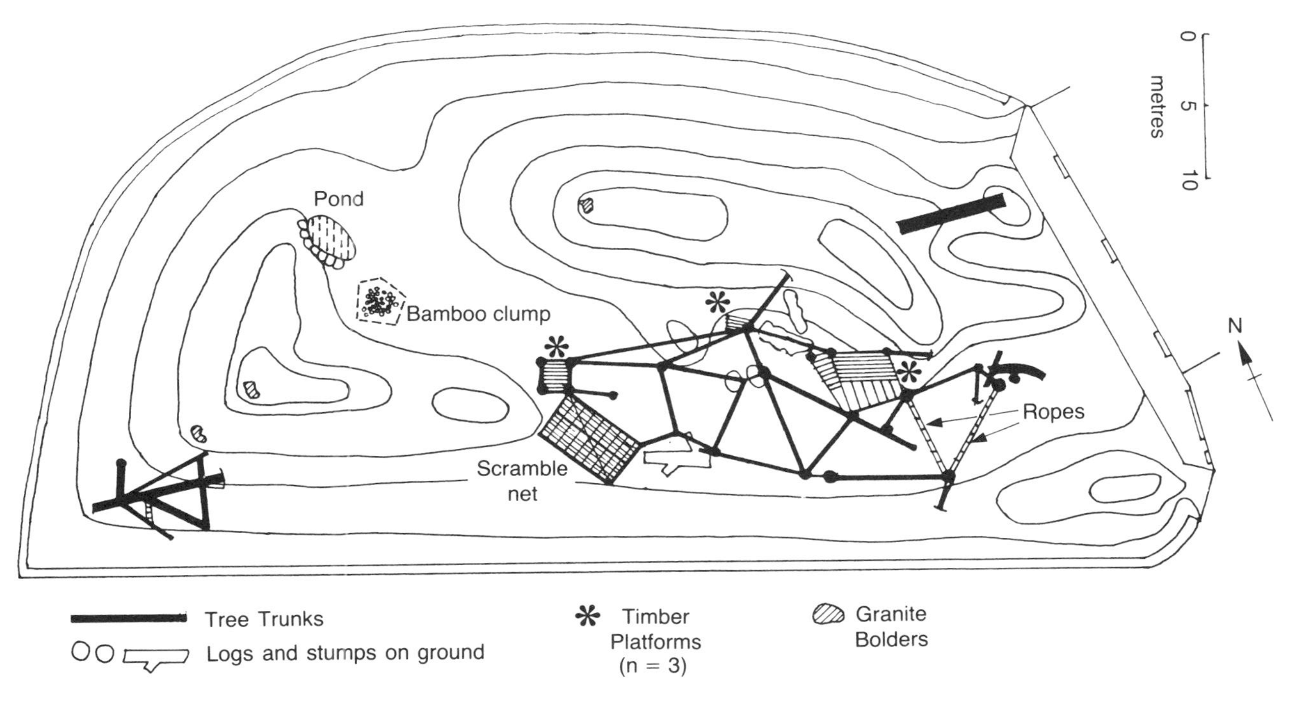

Figure 12.3 Map of outside area of gorilla complex area at Jersey, showing location of hills, apparatus and trees.

Figure 12.4 A group of lowland gorillas feeding and foraging in the outside gorilla area at Jersey. (Photo: Phillip Coffey.)

gave the animals a marked increase in space with a nearly two-fold increase in area and volume available.

Examination of Bowen's (1980) data from the period immediately prior to the move, and over the 8 weeks that followed, indicated an initial increase in locomotor activity and stereotyped behaviour and a decline in social play. During the first few days, the most obvious response of the immature gorillas to the new and unfamiliar surroundings seems to have been to rush about and explore, though, like other stressed animals, they then rocked or shuffled about and were less willing to indulge in the usual play-wrestling and chase games. These adverse effects were short-lived, for over the following weeks the social behaviour recovered, and stereotyped behaviours declined to pre-move levels or below. More positive long-lasting effects were observed in increased levels of manipulatory activity and foraging behaviour. The latter was probably achieved by the more frequent provision of small items like sunflower seeds, peanuts and raisins, and increased amounts of fresh forage. Though greater foraging could have been brought about before, it was undoubtedly facilitated by the closer contact by the keeper and observation that the new food preparation and access to the den areas allowed. The provision of extra manipulanda in the form of rubber rings and balls after the move also contributed to the increased levels of manual activity.

Possibly as a consequence of being confined in relatively small cages on a busy site, with little privacy and a high visitor flow, these young and growing animals appeared to have been stressed in some degree before the move. Two of them plucked bald patches on the belly, arms and groin, and all three rocked to some extent when tired or distressed. After the move the situation improved; only one animal was observed to pluck hair, and that infrequently.

Overall, it appeared that the subadult gorillas responded positively to the change and were interested in taking full advantage of the new facilities. Access to the outside area was given in April 1981. As Figure 12.3 shows, this area contains two hills, isolated boulders, a shallow pond, a stand of bamboo, several trees, logs, platforms and other wooden and rope climbing equipment. It is grass covered, with brambles, nettles, sorrel and clover growing well. On being let out into this large, well-furnished enclosure, the subadults showed only momentary hesitation and then explored the area thoroughly (Bowen, 1981). Considerable time and effort were devoted to playing around, investigating, picking and eating the different types of vegetation, particularly the bamboo. In the first few days they showed little interest in the climbing equipment, though subsequently it was in frequent use. In general, these subadults appeared to adjust rapidly to the new circumstances and routine, and were clearly on home ground when the next group of animals from the old building were introduced to the area.

Some of the adult gorillas showed a similar reluctance on being let out into the enclosure for the first time. This was most evident in the silverback male, who was zoo-born and had only ever been familiar with a substrate of concrete. Though tending to explore the periphery by means of the narrow concrete wall and gully which bounded the grassed area, he traversed most of the terrain, and even on the first day was seen using the highest point as a look-out. The adult females, juveniles and other individuals that have subsequently been introduced to the group have, without hesitation, enjoyed the grass and other plants, sunny slopes, shelters, logs and other equipment available.

In the last year or so, the main group has consisted of the adult male, three adult females, two juveniles and an infant. However, the composition continues to change, as happens with wild groups. An additional, nonparous, female has gradually been introduced, and now lives with these individuals, and more births are expected.

Use of the outside area

Because of the radical differences in facilities in the space available, and in the possible social groupings for the adult gorillas and offspring, no comparable behavioural data are available for activity in the previous gorilla accommodation.

By the summer of 1986 all the individuals were totally familiar with the large enclosure, and quantitative time-scan data on activity and use of the outside area were collected. These showed the captive, adult lowland gorillas to be relatively sedentary, at least at this time of the year. The adult male spent nearly three-quarters of the time resting and the adult females followed with between 54% and 70% of the time spent similarly. As expected, the juveniles were more active, resting for only 20% of the time. Though they spent 15% and 17% of the time feeding and foraging, this was somewhat less than their mothers. The two highest estimates of feeding and foraging time, of 20% and 24%, were in fact for these multiparous females that were still lactating and feeding the juveniles. This contrasted with the 14% recorded for the silverback. The figure for the female with an infant of less than 2 months was intermediate.

While the adults were resting the juveniles were active, social activity predominated (28% of the time) and consisted largely of rough and tumble play with each other, though on a few occasions this was with the silverback male. Manipulation of objects was frequent (13% adults and 20% juveniles) and often involved free objects, particularly items of vegetation, in nest-building and draping activity. Locomotor activity was also common, often in the form of solitary locomotor play involving climbing, running and rolling down the grassy slopes.

The impression is that the adults are now considerably more active socially, manually and in gross motor terms than in the old complex. Estimates of the time spent in locomotor activity ranged from 5% to 12% of the time. This mostly consisted in walking around the area looking out for leftover food items, avoiding other individuals, catching up with juveniles, and in moving from sun to shade or vice versa.

Figure 12.5 shows the use of the different parts of the enclosure by the individual group members, sampled approximately 130 times at 10-min intervals. All the group members had a preference for the southerly side of the area, where most of the apparatus is located. This is also where they can be partially sheltered from the wind, rain or sun, and out of sight behind large horizontal and vertical logs. The other half of the enclosure is more open. All the adults clearly had preferred vantage and resting places, while the juveniles had favoured play

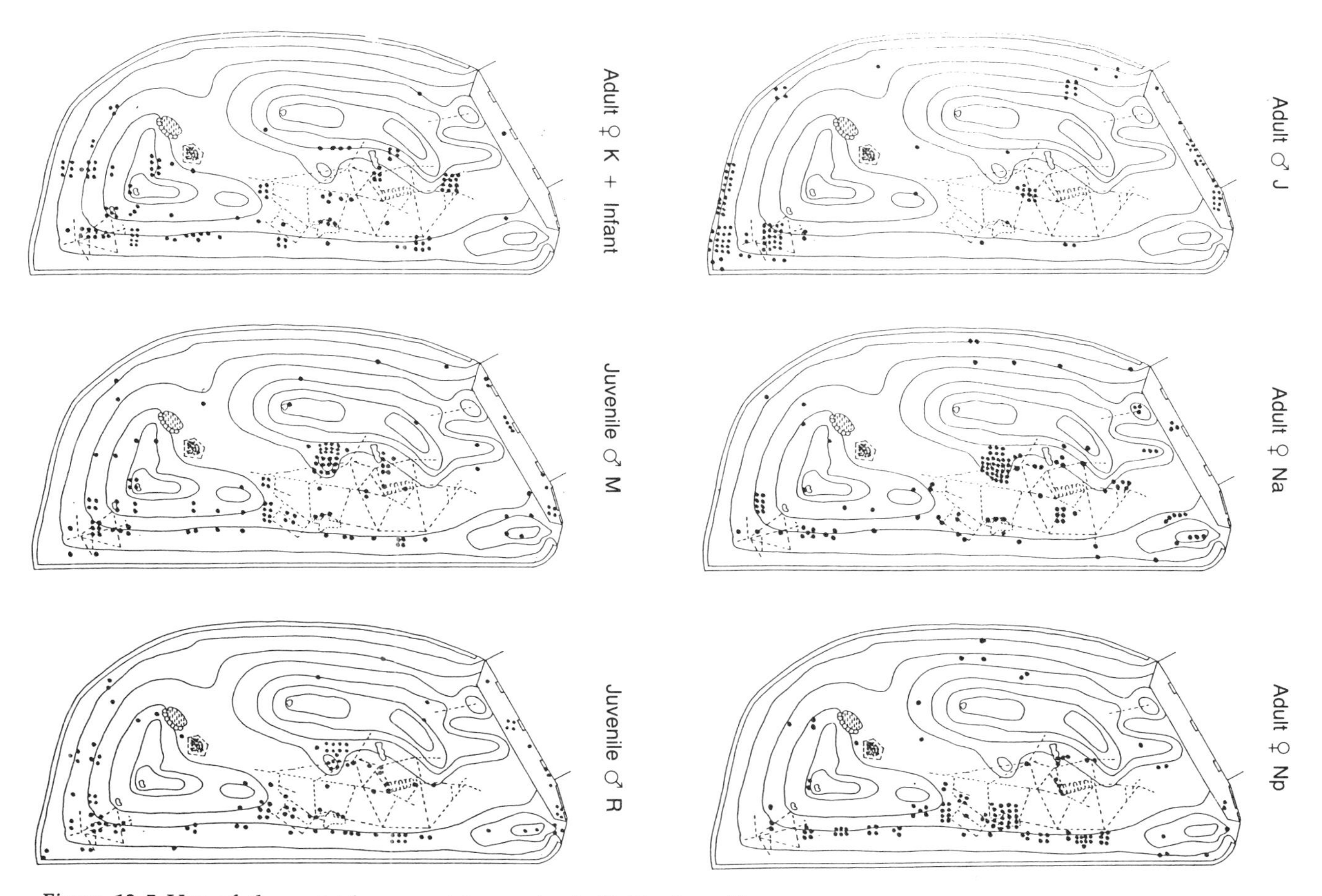

Figure 12.5 Use of the outside area at Jersey by individual gorillas.

areas. The adult male had three distinctly preferred locations and, in order, these were: the ditch at the bottom end of the enclosure, the concrete apron in front of the house, and the area of low tree stumps in the middle of the apparatus area. The other adult gorillas overlapped little with this, although the juveniles, as with wild youngsters (Harcourt, 1979), at times seemed highly attracted to the silverback and indulged in both solitary and vigorous social play in close proximity to him. This occurred in the ditch and apron areas.

The adult female with the young infant also frequented the areas used by the male, spending time in his view, though not immediately adjacent to him. That this should occur is not surprising in view of the finding among wild gorillas (Harcourt, 1979) that the younger the infant a mother has, the closer the proximity that is maintained with the silverback. The other adult females with juvenile offspring behaved differently from this. One would intermittently seek closer proximity with the male, and she was nearly always observed on the upper levels of the area, preferring to rest on a small, fairly central hillock. The other rested lower down, usually quite close to climbing apparatus and, in general, sought to avoid the male. In the choice of location both females were undoubtedly influenced by their offspring's enthusiasm for the climbing equipment and for romping with each other.

The use of different parts of the enclosure by the juveniles was similar, and in respects like that of their mothers, with one using the low central hillock more, and the other the lower area that runs parallel to the long straight ditch. With regard to the 'recreational value' of different habitat features, young gorillas use the hillocks and logs for 'king of the castle' and chase games with a central obstacle. Ropes and nets are used for social and solitary locomotor play and the concrete apron has a suitable surface for noisy games involving much slapping of hands and feet. Slopes function as slides and places to roll. The vegetation is an endless source of manipulable materials and logs are used not only for climbing but also as foraging sites, since they can be de-barked and the softer rotting parts dug out and consumed.

The juveniles were led to some extent into the last-mentioned activities by the adult females, which showed unexpected enthusiasm for such behaviour. As with the logs, the use of the other environmental features by the females is generally associated with feeding and foraging behaviour. They feed on and forage widely amongst the growing vegetation, using the contours of the land for paths, and also search for the small activity foods that are distributed in the area daily.

On cooler days, sunny slopes or hillocks are used for rest; on warmer days, a different aspect or some of the large logs provide shade and a place to lean. The adult male range over the area less widely, foraging infrequently, and still prefering to rest on the drier concrete areas. The site chosen seems to depend on the time of day and whether warmth or shade is required. Despite more limited use of the area, observations of animal movements suggests that, as in the wild, the silverback is the individual which the other gorillas follow and whose movements they coordinated with their own. In cooler conditions, the adult male may be more restive and less able to settle, and thus seems to provoke more changes of location by the other group members. On an early morning with heavy dew on the grass the entire group has been seen resting up on the timbered climbing area around the silverback male.

From the animals' responses to the facilities provided, it does seem that many of the needs of these captive gorillas have been looked after in the present gorilla complex. They can be maintained in a large group; they are more active; there are few incidents of aggression; and a number of births have successfully taken place, with other individuals present, including one birth that took place outside. The overall size and the landscaping of the outside area have been invaluable in the integration of the unrelated subadults, with the natural visual barriers of grassed slopes and gullies providing effective temporary boundaries.

The outside enclosure, with all its many advantages, is more obviously suitable for the gorillas than the inside accommodation. However, with the addition of more ropes, bedding material, a net and an everchanging supply of perishable items that can be manipulated, the range of behavioural needs that can be satisfied indoors has increased. As S.F. Wilson (1982) emphasized in discussing environments for great apes, space is not enough: making a cage 'better' may be possible by simply increasing the number and types of objects it contains, rather than its volume. On the other hand, she found that activity level was linked to the number of animals in an enclosure, and since more space is needed to maintain a large group, these factors tend to go together.

Future plans at Jersey include developing the facility in order to maintain another group of lowland gorillas. The aim will be to expand the breeding capability at the Trust. This will allow better genetic management as part of the cooperative plan for captive breeding of this species in Great Britain.

12.4 CONCLUDING REMARKS

There is a constant need to improve and up-date captive environments in the light of knowledge gained from studying primates in the wild, and as a consequence of the experience gained in the process of captive management. Captivity inevitably represents a compromise, but the aim should be to manage animals as naturally as possible, with the benefits of modern veterinary care, suitable housing and good diet.

Nothing is ideal, but some captive environments, from the perspective in which an animal's needs are given priority, are clearly preferable to others. In the wild, primates such as the golden lion tamarin and the lowland gorilla live in a rich and complex habitat in which they range freely, rest, forage and feed. This is not to say that such a life is without hardship or danger, for they are threatened in many ways and from many different directions. In this situation they mate, give birth and rear their young, which in turn learn the practical and social skills required for a successful life in the forest. In captivity, the aim must be to maximize the possibilities for the exercise of these skills in as natural a way as possible. If there is to be any hope of successful reintroduction, currently and in the future, and any chance of retaining the qualities that make these primates what they are, zoological collections must make greater efforts to take account of the psychological and physical needs of the animals entrusted to them.

ACKNOWLEDGEMENTS

Thanks are due to the following for assistance and comments: Ian Stevenson, Eliza Green, Hickson Ferguson, Kirstin Sundsith, Bryan Carroll, Richard Johnstone Scott and Cibele Carvalho.

13 Environmental challenges in groups of capuchins

DOROTHY M. FRAGASZY and LEAH E. ADAMS-CURTIS

13.1 INTRODUCTION

We are interested in the behaviour of socially housed capuchin monkeys (*Cebus apella*) in captivity, particularly in their instrumental behaviour and the social context in which it occurs. We have found it useful to incorporate novel objects and tasks in our programme to provide environmental change.

The intentional provision of changes to the physical environment is a means of probing behavioural organization and adaptability in captive animals. Change can be viewed as a perturbation of the organism-environment system. Behaviour associated with change informs us about such underlying characteristics of behavioural organization as tendencies to approach or withdraw (Schneirla, 1946), modes of investigation (Glickman and Sroges, 1966; Visalberghi and Mason, 1983), the integration of social predispositions with nonsocial activities (Fragaszy and Mason, 1978; Box, 1984), or attention span and rate of habituation (Wood *et al.*, 1979). These characteristics are the proximate sources of species-typical behaviour, and, on a longer timescale, of behavioural ecology (Klopfer, 1969; Fragaszy, 1985).

Studies of environmental challenge in captivity can and should have reasonable ecological validity, by which we mean that the independent variables in the captive environment tap one or more dimensions of the species' natural environment, and that the dependent variables are relevant to coping with the real-world equivalent of the independent variables. These manipulations do not require a one-to-one correspondence with natural change, but should reflect types of changes that may be encountered. In other words, ecological validity is a matter of modelling features of a natural environment and observing behaviour with an appreciation for its contributions to coping with the modelled challenge.

For example, one can provide in captivity foods of a sort never encountered in nature, and learn much about patterns of food selection and consumption which are relevant to feeding behaviour in nature (Fragaszy and Mason, 1983). Experimental manipulations in the laboratory can be arranged to produce change at different rates, times, places, and in any number of dimensions, as with light, heat, air currents, odours, food, objects, and to permit behaviour that affects changes in these dimensions at varying rates or times. In these ways, experimental manipulations which involve environmental change can model real-world phenomena. The other half of ecological validity requires a working knowledge of a species' behaviour in its natural environment. Lastly, it is useful to maintain a focus on the process of adjustment to change. The process is likely to reflect behavioural characteristics of greater generality than the particular end-points of adjustment to specific (natural or unnatural) perturbations.

13.2 BEHAVIOUR IN NATURE AND BEHAVIOUR TOWARDS OBJECTS IN CAPTIVITY

Several characteristics of a species' behavioural ecology bear on the study of behaviour which is directed towards objects in captivity. Feeding habits serve as a good illustration.

Wild primates interact with objects primarily during foraging. In captivity, once their safety has been acknowledged, contacts with objects also typically involve behaviour which occurs during the capture or processing of food. Hence, the interpretation of such behaviour is enhanced by recognition of the role of behaviour in natural foraging. We shall use a comparison of squirrel monkeys (*Saimiri sciureus*) and capuchin monkeys (*Cebus apella*) to illustrate the point.

Wild squirrel monkeys forage for much of their day for small insects on the surfaces of vegetation, and for fruits (Boinski, 1986). In both natural and captive settings, they are vigilant; they inspect surfaces by sniffing, licking and touching. They are attracted to novel spaces and surfaces; and, in general, they behave in ways which maximize the probability that they will detect small objects on newly encountered surfaces (Fragaszy, 1979, 1981, 1985). However, in comparison to capuchins, squirrel monkeys are not forceful or persistent in their investigations of individual objects.

In nature, and in addition to ripe fruits and other readily available foods, capuchins exploit hard-shelled fruits, inner fibres of toughly encased palm fronds, snails with thick shells, and many other foods which require long attention and persistent manipulation (e.g. Izawa, 1979; Fragaszy, 1986; Robinson, 1986). Capuchins in nature and in captivity are persistent manipulators of objects, and behave in ways which maximize the probability of opening a surface or discovering hidden, embedded objects. This involves extended interest in a single object and performance of several different acts with the same object (Fragaszy, 1986; Fragaszy and Adams-Curtis, 1987).

The contrasting ways these two species interact with objects are apparent in captivity, even when the objects are not edible (Fragaszy and Mason, 1978; Fragaszy, unpublished data). The actions directed towards objects, the duration of interest, and the energy devoted to their destruction, all vary consistently between them. Capuchin monkeys typically pick up, handle, visually inspect, bite, pull, hit and rub any nonfrightening object they encounter. Squirrel monkeys typically briefly inspect the same object visually, sniff it, perhaps touch it, and then go on to other activities.

Behaviour towards objects in captivity is correlated with other aspects of behaviour in nature, in addition to foraging. Social affiliation, for example, influences the organization of investigatory behaviour. Animals living in cohesive social units in nature, such as another cebid, the monogamous titi monkey (*Callicebus moloch*), coordinate their inspection of objects with the group-mate (Fragaszy and Mason, 1978; Le Poivre and Pallaud, 1985). Those living in larger social groups, such as squirrel monkeys, show less coordination of activity (Fragaszy and Mason, 1978; Fragaszy, 1979; see also Chapter 17). M. Andrews (1986) has also shown that the titis' tendency to remain near the pair-mate and to use more familiar pathways produced a significant difference between them and squirrel monkeys in the efficiency of exploiting new food sources in a novel environment. The point is, that what an individual does in nature reflects organizational tendencies which are potent in captivity. Perturbations of the captive environment result in behavioural adaptations, which in the wild are observed in the context of particular tasks (such as foraging). Behaviour seen in captivity is often most fully understood when its function in natural circumstances is considered.

Demographic characteristics typical of species in natural environments must also be kept in mind when studying behaviour in captive animals. In captivity, one must consider at least the size of the social unit, the composition of the group in terms of age and gender, and the stability of the social group (e.g. how recently it was formed). The relationship between these characteristics and the demography of wild populations is particularly relevant. Members of group-living species housed in pairs, for example, can behave quite differently from members of the same species housed in groups (Fragaszy and Mason, 1978; Vaitl *et al.*, 1978). Within groups, age and gender can be associated with pronounced individual differences in investigatory and change-producing behaviour. Because the behaviour of individuals affects the behaviour of their group-mates, groups of the same species with varying age-gender compositions can present quite different behavioural profiles in similar circumstances. For example, Visalberghi (1988) reported large gender differences in time spent contacting wooden blocks in groups of capuchins with even numbers of males and females. Juvenile and adult females were less investigative than males. But in groups of the same species, with no male subadults, and only a single adult male, juvenile and subadult females were highly investigative towards both novel and familiar objects (Fragaszy and Adams-Curtis, 1987 and unpublished data). Also, adult and juvenile females in the latter groups were successful at using tools in four different tasks, as were adult and juvenile males (Westergaard and Fragaszy, 1987a; Fragaszy and Visalberghi, 1989).

13.3 PRESENTING OBJECTS IN A CAPTIVE ENVIRONMENT

We are especially concerned with the provision of objects as a technique for studying behaviour associated with change. Often 'novel' objects are used for this purpose, although this is not at all necessary (see below). We conceive of novelty as a scaled property. Objects can range from extremely novel (many properties unknown), through less novel (properties partially known), to completely familiar (properties fully known). Descriptions of objects or situations as 'novel', reflect decisions about the relative proportion of properties which are unknown to the animals under study. As a practical matter, most investigators use the duration of exposure as a criterion for novelty. Yet, there is no agreed-upon standard for how long an object or condition must be present for it to be familiar rather than novel. Nor is there any agreement on how to assess the number of dimensions in which novelty can exist. Unfortunately, this state of affairs is unlikely to be remedied in the near future. It is simply too time-consuming to define novelty empirically for each case when, as a practical matter, an operational definition will suffice. Nevertheless, comparison across studies is constrained by differences in operational definitions.

Novelty may or may not accompany change. For individuals which have previously experienced a cycle of seasons, the seasons in a new cycle are not completely novel. Similarly, interaction with an object may produce a change in that object, but the new features are not necessarily novel. For example, an object may be bent repeatedly one way and then another; neither position is novel. Further, we often assume that novelty engenders interest, as defined by continued attention, and that familiarity leads to disinterest. But it is clear that non-novel, self-produced change can also maintain interest. We have only to think of repetitive operant performance, in which animals engage in the same acts with the same consequences for long periods, to see that this is so. Operant performance is sometimes continued even when a food item is delivered 'free', without any act required. This is a phenomenon called contra-freeloading (Osborne, 1977). Again, self-produced repetitive changes in body position or balance occur frequently in motor play and 'mastery' play (Bruner, 1974; Bower, 1982).

The ability to produce repetitive environmental changes by one's own actions – as, for example, by locating food by searching through bedding – is a key ingredient in the provision of comfortable captive environments. From the animal's point of view, the nature of change, and the exercise of control over the change are more significant than the degree of novelty. The frequent use of novel objects as environmental challenges reflects our limited abilities to provide inherently interesting situations for our animals, at least as much as it reflects the significance of novelty to their interest (see also Chapters 14 and 19).

Presenting objects, whether familiar or novel, to animals in a captive environment is a benign form of environmental challenge. It is benign in the sense that the individual is not restrained, and that the distance from, and the interaction with, the object(s) are controlled by the individual. It is a challenge in the sense that the individual can, and sometimes must, alter its activity to accommodate the change. Many aspects of behaviour can be studied with ecological validity in captivity through environmental challenge in this form. We have ordered our discussion of studies in this area by the behavioural outcome of the procedure of placing objects into a captive environment.

The first outcome is the alteration of temporal and spatial patterns of behaviour–activity levels, time budgets, and movement patterns. The time budgets and movements of captive primates often differ from their wild counterparts in several ways. Some of these are obvious. For example, captive primates do not range over distances to obtain food. Others are less obvious, such as the frequency of certain social behaviour. These may be greatly increased in captivity, as when individuals groom themselves or another to the point of producing baldness. The expression of species-typical activity levels is often a practical goal in studies manipulating the captive environment. McKenzie *et al.* (1986) report a study of this kind in which the addition of a specific kind of material altered behaviour in the intended fashion. By contrast with the situation in nature, common marmosets and cotton-topped tamarins made infrequent visits to the ground in a captive setting. However, by adding a covering of woodchips on the floor, the number of visits to the floor of the cage was increased. A simple manipulation of the physical environment can alter the behaviour of these animals to be more like the behaviour of their wild counterparts.

Second, the introduction of objects which vary in one or more dimensions can serve as a means of determining which dimensions influence the occurrence of a specific behaviour. An excellent example of this approach is Caine's (1984) study of vigilance in red-bellied tamarins (*Saguinus labiatus*). Vigilant scanning is a common behaviour in these animals, both in captivity and in the wild. Caine introduced novel objects as a method of examining the function of vigilant scanning. She found that objects that were intended to induce fear, such as a stuffed cat, increased scanning significantly more than benign objects. This suggested that scanning is sensitive to the level of perceived safety or risk. Furthermore, when items were introduced during periods of frequent scanning, they were discovered more rapidly than objects introduced during a period of infrequent scanning. Thus, high rates of scanning enhance detection of changes in the environment. However, by comparing rates of scanning after presentation of benign and fear-inducing novel objects, Caine showed that novelty by itself did not increase the occurrence of scanning. Therefore, scanning is not simply a function of change in the environment, but is instead dependent on the types of changes occurring in the environment. Caine's study strengthens the argument that scanning functions as a method of detecting predators in these monkeys.

The third category of behavioural outcome that can be inferred from presenting objects to captive individuals involves the stimulation of a wide range of natural behaviour, which, in turn, contributes to our appreciation of the adaptive capabilities of individuals and their species. In this domain, studies of captive animals may supplement field studies. The full range of exploratory and manipulative behaviour, for example, is easier to study in captive animals than in wild ones, for two reasons. First, it may be easier to make observations and, second, the animals may have more opportunities to express various behaviours. Captive monkeys do not devote as much of their time to food-gathering and predator-avoidance as their wild counterparts, and they sometimes have access to objects and surfaces unlike those they encounter in nature. Thus, they can tinker with problem-solving tasks (Kohler, 1927), explore new ways of using objects (Westergaard and Fragaszy, 1987a) and, in general, produce behaviour rarely or never seen in nature (Kummer and Goodall, 1985). Kummer and Goodall recount many examples of behavioural innovations seen in captive monkeys and apes.

The fourth category of behavioural outcome is the overt expression of social relationships and social dynamics, which are otherwise subtly or infrequently seen. The social context of responsiveness to change is a topic of increasing concern in primatology, as we recognize that complex social interactions are features of primate life which influence all others (Hinde, 1983b; Smuts *et al.*, 1987; see also Chapter 3, section 3.5). Presenting situations which induce or heighten the potential for direct competition is a common feature of studies concerned with social dominance (cf. Bernstein, 1981). The technique has equal usefulness in studying cohesion and affiliative social processes. A classic example in this area is E.W. Menzel's (1974) work with juvenile chimpanzees, showing that exploration of objects placed in a large enclosure by members of a group occurred in an affiliative social context.

Social learning, a related phenomenon, can be studied by presenting objects to groups of animals. There is increasing interest by comparative psychologists in the social transmission of information of behaviour (Box, 1984; Galef, 1988). Studies with group-living primates which have concerned the influence of social context on learning and response to change, include Menzel and Juno's (1982) work with saddle-back tamarins (*Saguinus fuscicollis*) showing 'one-trial learning' in a food-acquisition task; Le Poivre and Pallaud's (1985) work with guinea baboons (*Papio papio*) on social facilitation of food-finding; Cambefort's (1981) studies of social transmission of food choices in groups of chacma baboons (*Papio ursinus*) and vervet (*Cercopithecus aethiops*) monkeys, as well as studies in our laboratory on the use of tools in capuchins (Westergaard and Fragaszy, 1987a; Fragaszy and Visalberghi, 1989).

To summarize the ground covered so far: objects presented to captive animals can (1) influence activity levels; (2) allow specification of environmental features correlated with the performance of specific behaviour; (3) promote the performance of a wide array of behaviour, some of which may escape notice in nature; and (4) bring social dynamics into greater relief. The foraging predilections, social organization and group demographics in nature all correlate with responsiveness to objects, and with behavioural organization more generally, in captivity.

13.4 STUDIES IN OUR OWN LABORATORY

We turn now to our own work, in which capuchin monkeys living in social groups have been observed for several years. The studies reflect our interest in sensorimotor skills and social influences on behaviour. We believe they also illustrate the points made above which concern the opportunities to attain ecological validity in studies of captive animals responding to changing, or changeable features of their environment.

Capuchins provide an excellent opportunity to study manipulation, and the social context of manipulation in a slowly developing, dexterous, and adaptable primate genus. Wild capuchins are noted for their persistent manipulative activities during foraging (Izawa, 1979; Fragaszy, 1986), and the use of tools has recently been observed in nature (Boinski, 1988). Captive capuchins engage in manipulative behaviour unusual in other monkeys, including the common use of an object to act on another object or on a substrate (Torigoe, 1985; Visalberghi, 1988; Fragaszy and Adams-Curtis, 1987). Unlike all other New World primates and most Old World monkeys, they readily use, and occasionally even manufacture tools (Kluver, 1937; Westergaard and Fragaszy, 1985, 1987b; Visalberghi, 1986; Costello, 1987). Tool-use and manufacture in capuchins have most commonly been observed in food-getting tasks, but have also been seen in self-defence (e.g. Cooper and Harlow, 1961), in self-treatment of wounds (Westergaard and Fragaszy, 1987b) and in the treatment of another's wound (Ritchie and Fragaszy, 1988). As a sensorimotor correlate of these instrumental abilities, capuchins exhibit precision grips that require independent control of digits (Costello and Fragaszy, 1988), which contravenes the conventional wisdom that New World monkeys do not have such control. Further, capuchins are often keenly attentive to each other's manipulative behaviour (Fragaszy and Adams-Curtis, 1987). The influence of social factors on the exploration of objects, and on the spread of manipulative behaviour, is one of our primary current interests.

Our capuchins are housed in two breeding groups of around 15 monkeys in each. One group had been formed 4 years before the present studies began, and the other, 2 years before. Group membership remained relatively unchanged over that period except for natural recruitment by births. The composition of the groups during the period for which data are presented is given in Table 13.1. They approximated to an average to large group size with typical age and gender ratios for wild capuchins (cf. Terborgh, 1983). A full complement of ages allows us to assess differences in manipulation with

Table 13.1 Participation of capuchins in manipulation studies

Name	*Sex*	*Age class**	*Normative*	*Novel objects*	*Puzzle*	*Tool task*
Group 1						
Ike	M	Adult	X	X	X	X
Mercedes	F	Adult	X	X	X	X
Alice	F	Adult	X	X	X	X
Beth	F	Adult	X	X	X	X
Fanny	F	Adult	X	X	X	X
Jill	F	Adult	X	X	X	X
Quincy	F	J3	X	X	X	X
Patti	F	J3	X	X	X	X
Rita	F	J3	X	X	X	X
Tere	F	J2	X	X	X	X
Ulysses	M	J2	X	X	X	X
Willy	M	J1	X	X	X	X
Viola	F	Infant	X	X		X
Louisa	F	Infant	X	X		X
Dixie	F	Infant	X	X		X
Group 2						
Louis	M	Adult	X			
Sherlock	M	Adult	X	X		X
Nina	F	Adult	X	X	X	X
Hansje	F	Adult	X	X	X	X
Greta	F	Adult	X	X	X	X
Celia	F	Adult	X	X	X	X
Diana	F	Adult	X			
Katie	F	Adult	X	X	X	
Erna	F	Adult	X	X	X	X
Osa	F	J3	X	X	X	X
Xavier	M	J1	X	X	X	X
Yves	M	Infant	X	X		X
Zola	F	Infant	X	X		X
Bruce	M	Infant	X			X
Felicia	F	Infant	X			X
Andrea	F	Infant	X			X

*Adults = 6 years and older; J3 = 36–48 months; J2 = 24–35 months; J1 = 12–23 months; Infant = 0–11 months.

cross-sectional samples, and we study the ontogeny of behaviour of young monkeys, using longitudinal designs. The match between our captive groups and wild groups of the same species in size and composition, and the established nature of the groups, satisfy some of our criteria for ecologically valid studies.

Figure 13.1 Housing for the capuchin group at Washington State University. The straw on the floor of the living area provides an endless focus of manipulative activity for the monkeys.

The environment for our capuchins has been developed with the intention of meeting other criteria of ecological validity. Although we have limited space, we have tried to make the space available maximally useful to the animals as places to move around on, hang, swing and as locomotor and postural surfaces, with objects to manipulate. The groups' living areas have straw covering the floor (Figure 13.1). Small food items, such as seeds and diced vegetables, are regularly strewn in the straw bedding, which provides the monkeys with infinite opportunities to produce change. Each time the bedding is sifted, the configuration of the straw is changed and new surfaces are exposed. Pieces of straw also provide readily available portable objects. Because the straw is abundant, there is no competition for access to it. It also makes a large surface area of the cage warmer, softer, drier, cleaner, and generally more usable for the monkeys than the concrete alternative.

We regard straw bedding as our most important single means of providing benign environmental change and challenge to our groups, but additional enrichments are also provided. A tyre swing, plastic hangers suspended from the ceiling, a climbing rope, and basketball nets provide opportunities for gross motor behaviour (i.e. swinging from hanger to hanger). They also provide substrates to be explored, and also serve as items to push, pull and bite. Hanging rubber panels and interior walls allow the monkeys to segregate themselves from other group members. The intermittent introduction of tree cuttings and the presence of indestructible toys with moving parts (e.g. metal chain strung with washers and bolts), provide further opportunities for manipulation. In this enriched environment, our animals engage in manipulative behaviour, which occurs in from 34% (adults) to 65% (juveniles) of the time periods sampled (Fragaszy and Adams-Curtis, 1987). These figures are roughly comparable to those for time devoted to foraging in wild capuchins, which range from 38% to 66% (Terborgh, 1983; Fragaszy, 1986; Robinson, 1986). Thus we have met another criterion of ecological validity: our groups (on at least a gross level) have similar activity levels as their wild conspecifics (see also Chapter 19).

13.5 BASELINE DATA

We have made frequent observations of behaviour in these conditions in order to obtain baseline information on manipulative activity (Fragaszy and Adams-Curtis, 1987). We obtained data during repeated sessions per animal in which the focal animal was observed for 5 s, followed by 5 s in which all manipulative acts observed in the preceding 5 s were recorded. After 1 min the focal sampling was interrupted while each of the focal animal's neighbours (within 1 m) were observed and the manipulative behaviour of the neighbours was recorded (called the neighbour sample). Then focal observations were resumed for another minute. After 10 cycles of 1 min observations, another monkey was observed in the same manner. Manipulations were identified by the acts and the targets toward which acts were directed. The categories of behaviour which were recorded are given in Table 13.2. We made observations in 834 sessions during the first year of data collection.

Our capuchins devoted much time to manipulation. Over all animals, acts directed towards objects occurred in 50% of the recorded samples. Of these acts 15% involved two targets, such as banging chow on the wall or poking straw through the wire mesh. We divided the more than 30 000 scored acts into five general categories on a loosely functional basis. Food directed acts were the most common,

Table 13.2 Vocabulary (listed from most common to least common) used to record manipulations by capuchins

Acts (14)	*Targets* (10)
Bite, chew	Food
Hold	Enclosure
Lick, sniff, mouth	Browse, straw
Scratch, pick, tap	Moveable toys
Rub, roll	Unknown
Bang, slap, smear	Cage-mate
Handle	Outside
Insert, extract	Water
Sift	Faeces
Push, pull, tear	Self
Look at	
Drink, soak	
Reach towards (infants < 4 months only)	
Unknown	

occurring on 32% of samples. Exploratory acts (simple acts generally directed at a surface, including lick/sniff/mouth and scratch/pick/tap directed toward the enclosure or toys), browse-directed acts, and acts targeting a cagemate (e.g. picking up another animal's tail) made up 18%, 18% and 2% of the total number of acts, respectively. The remainder (29% of total acts) were classed as 'other', and included such behaviours as hitting the enclosure or pulling on the hangers. Many unusual combinations of acts and targets were observed in these sessions, which confirmed the impression that capuchins are creative manipulators.

Juveniles, infants and adults devoted differing proportions of manipulative activity to these categories. Acts that fell into the exploratory category were the most common in infants (0–12 months), making up 42% of their total acts. For juveniles (1–3 years), 'other' was the most frequent category, with 35% of their acts falling into this category. On the other hand, 57% of the adults' (6 years or older) acts consisted of food-directed behaviour (Figure 13.2). Clearly, age groups differ in the variability of manipulative behaviour, and in how closely routine activity is tied to mundane aspects of life such as eating.

Age differences were also reflected in the rate of manipulative activity and the number of different acts performed. Juveniles performed at least one object-directed act in 65% of all samples. Infants were active in 47% of all samples, whereas adults were active during only 34%. Juveniles also engaged in the widest variety of act and target combinations. On average, each juvenile produced 91

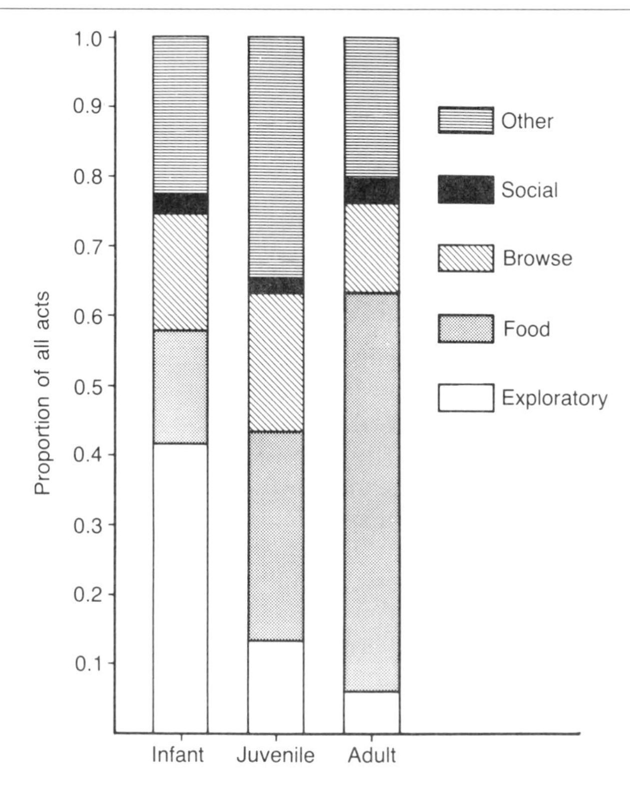

Figure 13.2 The proportional distribution of object-directed acts for each age group of capuchins during baseline observations. Infants were aged between 0 and 12 months old, juveniles were between 13 and 36 months old.

different act-target combinations (mean of 41 observations per juvenile). Adults performed an average of only 31 different combinations (mean of 17 observations per adult). Infants were intermediate to the adults and juveniles on this measure, and performed an average of 66 different combinations (mean of 39 observations per infant). This same pattern of age differences occurred when we considered the production of rare acts, defined as act-target combinations occurring less than 50 times in a sample of 31 620 acts. Of the different act–target combinations performed by juveniles, 69% were rare. The figure for infants was 63%, and for adults only 43%. Juveniles also performed the greatest proportion of two-target acts (i.e. acts in which an object is brought into combination with another object or a substrate): 19% of their acts were directed towards two targets. Only 8% of the adults' acts involved two targets, and 12% of the infants' acts did so.

By several measures, then, juveniles were clearly the most active and innovative age group. The infants' rate and variety of activity were less than that of the juveniles. Adults, as a group, were the least active and the least innovative. However, adults are quite capable of unusual exploratory behaviour when the task calls for it (see below).

In these noncompetitive conditions, capuchins interacted with objects amicably. Monkeys were within 1 m of others engaged in manipulating objects for one-third of all neighbour samples. Infants were near others so engaged in significantly more samples than were either juveniles or adults (infants: 47%; juveniles: 28%; adults: 25%). This reflected differential levels of affiliation and tolerance across age groups. Nonetheless, even the adult value is quite high compared to what one expects of these monkeys based on their reputation as competitive, aggressive characters (e.g. Janson, 1985).

We have developed our work by adding specific objects and tasks into this routine baseline situation. We have frequently imposed selected environmental challenges in the form of objects, in order to stimulate a wider range of manipulative behaviour than we observe without such challenges. The introduction of objects also stimulates the expression of social and cognitive characteristics, which are obscured in less challenging situations. For example, we introduced novel objects in order to determine the extent to which different or more varied patterns of manipulation would be stimulated, as well as to examine differences among individuals in the social context of their exploration (Fragaszy and Adams-Curtis, unpublished data).

Ten objects, one a week, were introduced to each group (Table 13.3). The objects varied along several dimensions, including size, number of parts, types of movement and the presence of food. There was no attempt to standardize the objects; they were chosen using our judgement of what the monkeys would enjoy manipulating. Some objects had a great deal of potential to produce noise (e.g. the metal salt shakers); others were designed to produce visual effects (e.g. the colour wheels) or to present a manipulative challenge (e.g. the Tygon tubing tied in knots). The objects were not indestructible, but instead were chosen to encourage a variety of manipulations, including tearing and biting.

In order to capture the initial flurry of activity which novel objects provoke, we completed a qualitative narrative describing the behaviour that occurred during the initial 5 min of presentation. A summary narrative was also completed at the end of each observation. The most interesting finding drawn from these concerned the behaviour of inserting straw and sticks into the crevices of objects and surfaces, which was seen frequently in connection with the novel objects.

The insertion of objects into substrates first appeared in our groups

Table 13.3 Novel objects presented to groups of capuchins (all measurements are rounded to the nearest cm)

Week	*Objects*	*Food available*	*Description*
1	Metal salt shakers	Sesame seeds	Metal salt shakers with handles, 5 cm in diameter and 6 cm tall; holes small enough to make it difficult to remove sesame seeds
2	Tygon tubing	Fruit Loops cereal	About 1 m of Tygon tubing (clear flexible plastic 2 cm in diameter) stuffed with cereal and tied in knots
3	Tethered ball	None	Pink rubber balls, 8 cm in diameter suspended from about 30 cm of metal cable
4	Mophead	Lifesavers candy	Dustmop head of cotton string with Lifesavers candy and coloured cardboard disks tied in the strands of the mop
5	Velcro	None	6–7 strips of Velcro, 2–5 cm long, sewn on to heavy pink cloth and glued to a board (30 × 30 cm); smaller pieces of Velcro 'Velcroed' to the attached pieces
6	Zippered pocket	Sesame seeds	Metal zipper, 18 cm long, attached to pink cloth and sewn on to board to form pocket
7	Hardware on tether	None	(3) 13-cm eyebolts, (2) 3-cm diameter washers and (4) 5-cm metal fasteners strung on metal cable
8	Colour wheel	None	Black and white pattern painted on 13-cm disk attached to turntable (8 cm); when spun, the wheel shows rings of colours
9	Slide toy	None	Plexiglass pieces, base 4 × 15 cm; slide (3 × 18 cm) fits in groove of base and moves about 5 cm in either direction; attached to board (30 × 30 cm); screws on either end prevent removal of slide
10	Marble rattle	None	Plexiglass tube, 15 cm long and 3 cm in diameter, sealed on either end and filled with five marbles; attached to turntable

following a study in which a tool-using task was presented 4 years ago (Westergaard and Fragaszy, 1987a). Many of these capuchins can obtain food (syrup or yogurt) by dipping straw or sticks into a well, and this activity has been acquired by several of the infants born into the group. Furthermore, infants that have not had experience with any tool-use problem also probe into the crevices and holes in the home cage. Thus, the use of probes in our groups is not context-dependent, but is expressed instead in a wide variety of circumstances, including the exploration of novel objects. As far as we can determine, this same behaviour is not prominent in other groups of capuchins which are not experienced with using sticks as probing tools. Nor do wild capuchins probe into holes with sticks or other objects, even though they do insert their hands into holes to retrieve food and water (Izawa, 1979; Fragaszy, 1986; S. Boinski, personal communication), as our captive monkeys do. This is, however, an open question until other capuchins with ready access to straw or sticks, and naive to their use as tools, are observed in detail.

Other than certain acts by chimpanzees (McGrew, 1977; McGrew *et al.*, 1979; Boesch and Boesch, 1984; Nishida, 1987), we know of no other manipulations requiring a specific spatial orientation of an object to a substrate that may be socially transmitted. The occurrence of the phenomenon in the exploration of novel objects is therefore of great interest to us.

A second finding of these observations bears on the relation between age and social status of the group members, and access to objects under conditions of greater social challenge than exists during the baseline observations. Because the number and distribution of the objects within the cage were restricted, there was competition for access to them, especially when they were first introduced. As a result, the relations among access to objects, the manipulative propensities, age and the social status of the monkeys were thrown into greater relief.

Access to objects clearly varied among individuals in our groups. Adult males enjoyed uncontested access to all the objects. Furthermore, avoidance of the males by others varied, most consistently as a function of the age of the others. Infants less than 1 year old were always allowed access (i.e. the presence of infants was unaffected by the presence of others). If there was food available (4 of the 10 objects) access by the juveniles, which were aged between 1 and 3 years, was often dependent on the relative rank of the mother, although juveniles approached when the adult males were present. Certain females approached the objects while the male and juveniles were there. Other adult females approached only when the objects were unattended by other animals. They exhibited pronounced avoidance of males, other females, and even most juveniles. Hence, age and social status affected the expression of nonsocial behaviour – in this case, the subjects' manipulative activity – through their efforts to gain access to specific resources. Investigators seeking to document each individual's range of behaviour should keep this in mind, and provide situations conducive to access by all members of the group.

13.6 THE INTRODUCTION OF A SEQUENTIAL PUZZLE

In another of our studies we presented a sequential puzzle to the monkeys in the home cage (Adams-Curtis and Fragaszy, unpublished manuscript). In this case, we were interested in combining two aspects of the study of primate behaviour which generally have been pursued independently: namely, problem-solving, traditionally studied with single animals, and the influence of social context, most commonly studied in terms of access to limited resources or other competitive situations (as in the previous example). In this study, we were interested in positive aspects of social influence, those which could contribute to social learning, in Box (1984) and Galef's (1988) sense of that term, as well as competitive aspects.

A three-step sequential puzzle was presented to each group of capuchins in a limited test-box area of the home cage for 20 min at a time (Figure 13.3). The puzzle was presented for a total of 24 times to each group over 12 days. On 12 occasions a food item was hidden beneath the final portion. The puzzle consisted of three parts. In order to solve the problem an eyebolt had to be removed from a hasp, then the hasp lifted, and finally a hinge opened. The task is similar to puzzles used many years ago by Harlow and his colleagues in the study of investigatory behaviour of monkeys (e.g. Davis *et al.*, 1950). The amount of time near the puzzle and the amount of time spent manipulating it were scored for each monkey. All solutions of each part separately were also recorded.

Our first surprise was that only one subject, Q, out of 20 successfully solved the puzzle repeatedly (over 300 times in 24 sessions). Given the manipulative propensities of capuchins (Fragaszy and Adams-Curtis, 1987), their ability to use tools (Kluver, 1937; Westergaard and Fragaszy, 1987a; Boinski, 1988), and the abilities of other species to solve problems that involve sequences (e.g. Davis *et al.*, 1950; Harlow *et al.*, 1956; Visalberghi and Mason, 1983; Fragaszy and Adams-Curtis, unpublished data), it was expected that a number of the monkeys would quickly learn to solve the puzzle. We suspect the apparent limitation of skill in the group can be attributed to some consequence

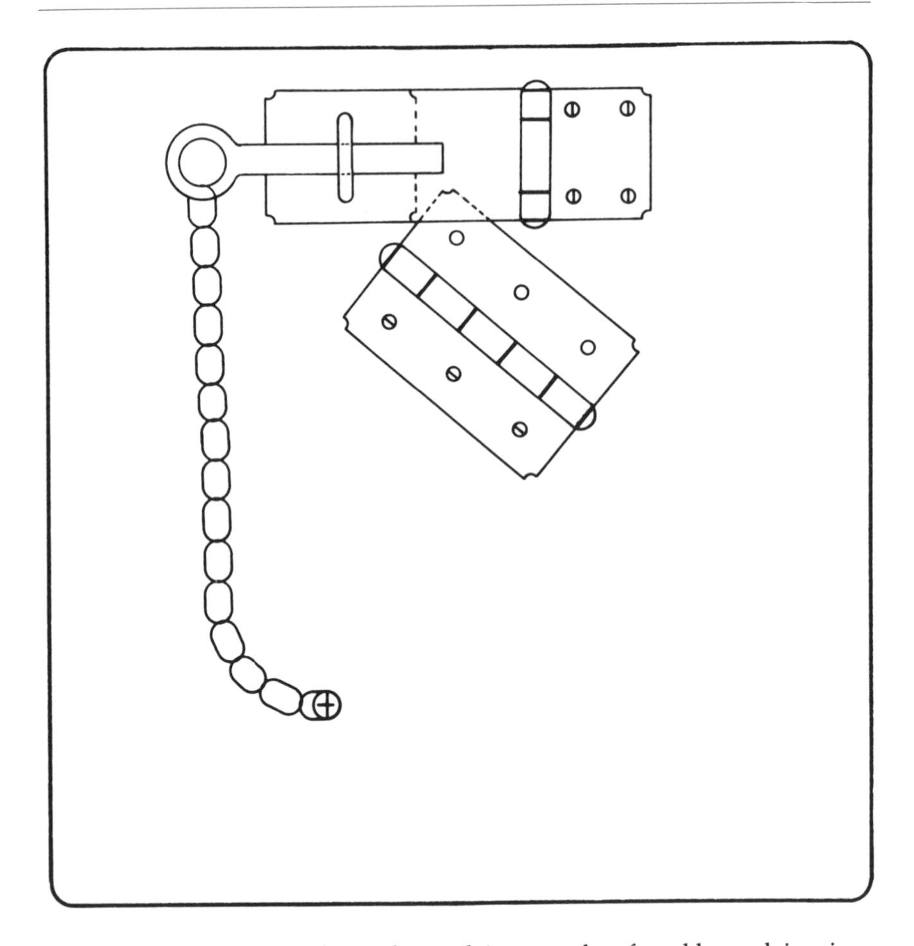

Figure 13.3 The sequential puzzle used in a study of problem-solving in a social context by capuchin monkeys. To open the puzzle, the eyebolt had to be removed from the hasp. The hasp then had to be opened and the hinge lifted to obtain a piece of fruit which was hidden underneath.

of the social context. Tests of single animals are necessary to confirm our suspicion. But our interest at this point is in the nature of social influences bearing on interaction with the puzzle in the group setting, as much as on the problem-solving abilities of individuals.

We did not observe direct competition for access to the puzzle.

There was no flurry of interest in it, as there was when objects were introduced into the main part of the cage, and as described above. Nor did antagonistic acts occur in the test area. However, we did see evidence of avoidance. In the sessions in which Q's mother (A), a socially powerful female, was present 10% of the time, the juveniles occupied the test area only 9% of the presentation time. Furthermore, they shared the box with Q only 25% of the time that A was present. When A was present the least (less than 4% of presentation time), the juveniles were present more than 20% of the time, and shared the box with Q much more (about half of the time Q was present). In other words, the juveniles were attracted to Q in the test area, but avoided A. In the study of social context, it is tempting to focus on a one-to-one correlation between the behaviour of particular animals. This study showed the social context has more dimensions than a simple model would suggest.

We also thought that positive social influence could include enhancement of interest in the puzzle, facilitation of common behaviour (such as slapping), or perhaps even imitation of specific sequences of acts. Although imitation and social facilitation were not observed, social enhancement of interest in the test area and the puzzle was evident. In the group with the larger population of juveniles (six) and the one successful monkey, the juveniles were in the test box on a significantly greater proportion of the presentation time than were the adults. Juveniles also spent a greater proportion of time actually manipulating the puzzle than did adults. Furthermore, the juveniles in this group were frequently present with other juveniles. However, adults were very rarely present with another adult. This result indicates that social enhancement of interest in the puzzle probably occurred among the juveniles in the larger group. Enhancement of interest was particularly related to the behaviour of the successful animal, but this was so even before she learned to solve the puzzle, and when there was no food associated with solving the task.

Since social enhancement can aid the spread of information in a group, the finding that it occurs differentially across age groups and living groups has potential significance for understanding social learning in this species.

13.7 PRESENTATION OF A TOOL-USING TASK

We have presented several tool-using tasks to our groups over the last few years. We discuss here one of these tasks which serves as an example of what can be gained from presenting tasks of this type to capuchins. Although detailed comparative data are not available, our studies and recent studies by Westergaard (1987) suggest there is greater continuity of tool-using abilities across simian primates than we previously thought. Furthermore, we have had good success with capuchins spontaneously adopting successful tool-using behaviour in a variety of tasks. In this particular case, however, we were interested in several familiar themes: the range of manipulative skills, the social context of exploration, and the social influences on acquisition and expression of specific behaviour.

In a previous study, Visalberghi observed two other capuchins (also *Cebus apella*) learn to use a wood block as a pounding implement to open walnuts (Visalberghi, 1986). We wanted to replicate those observations in our groups, and to focus particularly on the social context in which nut-cracking behaviour developed. To this end, we presented an apparatus which permitted many animals (potentially, the whole group), to observe others closely at the apparatus, and for several to manipulate tools simultaneously (Fragaszy and Visalberghi, 1989).

The apparatus was a wooden platform (the nut board) on which 24 whole walnuts were glued into predrilled holes so that only half of the nut shell was exposed. Four metal objects (hexagonal threaded nut, 5 cm in outer diameter with an inner hole 2.5 cm in diameter) and two wooden blocks (5 × 5 × 7.5 cm) were tethered to the platform on cords about 50 cm long (Figure 13.4a). The nuts could be opened by pounding them with the metal tools or (less efficiently) the wooden blocks, or by biting them. This is a straightforward task in that all the parts are visible, and the behaviour required to use the tool is a common one in wild capuchins, as well as in our captive animals. There were enough tools and nuts for several animals to work at one time. Earlier, we had presented loose walnuts to the monkeys. Some monkeys were able to open these without tools by banging them directly on the wall and biting them. Thus the monkeys were familiar with walnuts as a source of food. The board was presented on 20 occasions, for half an hour at a time, to each group over a period of 5 months.

Interest in the nut board was immediate, widespread and sustained in every session. Overall, the apparatus was visited on 89% of the observation samples in group 1 and 69% in group 2. All the monkeys were present at the apparatus at least once, and commonly several animals were present at once. Thus, by providing multiple opportunities to manipulate the same objects, and room for several animals to be present, we were able to produce a rich social context during exploration of the apparatus. Interest during early sessions was facilitated by the fact that the adult male in each group was able to open nuts by biting them, and thus a few nuts were opened in this

Figure 13.4 The nut board used to study tool-using behaviour by capuchin monkeys. (a) Nuts were glued onto the board, which could be broken by hexagonal pieces of metal. (b) An adult female using a piece of straw to probe the nut board.

Table 13.4 Percentage of intervals at the nut board for each capuchin

Group 1		Group 2	
Subject	%	*Subject*	%
Adult males			
I	24	S	11
Adult females			
A	42	H*	29
B	1	G	9
M	1	C	1
F	1	N	1
J	1	E*	16
Juveniles			
P	2	O*	21
Q*	53	X	6
R	5		
T	1		
U	13		
W*	25		
Yearlings and Infants			
V	3	Y	4
L	1	Z	10
D	4	A	3
		B	1
		F	1

* Denotes a monkey that solved the problem by using the tool.

way. Females and juveniles could not open nuts by biting, although they tried to do so. Eventually, five animals learned to use the metal tools provided with the apparatus to pound open nuts. The first successful monkey in group 1, Q, pounded a nut open on session 6. The second successful animal in her group, W, first opened nuts on session 13. In group 2, E first pounded open a nut on the fourth session, 0 on the twelfth, and H on the sixteenth.

Although general interest in the apparatus was high, some individuals seemed less interested than others and, for some, access to the apparatus was limited as a consequence of social factors. The percentage of intervals present at the apparatus for each monkey is presented in Table 13.4. The percentage of samples that each animal spent at the apparatus varied widely, from 52.9% (Q) to < 0.1% (B). In six of eight mother-offspring pairs, correlations in the number of samples at the apparatus over sessions were insignificant (−0.20 to +0.27). However, two mother-offspring pairs did show significant positive correlations: A with her daughter Q (successful), and H (successful) with her daughter Z (her only offspring present in the group). Both A and H were socially powerful females in their respective groups. The correlation between A and her son U was insignificant (−0.20), as was the correlation between the siblings Q and U. Thus, kinship alone is no guarantor of equivalent access, although the kin of powerful females more frequently tolerate (as Q and A) or exploit (as Z and H) their mothers than do others.

There were few attempts to defend the opened food. Most often mutual exploitation was simply permitted. Moreover, there were few overt threats towards animals which approached the apparatus, and most of these were given as a coalition by the mother-daughter pair, Q and A. As with novel objects, avoidance was a much clearer indication of social relations in these groups than was overt threat or aggression.

In addition to direct attempts to open the nuts by biting or slapping, several monkeys also performed behaviour that involved the tools, as well as other objects placed in relation to the tools or to the nuts. For example, small objects were placed inside of and then removed from the centre of the metal tool, or inserted all the way through it if it was standing on edge. Straw and sticks were probed into crevices of the apparatus, and into opened shells (Figure 13.4b). Small objects were also placed inside the opened nuts. We saw many episodes of 'testing' variations across all sessions. Successful animals were responsible for an average of nine bouts each involving an object in combination with the nuts or tools, and these occurred both before and after they learned to pound nuts open. Only two of the unsuccessful monkeys performed similar behaviour during a total of five bouts. As the animals which were eventually successful did not differ consistently from their age peers in the frequency with which combinations of acts were performed in baseline conditions, it appears that successful monkeys differed from others in their interest in the task, not in their talents.

We had thought that a successful demonstration of the task would influence the behaviour of observers. This seemed more likely in this task than in the sequential puzzle task. In this case, the one act required to open the nuts was visible, and common in the repertoire of all the animals. But even under these conditions, direct imitation of a model did not occur. Nor was interest in the tools, or exploratory behaviour involving combinations of objects at the apparatus enhanced. In fact, the data suggested that relevant exploratory behaviour was performed independently and, in some cases, preferentially while alone.

Although animals were clearly not imitating the pounding behaviour, their own behaviour may have been influenced in other ways. We looked at the temporal relationship between exploratory behaviour at the nut board by animals which were second or later in being successful, and the pounding activities of those which were initially so, to see if this might have been the case. We can take a particular case to illustrate the difficulty of disentangling the influences of the successful behaviour from other factors which were also operating during exploratory behaviour. Hence, in the sessions preceding his first successful pounding, W occasionally tapped or otherwise explored the nuts while Q was present or just after she had left. However, the temporal relation between their activities was almost certainly a consequence of the behaviour of a third animal, Ike (the adult male, and father of W, but not of Q). Q usually left the apparatus when Ike approached, and W often came to the board while Ike was present (and opened nuts by biting them).

In sum, there was no clear relationship between W's exploratory behaviour and Q's pounding, although this does not rule out the possibility that such a relationship may have in fact existed occasionally. The same is true for the second and third successful monkeys in group 2. There was no evidence that exploratory behaviour or pounding performed by animals in the process of learning to pound efficiently occurred more frequently when successful monkeys were active at the apparatus, or shortly after they had left than at other times. Overlaps at the apparatus with the previously successful animal and periods of potential 'modelling' occurred now and then for all subsequently successful animals before they learned the task.

When the behaviour of animals which did not pound open nuts is examined in relation to the presence of models of the behaviour, there was again no pattern which suggested social facilitation of particular acts, nor any effect on interest in the tool (Table 13.5). However, some aspects of their behaviour were affected by the successful activities. For instance, the presence of a successful monkey at the nut board appeared to serve as a cue for others that nuts may be available. Successful animals attracted other animals when the social conditions at the board permitted, but the activity of the successful monkey with tools seemed irrelevant to the behaviour of the others. As in the sequential puzzle task, great interest by one animal led certain others also to display interest in the same place, but not to engage in specific

Table 13.5 Mean frequency of acts per nonsolver (n = 6) at the nut board over blocks of four sessions, by social context

Block	*Social context*			
	Alone	*+ Others*	*+ Solver*	*+ Others and solver*
First				
At apparatus	17.0	38.0	NA	NA
Bite	0.7	7.5	NA	NA
Other	3.5	0.7	NA	NA
Last				
At apparatus	19.2	23.5	46.2	61.8
Bite	1.8	1.3	2.3	1.6
Other	0.0	0.0	0.0	0.0
Investigate/consume	8.7	9.8	27.7	39.5

Note: data are numbers of 10-s intervals of 480 total intervals (from Fragaszy and Visalberghi, 1989).

behaviour. In this case, in contrast to the puzzle task, adults were as involved as the juveniles. Why this should be so is not clear.

The contrast in the nature of social context during presentations of the puzzle task and the nut board deserve some comment. In the former, juveniles were the most frequent investigators. Interest in the apparatus and competition for access to the apparatus was low-key, and social enhancement was evident among juveniles, but not among adults. In the latter, competitive exclusion of adults and juveniles was clear, social vigilance at the apparatus was marked, and adults were as involved in the task as juveniles. Positive social influences in this case were muted compared to those seen in the puzzle task. The set of results suggest that juveniles are most susceptible to positive influence by others when they are least likely to be constrained by status-related competition and vigilance.

In summary, the two tasks discussed above were successful in generating a lot of activity by many individuals at a specific location. They served our purpose of providing benign challenges, eliciting displays of affiliation, tolerance and avoidance which are rarely observed in calmer conditions. They also elicited a variety of manipulations of interest to us because of their relationship to problem-solving abilities. Lastly, they provided a means of observing social influences on the acquisition of novel instrumental skills. The group context, while very rich and ecologically valid, nevertheless produces complex situations in the interpretation of data on social learning (cf. Box, 1984; Galef, 1988). We view studies of this sort as a necessary first step in an empirical programme which must also include studies of single animals and pairs of animals.

13.8 CONCLUDING REMARKS

Our aim has been to emphasize the usefulness of studying behavioural responsiveness to objects in groups of capuchins as a means of probing the general organization of behaviour and social processes. We want to emphasize the benign, often enriching character, of this sort of environmental perturbation. Varied approaches, normative observations, presentation of novel objects, and presentation of problem-solving tasks, have been described. These approaches, and others not mentioned here, provide threads of information which we can weave into our understanding of behaviour as a means of coping with varying environments in which the social context of behaviour is evidently important. Studies of single animals or pairs of animals are also useful at times, and sometimes even essential. The group context contains too many interacting elements to permit isolation of a few. Particularly when one gets involved in questions of learning mechanisms, controlled manipulations must be arranged.

Although most primate genera are not as manipulative as capuchins, the general approach outlined above is relevant to other genera. As we have tried to show, one can choose experimental manipulations and behavioural variables to fit the behavioural ecology of the species under study, or which can help one understand little-known aspects of behavioural adaptability (e.g. Ehrlich and Musicant, 1977). If we do this wisely, studies of responsiveness to objects will contribute to the vigorous study of behavioural adaptability.

ACKNOWLEDGEMENTS

The research reported in this chapter was supported by grant 8503603 from the National Science Foundation, grant MH 41543 from the National Institutes of Health, and a Research Scientist Development Award from the National Institutes of Mental Health, USA, to D. Fragaszy. A bilateral Italy-USA collaborative grant with E. Visalberghi from the Consiglio Nazionale delle Richerche partially supported the study of pounding behaviour. W. London of the NIH lent us the adult core of the breeding groups. We thank the editor for insightful comments on an earlier version of this chapter, and Ruth Day for patiently retyping our many revisions. We also thank the numerous students who assisted in collecting the data presented here.

14 Environmental enrichment for single housed common marmosets

LEAH SCOTT

14.1 INTRODUCTION

There is increasing concern about the needs of captive nonhuman primates (see Chapter 19). In the UK, for example, a Home Office Code of Practice for the care of animals used in scientific procedures (1989) suggests that special consideration be given to the social and physical needs of primates used in research.

Often, the nature of research precludes group-housing, and the well-being of singly housed animals is an area of particular concern. Chamove (1989) has recently reviewed some of the measures adopted for enriching the environments of captive group-housed primates, and has suggested criteria for assessing the effectiveness of these measures. However, there are few data relating to environmental enrichment for monkeys caged individually. Harris (1988) and Bryant *et al.* (1988), for example, have studied the effects of enrichment techniques in singly housed species of Old World monkeys. However, there are no data available for the common marmoset (*Callithrix jacchus*), although, interestingly, a low incidence of abnormal behaviours has been reported (Ridley and Baker, 1982).

Common marmosets breed well in captivity; they are tractable and relatively simple and inexpensive to maintain. They are, therefore, used widely in behavioural (Williams, 1987) and biomedical research (Stellar, 1960; Burt and Plant, 1983). Further, the experimental protocols employed in such work often require the marmosets to be caged singly (cf. Ridley *et al.*, 1981).

At Porton Down, we have developed techniques to improve the social and physical conditions of marmosets. The issues of housing, interaction with other animals, interaction with humans and the use of play objects have all been addressed and will be described here.

14.2 CONDITIONS OF HOUSING

It is a common practice to keep marmosets in metal cages fitted with a nest box and wooden perches. Although the size of the cages, typically 75 cm high × 50 cm wide × 60 cm deep, may be adequate for locomotion, the overall designs are somewhat unimaginative. Moreover, most of the caging available commercially involves a two-tier system.

A consequence of the two-tier rack is that the level of illumination of cages in upper and lower tiers is different. In this laboratory, upper cages were found to be at least three times brighter than those in the lower tier. Informal observations suggested that the behaviour of animals housed in the upper tier differed from that of marmosets housed at the lower level. This was confirmed by a crude estimation of spontaneous activity. Locomotor activity was quantified by counting the number of 15 s periods in which activity occurred, over 10 min samples.

Figure 14.1 shows that animals in the lower tier were significantly less active than animals above them. The level of illumination which would be preferred by laboratory marmosets, if they were given a choice, is not known. It may be possible, however, to allow marmosets to exert some control over aspects of their environment, including illumination. A computer-controlled response panel, which has been designed specifically to monitor operant behaviour in the home cage, can also be used to allow the animal to control illumination, access to visual stimuli, and so on. The potential of this device is being investigated. Work in other species (cf. Line *et al.*, 1987) has supported such an approach.

Some enrichment techniques in use in our laboratory will now be considered.

14.3 ACCESS TO SOCIAL STIMULI

If single-housing is necessary, it is particularly important to ensure that animals are maintained in more or less continual aural and visual contact with conspecifics. Marmosets do not respond well to isolation and, in common with all simian primates, have strong social needs.

We have developed a system in which pairs of marmosets, of the same gender, have access to physical contact with one another at times when they are not involved in experiments. The idea is similar to the

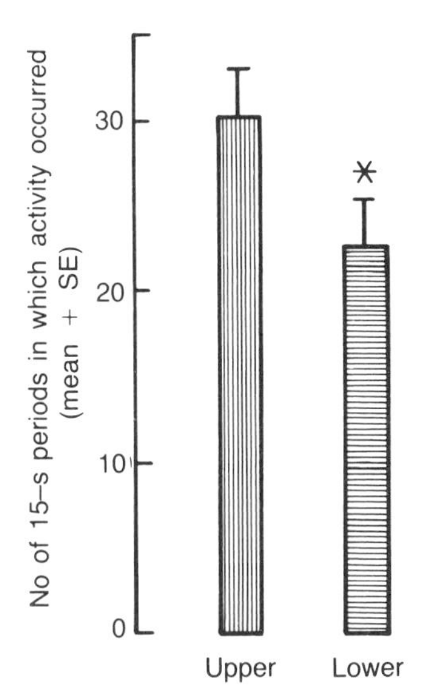

Figure 14.1 A comparison of locomotor activity for common marmosets in the upper tier of a rack and those housed in the lower tier, in which the monkeys were significantly ($P = 0.05$) less active.

successful introduction of periods of social visiting for female baboons (Jerome and Szostak, 1987) and rhesus monkeys (Hauser *et al.*, 1987).

In our laboratory, animals gain controlled access to one another by means of purpose-built rigid cage links, made of stainless steel mesh. These may be used to link horizontally or vertically adjacent cages (Figure 14.2). A link is connected to the cages by an opposed pin-lock system, which exploits an existing feature of the standard cage and allows rapid attachment and removal.

It is important, however, when pairs of marmosets are linked together for the first time, that they are observed continually to make sure that they are compatible. To date, only a small number of marmosets have exhibited marked levels of aggression when linked with others, and these had always been caged singly for more than a year. By contrast, other animals, which had been caged similarly for just as long, adapted very readily to close social contact. Pairs of animals are now linked routinely for 4 to 5 hours a day, when they are not involved in experiments or being fed. During this time they play, groom and look at other marmosets in the home room from the vantage point provided by the cage link. Indeed, when individual animals are provided with the opportunity, they spend a good deal of

Figure 14.2 Removable stainless-steel mesh cage links are used at Porton Down to allow pairs of marmosets of the same gender controlled access to social contact.

time in the link itself, and its popularity indicates that simple rigid extensions have considerable potential for improving cage design.

14.4 FORAGING TASKS

The value of stimulating captive animals to search for food and to

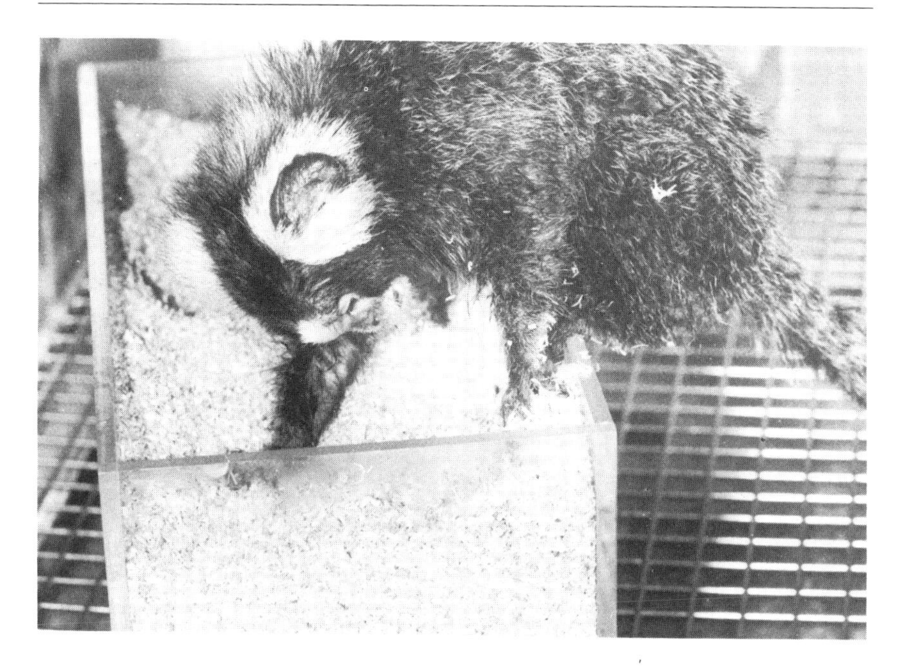

Figure 14.3 Perspex boxes filled with sawdust in which pieces of favourite food are distributed provide a useful foraging task for common marmosets housed individually.

spend a good deal of their otherwise 'purposeless' time engaged in such activity, has been discussed in several contexts and for various species (cf. Chamove *et al.*, 1982). At Porton Down, some simple techniques in which marmosets may 'work' for food are used. For example, Figure 14.3 shows a marmoset retrieving morsels of preferred foods, such as malt loaf, from a sawdust-filled Perspex box in the home cage. Relatively stable baselines of foraging have been generated, and some animals have been observed to forage for at least 6 h at a time (A.S. Chamove and E.A.M. Scott, in preparation).

Other techniques exploit the fact that, in nature, tree exudates play an important part in the diet of marmosets (Coimbra-Filho and Mittermeier, 1976). McGrew *et al.* (1986) described an artificial gum tree, consisting of a hollow wooden dowel filled with a solution of gum arabic, which marmosets could gnaw to extract the gum. In preliminary studies with a scaled-down version of the artificial gum tree, some practical difficulties were encountered. For example, the device was fairly difficult to build and maintain, other than on a one-off basis. A more simply constructed version is required. In the interim

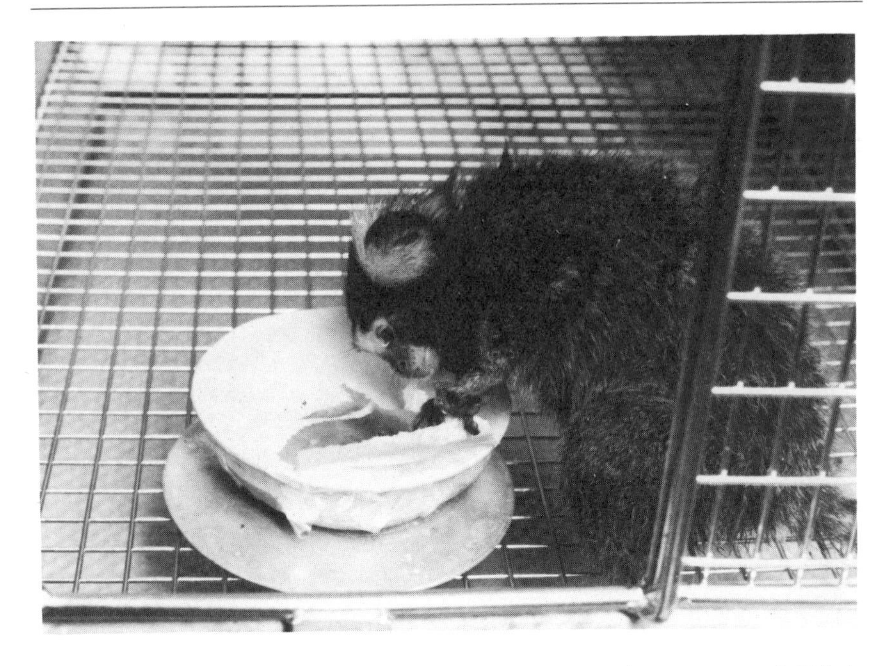

Figure 14.4 A common marmoset removing the paper from a covered dish in which pieces of malt loaf have been placed.

in our laboratory, marmosets have access to plain wooden dowels for gouging and scent-marking.

One very successful technique of 'foraging' that we have developed involves a standard food bowl, which contains a morsel of malt loaf. Access to the food is prevented by covering the bowl with thin paper. Marmosets learn very quickly to tear the paper and retrieve the food (Figure 14.4). This approach is very inexpensive and easy to implement, because it uses materials which are readily available.

14.5 OTHER EXAMPLES OF ENRICHMENT

Another example involves a manipulative problem-solving task suitable for marmosets (see Figure 14.5). The device, which is constructed in Perspex, attaches to the front of the home cage in a similar manner to the cage links described previously. In order to obtain a piece of malt loaf, the marmoset is required to move the vertical slider to allow the food to drop through a hole. A typical learning curve for such behaviour is shown in Figure 14.6. There is, undoubtedly, scope for the design of more complicated devices based upon this simple

Figure 14.5 A Perspex device that attaches to the front of the cage of a common marmoset and dispenses a piece of food through a hole when the monkey moves a vertical slider.

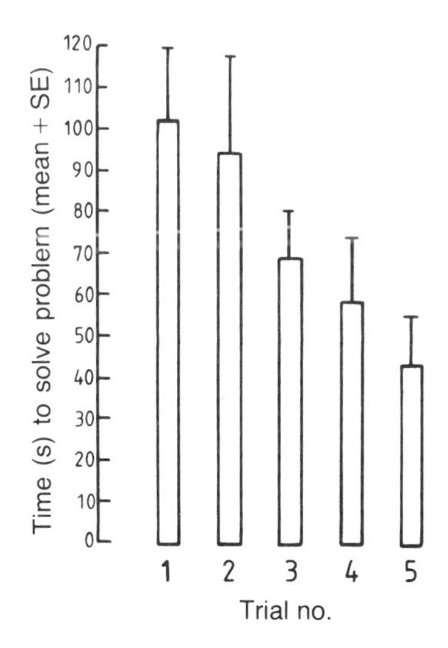

Figure 14.6 The improvement of six individual common marmosets in learning the task shown in Figure 14.5.

Figure 14.7 Common marmosets may be trained to retrieve small pieces of food from a moving belt that is attached to the front of their home cage.

situation, which would occupy the attention of marmosets for longer periods of time, and generate additional behavioural information.

A test of visually guided reaching, developed in our laboratory, also involves a food reward (D'Mello *et al.*, 1985). Figure 14.7 shows a marmoset trained to retrieve small pieces of apple from a moving belt, which replaces the front of the cage during test sessions. Animals involved in training and testing are never reluctant to take part. The procedure has an additional advantage in that other animals in the room, which are not involved in the experiment, are very attentive to this activity.

Moreover, experiments involving the use of apparatus attached to the home cage facilitate direct interaction between the marmoset and the experimenter. This approach has several advantages. Animals are not subjected to the additional stresses associated with remote and isolated test situations, and the opportunity is provided for direct observation of their condition.

Contact with humans may, in itself, be an important source of environmental stimulation. Visitors to the home rooms generate the interest of the marmosets. Visits are encouraged at times appropriate to the experimental protocols. Further, experience suggests that animals appear to adapt more readily to training when they have been habituated to human contact, but no data have yet been collected to support this observation. Again, the practice of identifying marmosets by names, rather than by numbers, has been found to be an important factor in establishing individual care and close observation of the animals by the technicians and the experimenters.

At various times, different items such as swings and suspended chains, have been added to standard cages (see Chapter 19, section 19.7), but the use of such devices by common marmosets has been found to be very variable. Some individuals use particular objects for long periods of time, while others are reluctant to touch the objects, even though they are habituated to their presence.

When play objects are associated with food, however, the behaviour of the marmosets is more predictable, and a much greater level of interest is maintained.

14.6 CONCLUDING REMARKS

The opportunity to interact socially with conspecifics, and the use of play objects associated with food, are key factors in approaches to environmental enrichment in this laboratory.

Rupniak and Iverson (1989) and Chamove (1989) suggest that there has been insufficient rigour in the assessment of the effectiveness of environmental enrichment measures. It is undoubtedly true that more data are required in many cases. Particular methodological difficulties arise in studies where there are insufficient baseline data with which to assess the influence of environmental change. Baseline data from our situations suggest that single-housed marmosets spend most of their time in relatively inactive pursuits, such as watching and sitting. The use of the techniques described previously changes that situation (see A.S. Chamove and E.A.M. Scott, in preparation).

A commonsense and pragmatic approach, which does not preclude the collection of data for appropriate empirical evaluation, has been adopted at Porton Down. When the work was begun it was felt that it would be unacceptable not to take steps to improve the environment of laboratory-housed marmosets before unequivocal scientific evidence of the benefits had been obtained.

The enrichment devices most likely to be used by experimenters have several design characteristics in common. They tend to be uncomplicated and inexpensive to construct, and their operation and maintenance should be safe and simple. These features help to ensure that the devices are used regularly. The cage links, for example, do not involve any structural modification to the cages, and can be attached and detached in less than 15 s. Moreover, the introduction of enrichment techniques as part of daily/weekly routines reinforce their importance in animal care, and animal technicians are keen to become involved in their use.

The techniques which we have developed are not only valuable in improving the environment of laboratory-housed marmosets, but also for the collection of additional information about aspects of marmoset behaviour. The process of allowing the animals to demonstrate more components of their behavioural repertoire broadens the baseline of spontaneous and simple conditioned behaviour, upon which experimental treatments such as the influence of drugs may be demonstrated. Direct observation of marmoset behaviour has been used as a baseline to study the acute and chronic effects of drugs (Ridley *et al.*, 1979; Scraggs and Ridley, 1979). Environmental enrichment, which would increase the number of observable categories of behaviour, might provide more sensitive baselines upon which to demonstrate the effects of experimental procedures.

Some work on zoo animals suggests that failure to use enrichment devices has been indicative of health problems, which would not have been detected at such an early stage by simple observation of the animals (Spinelli and Markowitz, 1987). If environmental enrichment measures are to be used in this way, methods for the automatic data collection would be useful.

For the present, although there is an increasing number of attempts to enrich captive environments, it is also the case that there is no information about the effects which withdrawal of enrichment has upon the animals.

ACKNOWLEDGEMENTS

I am grateful to M. Baskerville, R.N. Hollands, J.W. McBlane and S.S. Miles for their encouragement and practical assistance, and to G.D. D'Mello for his comments on the manuscript.

15

Responses to novel social stimuli in callitrichid monkeys: a comparative perspective

JEFFREY A. FRENCH and BETTY J. INGLETT

15.1 INTRODUCTION

Among the unfamiliar stimuli that primates confront during the course of their lifetime, social stimuli are arguably the most critical. The potential for novelty inherent in social interactions is extensive when one considers the myriad of ways in which social partners may interact. An encounter with an unfamiliar conspecific involves unpredictability and novelty due to a lack of information regarding the behaviour of a new social partner. Moreover, novelty in a social context is not limited to interactions between unfamiliar individuals; it can be generated in an interaction between two well-acquainted individuals simply by modifying the sequences of behavioural elements in a social exchange. A behavioural pattern of low probability inserted into a well-established behavioural sequence can alter the nature and outcome of even the most familiar social interactions. For instance, play behaviour in primates is characterized by such behavioural flexibility, and some consider behavioural novelty to be the critical feature of primate play (Baldwin, 1986).

Interactions with, and responses to, novel social partners have many short-term consequences. For instance, an aggressive encounter with an unfamiliar conspecific carries with it the risk of injury due to a lack of knowledge about the strength and combative strategies of the partner. Males in many species respond differently to familiar neighbours than they do to unfamiliar, novel males ('dear enemy' or 'neighbour' effect). Unfamiliar social partners often elicit a more elaborate response

Figure 15.1 Cotton-top tamarins (*Saguinus o. oedipus*) housed in natal group. (Photo: by J. French.)

than familiar ones (Marler, 1976). In tests of this hypothesis in captive capuchin monkeys (*Cebus apella*), Becker and Berkson (1979) noted that males directed more attacks towards an unfamiliar male than towards a male from a neighbouring group, or from their own group.

In addition to the immediate consequences of responses to novel social stimuli, there are several social interactions that have considerable impact on the lifetime reproductive success of individuals. Examples include intergroup agonistic exchanges, male–male competition for mates or other reproductive resources, and female choice of reproductive partners. Given the importance of these interactions for reproductive success, one would expect natural selection to exert strong selective pressures on the nature of interactions with unfamiliar individuals. As a consequence, it can be expected that differences in the nature of these interactions, among and within species, will be explicable in light of these pressures.

In this context we describe the social responsiveness of two callitrichid primates, the cotton-top tamarin (*Saguinus o. oedipus*) and the golden lion tamarin (*Leontopithecus r. rosalia*), to unfamiliar conspecifics (see Figures 15.1 and 15.2). We highlight differences between them as an integral part of each species' complex of adaptive

Figure 15.2 Golden lion tamarin (*Leontopithecus r. rosalia*) natal group. (Photo: G. Anzenberger.)

behaviour, but first we begin with some general information about the social biology of tamarins.

15.2 SOCIAL BIOLOGY OF TAMARINS AND MARMOSETS

There are approximately 20–37 recognized species of marmosets and tamarins (family Callitrichidae), depending upon which taxonomist one trusts (Napier and Napier, 1967; Hershkovitz, 1977; Rosenberger, 1983; Ford, 1986). Callitrichids are relatively homogenous with respect to the standard morphological measures used for taxonomy (e.g. dental patterns and clawed digits). Several other characteristics, such as the occurrence of twin births, small body size, large vocal repertoires and pronounced territorial behaviour, also occur throughout the family.

Historically, the predominant interpretation of the callitrichid mating system has been one of monogamy with an extended residence of offspring in natal family groups (Epple, 1975; Kleiman, 1977). However, recent field and laboratory reports suggest that no unitary characterization of callitrichid social systems adequately describes the diversity that actually exists. Hence, mating systems of these primates can be characterized as variants on a basic theme, that of communal rearing, a system in which variable numbers of adult males and females are reproductively active, but most group members provide parental care regardless of their breeding status. This system is rare among primates (Kleiman, 1977). Demographic analyses of callitrichid groups have shown that they may contain more than one adult of both sexes (Sussman and Garber, 1987), and in some cases the adults are known to be unrelated to the core breeding adults (Terborgh and Goldizen, 1985; Goldizen, 1986). Intergroup transfer has been documented with some frequency in wild, tagged groups (Neyman, 1978; Dawson, 1978; Garber *et al.*, 1984), indicating that social groups do not necessarily consist exclusively of related family members.

Recent reports have identified three variants on the communal rearing theme in callitrichids. First, classically monogamous social groups (one pair of breeding adults and their dependent offspring) have been reported in all species studied in the field to date (e.g. *Saguinus fuscicollis*: Terborgh and Goldizen, 1985; Goldizen, 1986, 1987; *S. mystax*: Garber *et al.*, 1984; *S. oedipus* ssp.: Dawson, 1978; Neyman, 1978; *Leontopithecus rosalia*: Kleiman *et al.*, 1986, personal communication; *Callithrix jacchus*: Hubrecht, 1984). Second, polyandrous mating relationships among marked individuals, in which the breeding female has been observed mating with more than one male group member, have been observed in at least one species (*S. fuscicollis*: Terborgh and Goldizen, 1985). Third, groups in which more than one breeding female is present have been reported in two species, *S. fuscicollis* (Terborgh and Goldizen, 1985) and *L. rosalia* (Kleiman *et al.*, 1986). These polygynous social groups are rare, and seem to occur when group sizes are large.

The reason that tamarins and marmosets display communal rearing appears to be due in large part to the cost of reproduction in these species. Allometric analyses of maternal/fetal weight ratios among primates indicate that callitrichid females give birth to the proportionately largest infants of all primates, with neonate weight in some cases equalling 15–20% of maternal weight (Leutenegger, 1973; Snowdon and Suomi, 1982). In addition, twin births are the rule, occurring in about 80% of births, and nursing continues for 3–4 months. Again, in lion tamarins, in common marmosets, and perhaps in many other species, females can, and do, become pregnant again immediately post-partum. In addition, there appears to be little lactational suppression of fertility in marmosets and tamarins (French, 1983). The tremendous costs of callitrichid reproduction are further emphasized by data which suggest that a solitary female or a single female/male pair are incapable of successfully rearing a set of offspring (Goldizen, 1986, 1987; Snowdon *et al.*, 1985). High reproductive costs and communal rearing are both likely to contribute to reproductive competition in these primates, particularly between females. The role of female–female reproductive competition in shaping the way in which callitrichids respond to unfamiliar conspecifics is discussed later.

15.3 RESPONSES TO UNFAMILIAR INTRUDERS – EXPERIMENTAL STUDIES

Experimental paradigm

Our experiments concerned the responses of breeding adults in family groups (hereafter referred to as 'residents') to the presence of unfamiliar conspecifics (hereafter referred to as 'intruders'). The aim was to acquire an analogue of the common field situation in which an established social group encounters a lone individual which has either emigrated from a natal group, or who is a transient and prospectively investigating a new group. By varying the identity of the unfamiliar tamarin or intruder introduced to residents, we were able to address the following issues: (1) What is the nature of the response of breeding adult tamarins to unfamiliar conspecifics? (2) Does the age/sex demography of the social group influence the responses of residents to the intruder? (3) Are the responses of residents modified by characteristics of the unfamiliar animal such as its age, gender, or reproductive status? (See Chapter 16 for an additional discussion.)

Although specific details of the experimental protocols differed slightly for cotton-top and lion tamarins, the basic paradigm for assessing responsiveness to social novelty was similar. Intruder tamarins were removed from their own social groups and placed in a portable cage with small-gauge wire-mesh walls. For both species the cage was sufficiently large (approximately 45 × 45 × 60 cm) to allow free movement of the tamarin inside; it also permitted good visual access between the residents and the intruder (see Figure 15.3 for an intruder interaction in lion tamarins). Intruders were given time to become accustomed to the cage through a series of habituation sessions and, by all appearances, exhibited normal behaviour such as eating, vocalizing and grooming while inside it.

Experimental trials involved placing the intruder cage and its occupant inside the home cage of a resident pair. Two observers recorded the responses of the resident pair to the presentation of the intruder

Figure 15.3 Male and female residents investigating a lion tamarin within the intruder cage. (Photo: J. French.)

throughout each trial (20 min for cotton-top and 30 min for golden lion tamarins). In experiments with both species, residents were involved in multiple trials, and were presented with different age-gender classes of intruder. The identity of intruders was varied to prevent the development of familiarity and potentially idiosyncratic relationships.

In the experiments that involved cotton-top tamarins, residents were presented with both unfamiliar adult male and female intruders (French and Snowdon, 1981). The main question of interest involved a comparison of male and female resident responses to intruders of either gender. In evaluating the responses of golden lion tamarins to novel social partners, we also presented male and female intruders, but varied a second variable; the sociosexual status of the intruder (French and Inglett, 1989). Intruders were either breeding adults (animals that occupied the breeding role in their social groups) or nonbreeding adults (animals that were nonbreeding, adult offspring in their social groups). All nonbreeding intruders were at least 1.5 years of age, and hence were post-pubertal (French, 1987). In lion tamarins, then, we evaluated the additional question of whether responses to unfamiliar tamarins varied as a function of the sociosexual status of the intruder (see Chapter 16 for complementary information in common marmosets).

Three major classes of behaviour were recorded during trials. First, we monitored spatial relationships between residents and intruders, including approaches and the time residents spent in close proximity to the intruder. Second, several noncontact agonistic displays were recorded. These included visual and vocal threats, and scent-marking behaviour. Finally, attempts by the residents to aggress the intruder, by chasing, attacking or attempting to bite were recorded (for detailed behavioural definitions see French and Snowdon, 1981; French and Inglett, 1989). During the experiments with both species, extremely aggressive interactions did occur. The use of the intruder cage proved to be valuable for preventing injury to the intruding tamarin, but the cage did not limit the ability of residents to direct aggressive responses toward intruders.

Responses of cotton-top tamarins

The responses of cotton-top tamarins varied with the gender of the intruder, and male and female residents behaved in strikingly different ways. Figure 15.4 presents a summary of resident scent-marking scores. Females showed significantly higher rates of both the two measures of scent-marking – namely, suprapubic (S–P) and anogenital (A–G) marking – than the males under all conditions (see also French and Cleveland, 1984). Similar results were obtained in baseline conditions with no intruders present. Rates of female scent-marking were clearly increased by placing unfamiliar conspecific intruders in the home cage, and this was highest when the intruder was a female. However, increases in scent-marking over baseline rates were only statistically significant for S–P marking, regardless of the gender of the intruder. Rates of scent-marking in males were not affected by the introduction of intruders of either gender.

The dimorphic pattern of male and female responses to unfamiliar conspecifics in cotton-top tamarins persisted for agonistic displays. Figure 15.5 presents mean rates of visual threats, which is a summed score of the number of frowns, arch posture, upright bipedal stance, and tongue-flicks directed towards the intruder by residents. Each of these patterns is associated with threat and aggressive contexts in cotton-top tamarins (Moynihan, 1970; Wolters, 1978; French and Snowdon, 1981). Male residents clearly discriminated the gender of the intruder, and showed high levels of agonistic threat responses in the presence of male intruders. Male cotton-top residents infrequently threatened female intruders. In contrast, female residents showed

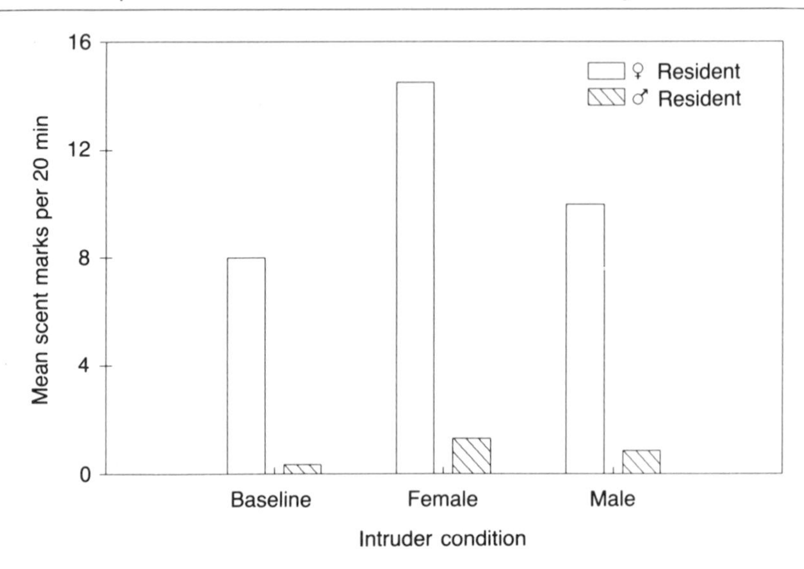

Figure 15.4 Mean rates of scent marking (summed scores for both anogenital and suprapubic marking) by resident male and female cotton-top tamarins under the following intruder conditions: baseline = no intruder; female = breeding female intruder; male = breeding male intruder. (After French and Snowdon, 1981.)

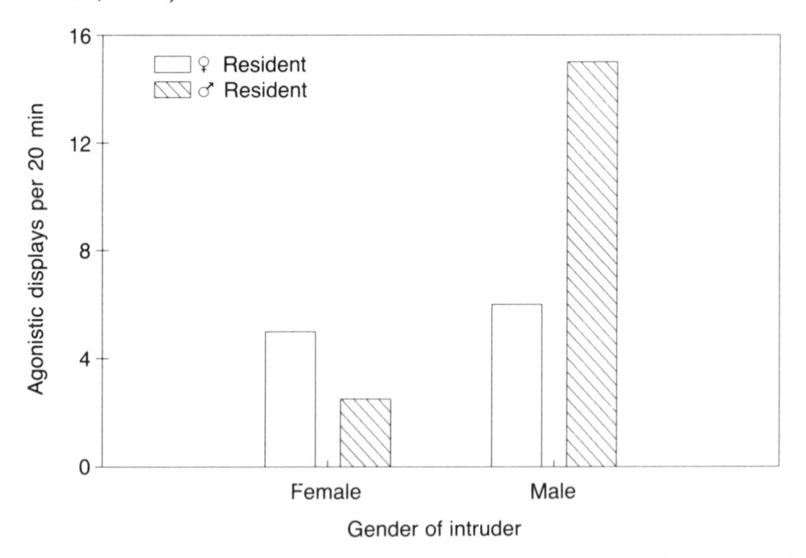

Figure 15.5 Mean rates of agonistic displays by male and female resident cotton-top tamarins in the presence of male and female intruders. Note the sexually dimorphic distribution of agonistic displays, performed primarily by males, and the gender-specificity of these displays. (After French and Snowdon, 1981.)

moderate levels of threat in the presence of unfamiliar intruders, and did not respond differentially to male and female intruders.

Only male cotton-tops were observed to attack intruders, and they exclusively attacked male intruders. These attacks were usually the terminal behaviour in a sequence of escalating agonistic displays. They were sufficiently severe to result in serious injury had the intruder not been located in a protective cage. Female residents exhibited a complete absence of aggressive behaviour toward either male or female intruders. Indeed, their responses in the presence of the intruders, with the exception of increases in scent-marking, could be characterized by benign indifference.

In summary, the response of breeding adult cotton-top tamarins to unfamiliar conspecifics was sexually dimorphic and, in males, gender-specific. Females showed increases in scent-marking, but were not involved to any extent in agonistic displays or attacks on intruders. In contrast, males showed no changes in scent-marking as a function of intruder gender, but were aggressively intolerant of conspecific males. These patterns are similar to the qualitative description of intergroup encounters and encounters of groups with transient individuals in free-ranging *S. oedipus* reported by Dawson (1978, 1979) and Neyman (1978). The concordance between these observations and those from the field lends credence to the ecological validity of the technique we used to evaluate the response of cotton-top tamarins to social novelty.

Responses of golden lion tamarins

As with cotton-tops, male and female lion tamarins showed dimorphic behavioural responsiveness, which was specific to the gender of the intruder. However, we found striking species differences between lion and cotton-top tamarins with respect to the gender of the resident that showed intolerance toward intruders.

Lion tamarins quickly discriminated the gender of an unfamiliar conspecific; they approached females more quickly than they did males (Figure 15.6). Female residents showed extremely rapid approaches to female intruders (within the first 15 s of the onset of a trial). Female residents also spent more time in close proximity to female than to male intruders. Female intruders also tended to be approached more frequently by residents than were male intruders.

Further, clear differences in agonistic displays and aggression were observed in response to intruders of different gender and sociosexual status. Figure 15.7 presents a cumulative score for noncontact agonistic displays, which included vocal threats (long calls and

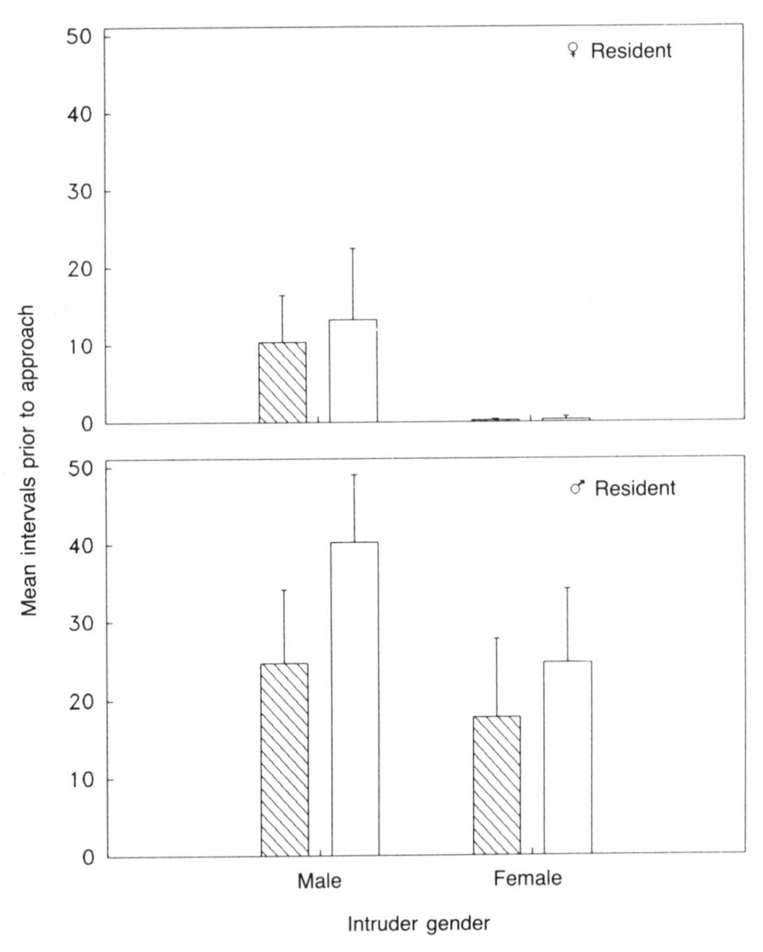

Figure 15.6 Mean latency to approach breeding and nonbreeding intruders by resident male and female golden lion tamarins. Resident females clearly discriminated the sex of the intruder, approaching female intruders within 15 s of the onset of a behavioural trial. (From French and Inglett, 1989.)

chatters: McLanahan and Green, 1978), visual displays (frontal stare, arch walk and 'look away': Rathbun, 1979) and scent-marking (both circumgenital and ventral: Kleiman and Mack, 1980). Females exhibited high levels of these threat displays; they directed them almost exclusively toward female intruders, and more so at breeding than nonbreeding females. No resident female was ever observed to attack a male intruder (of either status), and only three instances of resident

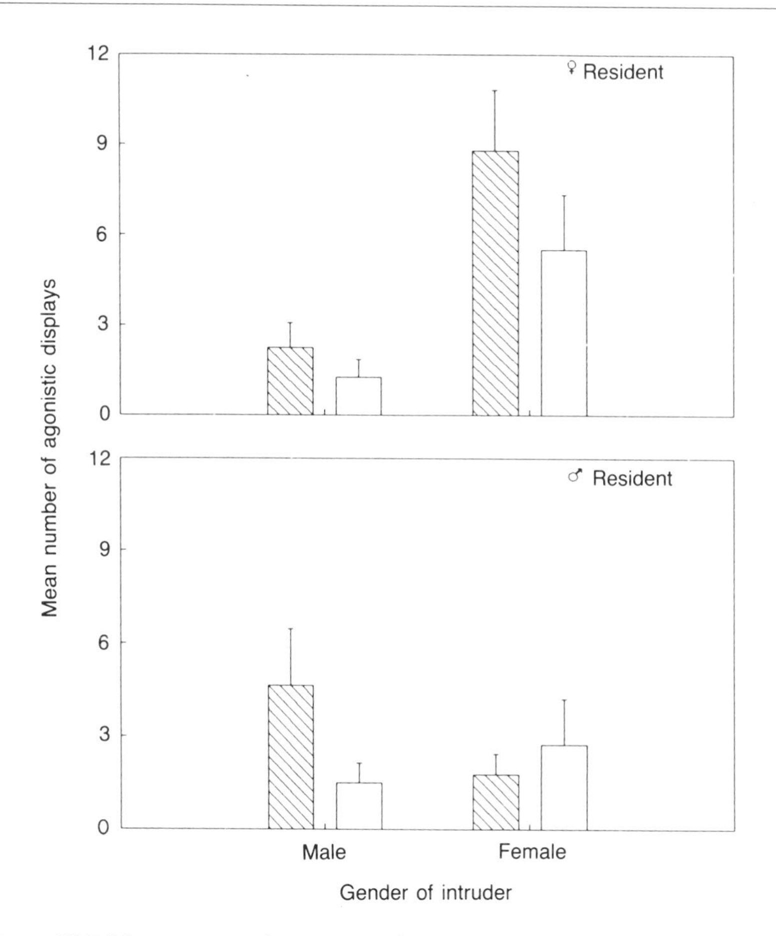

Figure 15.7 Mean rates of agonistic displays (vocal, visual and olfactory) performed by golden lion tamarin residents in the presence of intruders of different gender and breeding status. Female intruders received significantly more agonistic displays from female residents than in any other intruder condition. (From French and Inglett, 1989.)

male aggression were recorded during the course of the study. One resident male attacked a breeding male intruder twice during one trial, and the same male attacked a breeding female intruder once during an extremely aggressive interaction between the male's mate and a female intruder. Hence, resident aggression towards intruders was by far the most sexually dimorphic and gender specific type of behaviour exhibited by lion tamarins (Figure 15.8). Aggression occurred often during intruder trials and, with only three exceptions, all attacks were performed by female residents toward female intruders, with breeding

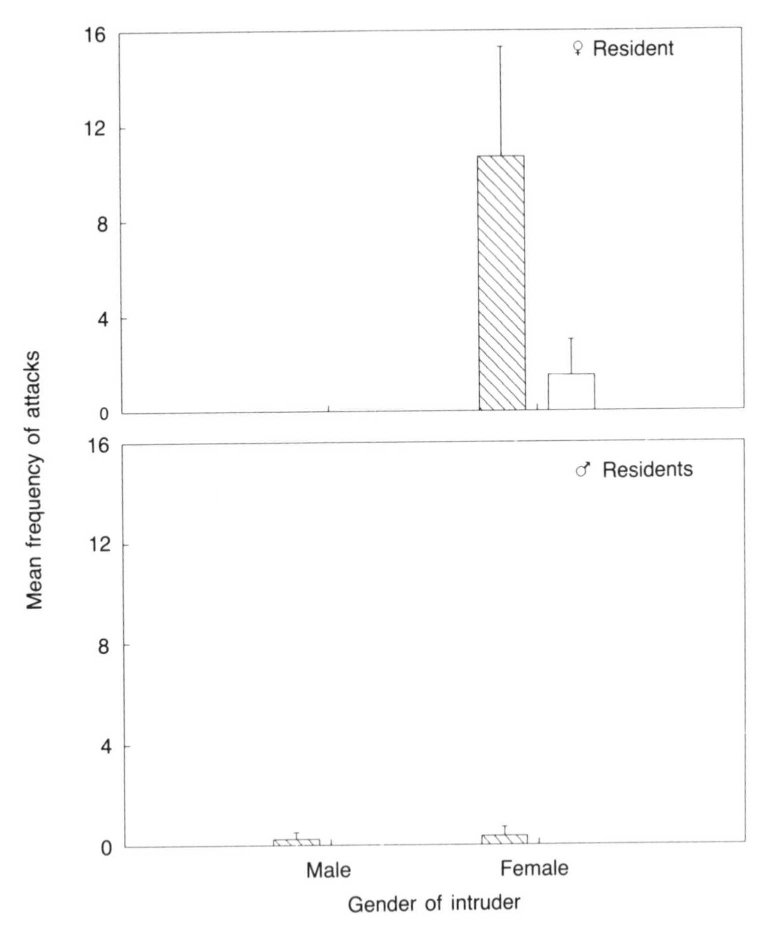

Figure 15.8 Mean rates of aggression performed by resident golden lion tamarins in the presence of intruders. Breeding female intruders received the highest levels of aggression from female residents, and little aggression was observed in other intruder conditions. (From French and Inglett, 1989.)

females receiving significantly more attacks from resident females than nonbreeding female intruders.

15.4 ACCOUNTING FOR SPECIES DIFFERENCES IN RESPONSES TO INTRUDERS

In the previous section, species differences in the responses of tamarins to unfamiliar conspecifics were described. We are now faced with the problem of accounting for these differences in an integrative fashion within the broader context of adaptive callitrichid social behaviour. To this end, and in this section, we draw upon two aspects of cotton-top and golden lion tamarin social biology; namely, the reproductive suppression of nonbreeding females in social groups, and evidence of species divergence in the dynamics of intragroup aggression.

Reproductive suppression

With only a handful of exceptions (see section 15.2 above), a characteristic of callitrichid social groups is that only one female produces offspring. Moreover, until recently the reproductive status of nonbreeding females (daughters and emigrant females) had been a matter of speculation. However, the development of sensitive and noninvasive techniques for assessing reproductive function (e.g. Hodges *et al.*, 1981; French and Stribley, 1985) has made it possible successfully to monitor ovarian and reproductive activity in social groups. These analyses have shown that nonbreeding female common marmosets (*Callithrix jacchus*) and saddle-back tamarins (*Saguinus fuscicollis*) are physiologically, reproductively inhibited (Abbott *et al.*, 1981; Abbott, 1984; Epple and Katz, 1984; Evans and Hodges, 1984). That is, they exhibit low and acyclic endocrine profiles that reflect ovulatory failure. Thus, the nonbreeding status of subordinate females in these species is obligatory, and the relevant stimulus for ovulatory suppression would appear to be the presence of the reproductively active female (Epple and Katz, 1984).

We have evaluated reproductive functioning in both cotton-top and golden lion subordinate females (French *et al.*, 1984; French and Stribley, 1987; French *et al.*, 1989), and a comparative analysis of reproductive suppression reveals further species differences. Nonbreeding female cotton-tops are clearly physiologically suppressed with respect to their ovarian activity. Figure 15.9 shows mean levels of oestrone and oestradiol excreted in the urine of five females while housed in either the presence of a dominant, breeding female ('family') or in the presence of an unrelated adult male with no other breeding females present ('mate'). Levels of urinary oestrogen excretion were clearly depressed when females were in the presence of a dominant breeding female.

The suppression of ovulatory activity does not appear to be a delay in puberty, for two reasons. First, daughters which were removed from their families began ovulatory cycling almost immediately (within 1–2 weeks), and females can conceive within 2 weeks after being removed from their family group and housed with a mature male

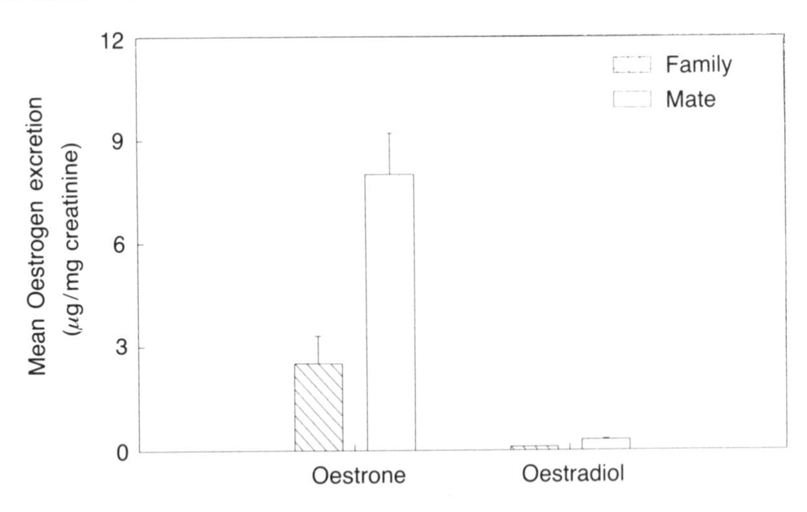

Figure 15.9 Excretion patterns of urinary oestrogen in female cotton-top tamarins while housed in the presence of a dominant, breeding female ('Family') and after pairing with an unrelated adult male in a separate cage ('Mate'). Urinary oestrogen concentrations are corrected by urinary creatinine concentrations to control for variable fluid intake and output. (After French *et al.*, 1984.)

(French *et al.*, 1984; Ziegler *et al.*, 1987c). Second, ovulatory activity in cotton-top females can be terminated when they are housed in the presence of another dominant, breeding female (French *et al.*, 1984).

Lion tamarin social groups are identical to other callitrichid species in that only one female in a group produces offspring. However, we have found differences with regard to the hormonal status of non-breeding females when compared with those in which ovarian cycling is suppressed (i.e. cotton-tops). We investigated reproductive functioning in seven females, all over the age of 1 year, which occupied a nonbreeding or subordinate role in their social group (six daughters in natal groups and one behaviourally subordinate female in a pair of females). The results of our evaluation indicated a complete absence of endocrine suppression in these nonbreeding golden lion females. Figure 15.10 presents urinary hormone profiles for two daughters and a subordinate female. The profiles for all the females are characterized by cyclic patterns of oestrogen excretion. The large peaks of oestrogen probably reflect a postovulatory surge of oestrone, and hence are confirmatory of ovulation (Ziegler *et al.*, 1987a; Eastman *et al.*, 1984). All seven females showed urinary oestrogen profiles that were indistinguishable from those of breeding adult females.

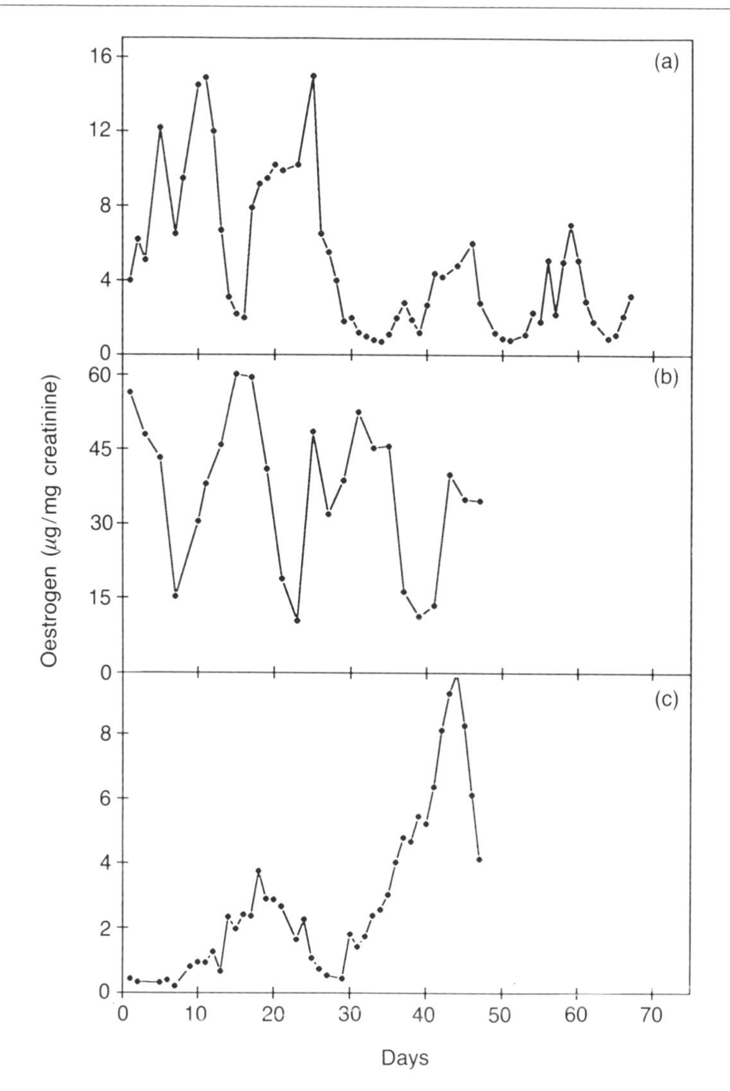

Figure 15.10 Patterns of urinary oestrogen excretion in three female golden lion tamarins: (a) profile from a 24-month-old daughter in a family group; (b) profiles from a 39-month-old female housed in her natal family group; (c) oestrogen cycles in a 120-month-old subordinate female in an isosexual peer group.

These results show that the mechanisms which produce reproductive inhibition in these two tamarin species are radically different. In cotton-top tamarins, reproductive suppression in all but the breeding female is accomplished by a physiological inhibition of ovulation. In

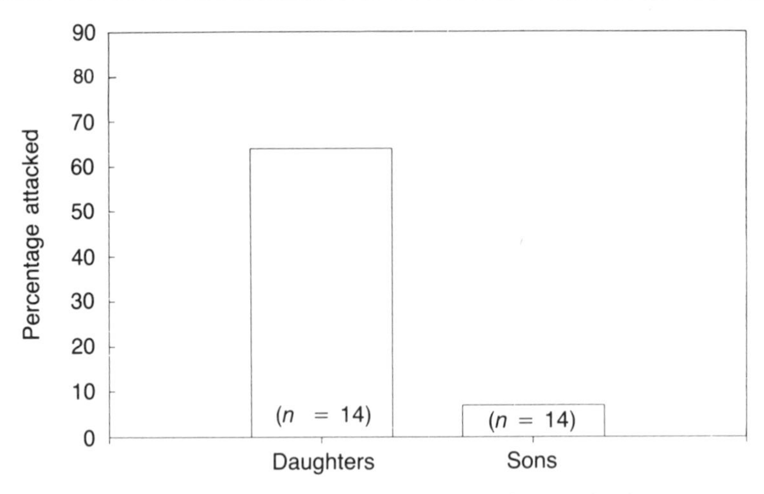

Figure 15.11 The proportion of offspring (out of a total of 14 sons and 14 daughters) that were targets of attack in cotton-top tamarin social groups. Daughters are significantly more likely to be attacked than sons. (Data modified from McGrew and McLuckie, 1986.)

lion tamarins, reproductive activity in nonbreeding females is prevented by a nonphysiological mechanism. Our results show that the complex cascade of changes underlying puberty and the onset of reproductive functioning in females (Terasawa *et al.*, 1983) can mature in lion tamarin daughters as early as 1.5 year of age. However, there is a clear dissociation between the endocrine markers of puberty and the behavioural manifestation of reproductive activity. In a long-term study of sexual behaviour in lion tamarin groups (Stribley *et al.*, 1989), mating activity was limited exclusively to the breeding male and female, so it is likely that some behavioural mechanism operates to prevent mating among nonbreeding individuals. We will return to this point later on, but first will consider intrafamily aggression as one such mechanism in callitrichids.

Intrafamily aggression

It is well established that strife within social groups is common in captive (and presumably free-ranging) marmosets and tamarins (e.g., Rothe, 1975; Wolters, 1978). When we considered intragroup social dynamics in these two tamarin species, we found further contrasts.

Several anecdotal reports of intrafamily conflict within cotton-top groups indicated that fights among sisters tended to be more frequent, and more severe, than fights among males (Hampton *et al.*, 1966;

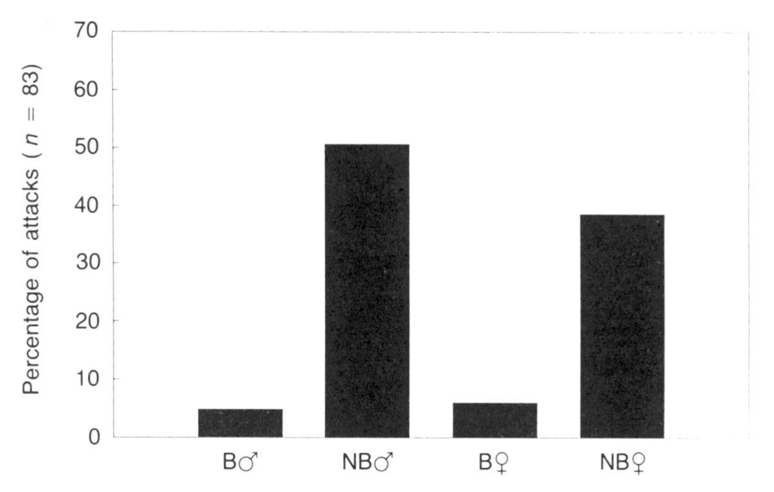

Figure 15.12 Distribution of targets of intrafamily aggression in golden lion tamarin family groups (B, breeding; NB, nonbreeding). Nonbreeding animals were more likely to be targets of attack in family groups than adult breeding animals of either sex. Juvenile males were the targets of more than half of all reported attacks in lion tamarin groups. (Data modified from Inglett *et al.*, 1989.)

Wolters, 1978). In a more systematic analysis, McGrew and McLuckie (1986) documented cases of aggression directed towards older offspring in several cotton-top family groups. The results of their analysis, presented in Figure 15.11, show that, although the number of sons and daughters in the families was identical, daughters were much more likely to be the targets of attack than sons. The initiators of attacks were also likely to be same-gender siblings, although aggression from mothers or other family members was not uncommon (see also Wolters, 1978). To the extent that intrafamily aggression represents a potential proximal stimulus for emigration, females appear to be more likely than males to emigrate. This hypothesis has received some, but not overwhelming, confirmation from field reports that juvenile females are observed slightly more often as emigrants, immigrants, and in transient status (Dawson, 1978; Neyman, 1978).

To provide comparable data for lion tamarins we (Inglett *et al.*, 1989) compiled well-documented cases of intrafamily aggression from 11 institutions around the world that house large colonies of *L. rosalia*. Our analysis focused on 83 instances of aggression within family groups. The vast majority of attacks (89%) were directed towards older juveniles or nonbreeding adults (Figure 15.12). In lion tamarin family groups, sons received the highest amount of aggression. Attacks on sons accounted for 51% of all attacks, and 57% of attacks on nonbreeding tamarins (Figure 15.12). Daughters were the targets of attack in 38% of the reported cases. The finding that nonbreeding individuals are more likely to be the targets of intrafamily aggression agrees with Kleiman's (1979) earlier analysis of family aggressive conflicts, although she reported a higher incidence of lethal attacks on daughters (three out of six instances of aggression involved the death of the attacked daughter). Again, if we are correct in our interpretation of intragroup aggression as a proximal stimulus for emigration from natal groups, the results of our survey indicate that there should be a bias toward the emigration of sons from natal family groups in free-ranging lion tamarins. We anxiously await field data to evaluate this hypothesis.

15.5 CONCLUDING REMARKS

We submit that the species differences we have identified in tamarin responsiveness to unfamiliar conspecifics can be interpreted in the light of species differences in the mechanisms of reproductive suppression, and the maintenance of reproductive exclusivity of the breeding female. Table 15.1 presents a summary of the information outlined in this chapter; consistent patterns emerge which allow an integrated interpretation.

Female cotton-top intruders that encounter established social groups are treated with indifference by female residents. We interpret this lack of response as a consequence of the fact that immigrating females are unlikely to pose a significant reproductive threat to resident females. The dominant female in a group possesses the capacity to inhibit ovulatory activity in all other females, effectively eliminating any potential reproductive competition. Immigrating females, should they join a social group, may represent additional nonbreeding helpers in the communal rearing system. In turn, daughters are much more likely than sons to be evicted from natal groups through intrafamily aggression. Sons are least likely to be successful in gaining entry to new groups because of high levels of male–male threat and aggression; they are also least likely to be evicted from their natal group. Sons may remain in natal groups until the male breeding position is vacated through death or departure (McGrew and McLuckie, 1986).

In contrast to cotton-tops, female lion tamarins respond very aggressively during confrontations with unfamiliar female intruders. In the absence of the capacity to control or suppress reproductive activity in other females, the breeding female faces a threat to her monopolization of reproductive resources such as food, and, perhaps more

Table 15.1 Comparison of cotton-top and golden lion tamarins: responses to intruders, reproductive suppression, and intrafamily aggression

Measure	*Cotton-top tamarins*	*Golden lion tamarins*
Response to male intruders	Intense male aggression (1,2)	Weak response (3)
Response to female intruders	Weak response (1)	Intense female aggression (3)
Reproductive suppression	Strong endocrine suppression (4,5)	Lack of ovulatory suppression (6,7)
Intrafamily aggression	Biased toward attacks on daughters (8)	Biased toward attacks on sons (9)
Suspected emigrating sex	Females (2,8,10)	(Males) (11)

Sources: (1) French and Snowdon, 1981; (2) Dawson, 1978; (3) French and Inglett, 1989; (4) French *et al.*, 1984; (5) Ziegler *et al.*, 1987c; (6) French and Stribley, 1987; (7) French, 1987; (8) McGrew and McLuckie, 1986; (9) Inglett *et al.*, 1989; (10) Neyman, 1977; (11) no field evidence currently available.

importantly, aid from helpers in the family group. High levels of intolerance for unrelated females may represent an important mechanism by which females maintain reproductive exclusivity in lion tamarin social groups. Patterns of intrafamily aggression complement this intolerance, as sons may be more likely to be forced from their natal group. However, due to the absence of male–male aggression sons are also more likely to gain access into unfamiliar groups. Females are least likely to be successful in joining unfamiliar groups for the reasons discussed above, and our analyses indicate that daughters are less likely than sons to be forcefully evicted from natal groups. However, a complicating factor in this scenario is the observation that female–female aggression in an intrafamily conflict tends to be more serious than male–male aggression. Thus, although daughters are slightly less likely to be targets of aggression, attacks on them are more severe than those on sons.

Our interpretation of Table 15.1 is that reproductive competition is a critical pressure on both cotton-top and golden lion tamarins, and that the manifestation of this competition differs between species and across genders. Intrasexual reproductive competition in cotton-top tamarins has been described in both males and females, but the mechanism used by each gender differs from the other. Male cotton-tops are overtly aggressive to potential reproductive competitors, while females are not. Female cotton-tops possess the capability to inhibit directly the reproductive capability of any other female in the social group. In golden lion tamarins, females use the same competitive mechanism as male cotton-top tamarins: breeding females are aggressively intolerant of other adult breeding females. There is a lack of overt aggressive competition and physiological suppression (French, 1987) in male golden lion tamarins. This lack of reproductive competition is surprising, and the reasons for its absence are not immediately apparent.

Although the specific manifestations of reproductive competition are markedly different, the ultimate outcome of the competition is identical for both tamarin species. Aggressive attacks of same-gender conspecifics by male cotton-top and female lion tamarins would probably effectively repel reproductive competitors from a group, while ovarian suppression by female cotton-tops would eliminate reproductive competitors within a group. For female tamarins in particular, either mechanism, aggression or suppression, produces the same outcome; namely, a reduction in competition for limited reproductive resources.

ACKNOWLEDGEMENTS

We would like to thank the many individuals who contributed to this project, including Anne Eglash for assistance with behavioural observations, Denise Hightower and Bill Oldenhuis for assistance with animal husbandry, and Judy Stribley and Theresa Dethlefs for assistance with urine collections and helpful comments on previous versions of the manuscript. Chuck Snowdon at the University of Wisconsin and Ken Deffenbacher at the University of Nebraska at Omaha facilitated the conduct of research at the respective institutions. This research has been supported by funds from the University Committee on Research at the University of Nebraska at Omaha, and by grants from the PHS, including HD 23139 awarded to Jeffrey A. French and MH 35215 awarded to Charles T. Snowdon.

16

Reproductive consequences of changing social status in female common marmosets

DAVID H. ABBOTT and LYNNE M. GEORGE

16.1 INTRODUCTION

One of the striking features of the biology of marmoset and tamarin monkeys (Callitrichidae) is a distinct and reliable relationship between social status and reproductive physiology. These monkeys demonstrate an extreme example of social control over female reproduction. As was noted in Chapter 15, only one female of a group breeds in the wild (Garber *et al.*, 1984; Terborgh and Goldizen, 1985; Stevenson and Rylands, 1988). The same reproductive monopoly is found in captivity (Epple, 1975; Abbott, 1984; French *et al.*, 1984; Ziegler *et al.*, 1987b). Moreover, laboratory studies have shown that it is the social status of a female that determines which female will breed in any one group. Hence, the socially dominant or rank 1 female becomes the breeding female (Epple, 1975; Abbott and Hearn, 1978) and subordinates, ranks 2 or below, become nonbreeding females. The system of reproductive inhibition is frequently related to the need of breeding females to have helpers to rear the proportionately large individuals of multiple births (see Abbott, 1987; Sussman and Garber, 1987).

A clear-cut reproductive division between the dominant female and her subordinates is achieved in a majority of callitrichid species by the suppression of ovulation (Abbott *et al.*, 1981; Epple and Katz, 1984; Evans and Hodges, 1984; French *et al.*, 1984). For example, in a study of common marmoset families (*Callithrix jacchus*; Abbott, 1984), one daughter ovulated in about 50% of the families studied, whilst in another study of common marmosets, no ovulatory daughters were found (Evans and Hodges, 1984). However, since the cycling daughters did not copulate, no pregnancies occurred. The exceptions to the anovulatory rule are rare and, where they exist, a further behavioural block to reproduction operates because the subordinates do not copulate. In family groups of golden lion tamarins (*Leontopithecus rosalia*), for instance, all mature daughters ovulate (French and Stribley, in press) but do not mate (see Chapter 15). In most marmoset and tamarin groups, then, social status completely determines whether or not a mature female will ovulate, which in turn means that strict social control is maintained over female reproductive physiology (Abbott *et al.*, 1988). The more usual situation in primate species is that subordinate females just produce or rear fewer offspring than dominant females because of the harassment they receive from their dominants (Abbott, 1987; Harcourt, 1987a; Dunbar, 1989). Ovulation is therefore not completely suppressed as in most marmosets and tamarins.

Our aim in this chapter is to capitalize on the definite relationship between social status and female reproductive function in marmosets and tamarins, and to illustrate how the social environment in a group can have profound and predictable effects on each individual female's reproductive physiology. These effects differ between females, depending upon each animal's social status in the group. The female common marmoset is given as a case in point to show how an individual female's reproductive physiology can be drastically altered by changing her social status. The results may serve as a good example of the physiological mechanisms which operate naturally in a female primate to translate the social stress of subordination into infertility.

16.2 THE ESTABLISHMENT OF SOCIAL STATUS

Over the 4 years of the present study, 28 standardized social groups of marmosets were established. They were made up of unrelated post-pubertal monkeys, normally with two to four females and two to four males. All the animals were born in captivity and were housed in the Institute of Zoology of The Zoological Society of London (Hearn, 1983). Despite the total number of groups, we actually maintained a maximum of eight groups at any one time. The turnover was due to an almost equal mixture of colony management practices and the instability of some groups beyond 6–10 months.

The social status of each female in a group was usually determined within 2–3 days of group formation. Dominant (rank 1) or subordinate (ranks 2 and below) status was assigned from quantitative analyses of behavioural recordings of aggressive and submissive interactions between female members of the group (Abbott and Hearn, 1978; Abbott, 1984). Each gender formed a discrete hierarchy and a marmoset's rank within its own gender was the key determinant of its reproductive performance (Abbott, 1978, 1979, 1984). The rank of an animal within its group as a whole was of lesser importance.

16.3 MONITORING OVARIAN FUNCTION

As female marmosets show no physical signs of approaching ovulation, and do not menstruate, ovarian function was monitored by the measurement of circulating progesterone concentrations in blood samples collected twice weekly (Hearn *et al.*, 1978; Abbott *et al.*, 1988). The frequency of this blood sampling was sufficient to cover the period of time when plasma progesterone concentrations were elevated, following ovulation, during the approximately 19–20-day luteal phase of the 28-day ovarian cycle (Harlow *et al.*, 1983). The day of ovulation was designated as the day before plasma progesterone concentrations exceeded 10 ng/ml. This plasma concentration represents the earliest detectable increase in progesterone consistently associated with a luteal phase (Harlow *et al.*, 1983). The luteal phase terminated when plasma progesterone concentrations returned to below 10 ng/ml. With a blood sampling frequency of twice a week, these time estimates for ovulation are only accurate to ± 1 day.

When offspring were born to the dominant female in the standardized social groups, she aggressed the unrelated subordinate females to an extent which warranted the removal of all unrelated female subordinates (Abbott and Hearn, 1978; Abbott, 1984). In order to avoid disbanding these established groups, pregnancies of dominant females were terminated artificially in all but two groups. To achieve this, dominant females were given an injection of a prostaglandin $F_{2\alpha}$ analogue, cloprostenol (Summers *et al.*, 1985), 14–30 days after ovulation. The prostaglandin treatment induced the demise of the corpora lutea and thus terminated the luteal phase of the cycle or early pregnancy. Ovulation recurred within 9–13 days of prostaglandin treatment (Summers *et al.*, 1985). In the two exempted groups, no artificial fertility control was applied.

Only the dominant females in our standardized social groups showed normal ovulatory cycles. Subordinate females either did not ovulate (39 out of 52 subordinates: 75%) or underwent inadequate cycles (one cycle: 9 females; two cycles: 3 females; four cycles: 1 female), where plasma progesterone concentrations were elevated about 10 ng/ml for only 1–14 days, in contrast to the 19–20 days of the normal cycle. We only consider here groups with acyclic subordinate

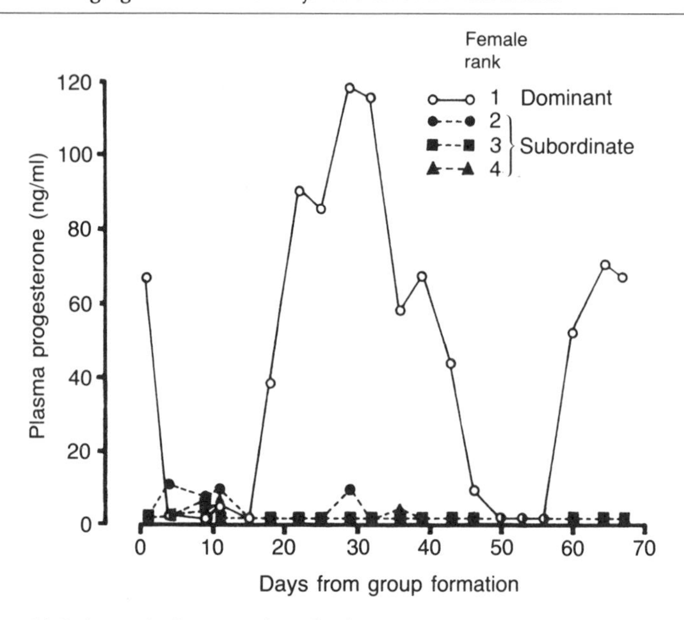

Figure 16.1 A typical example of plasma progesterone concentrations in a dominant female marmoset monkey and her three subordinates in one of the 22 groups. All subordinates were acyclic. The period of time shown covers the group's first 70 days and starts at the end of a luteal phase for the dominant female and continues through the next two ovarian cycles, finishing in the mid-luteal phase of the second cycle. The first cycle may have been a short-lived pregnancy because of the longer than usual 30-day elevation in plasma progesterone concentrations.

females. Figure 16.1 shows a typical example which, in fact, represents the majority of the standardized groups that were established (22 out of the 28, i.e. 82%).

Given that we were concerned to illustrate the control of ovulation by social status, manipulations of status and subsequent evaluation of endocrine parameters provided excellent empirical opportunities. Some instances are now described to show that imposing subordinate status on a female switches *off* her ovulatory cycles and that removing a female from subordinate status switches *on* her ovulatory cycles. This functional change is rapid, clear-cut and repeatable. We begin with cases in which females were removed from subordinate status.

16.4 REMOVAL OF THE DOMINANT FEMALE

In two groups, the dominant female died. This was not associated with any fighting or aggression within either group. The effect of losing the dominant female on the reproductive physiology of the

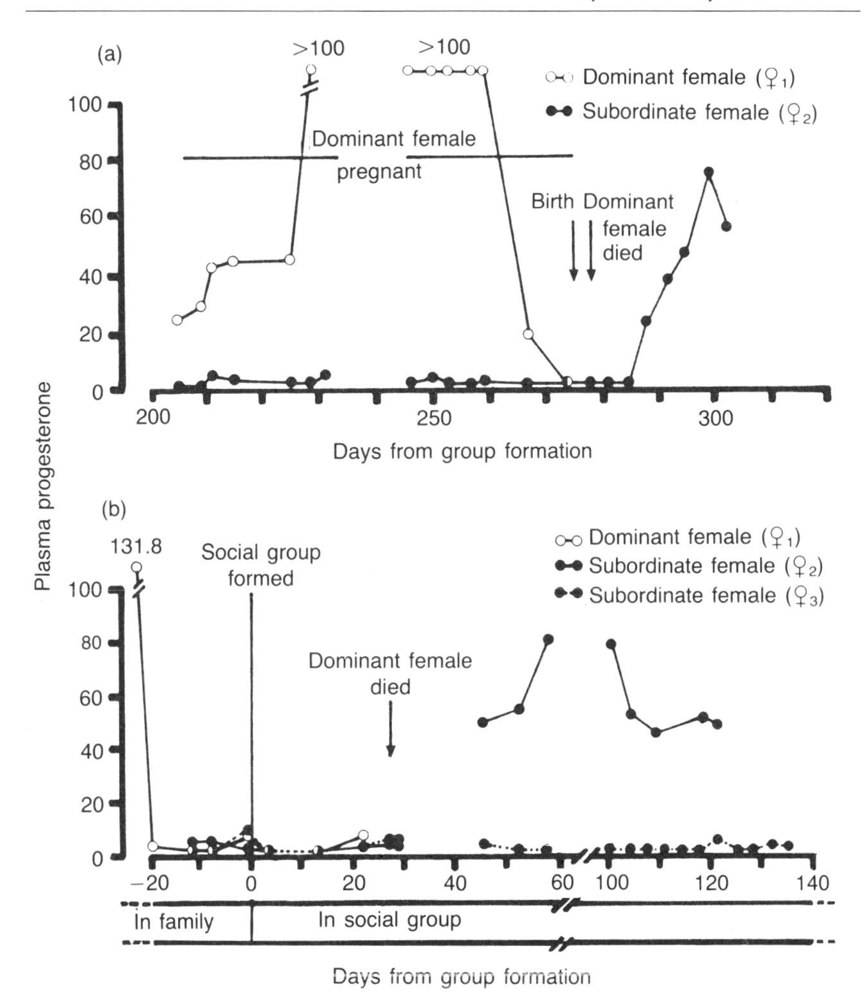

Figure 16.2 Plasma progesterone concentrations in dominant and subordinate female marmoset monkeys in (a) a social group where the dominant female died following a postpartum haemorrhage and (b) a social group where the dominant female died from a 'wasting syndrome' shortly after the group was established. (Reprinted with permission of the British Ecological Society from Abbott, 1989.)

remaining female subordinates is illustrated in Figure 16.2. In one of these groups (Figure 16.2a), the dominant female died approximately 9 months after the group was formed due to a postpartum haemorrhage. The only subordinate female in that group ovulated 9 days later and conception occurred during this cycle. In the second group (Figure 16.2b), the behaviourally identified dominant female died 28 days after group formation from a 'wasting syndrome' which produced a chronic loss of body weight and, probably as a consequence, she stopped ovulating.

Of the two subordinate females which remained, the highest ranking of them, the rank 2 female, ovulated and conceived within 12 days of the dominant female's demise. This was indicated by continually elevated plasma progesterone concentrations after the death of the dominant female, and by the subsequent parturition date (Figure 16.2b the postovulatory progesterone rise was missed because of an interruption of blood sampling). The subordinate rank 3 female did not ovulate.

In both these cases, in which previously subordinate females started ovulating, they conceived, produced live offspring at full term and raised the offspring successfully. Their ascension to the status of breeding female of their group was therefore complete.

16.5 THE DOMINANT FEMALE DEPOSED

Given the inhibitory effect of a dominant female on the ovulatory capacity of subordinate female marmosets, the question arose as to whether subordinate female marmosets would attempt to depose their dominant female and take over the breeding position; and, further, what would happen to the previously dominant female?

We can answer these questions only on the basis of one instance to date. A dominant female was deposed by a subordinate female in only one of our 28 groups. In this case, the group was established for about 9 months before the dominant female (226W) injured her back and legs, probably during an accidental fall in her cage which resulted in temporary impairments in her mobility. She had also been ovariectomized 3 years previously; however, ovariectomized females are completely effective dominant females (N = 5: D.H. Abbott and L.M. George, unpublished data). About a month following the injury to this dominant female, the intact subordinate female in the group (359W) attacked her ovariectomized dominant, and took over the rank 1 position. This rank change occurred within 3 days of the initial aggressive behaviour displayed by female 359W. The previous dominant female (226W) adopted the submissive role of a rank 2 female. The now-dominant intact female (359W) ovulated within 5 days of assuming the rank 1 position (Figure 16.3). The luteal phase was normal in all respects. This female was then removed from her group for other experimental purposes.

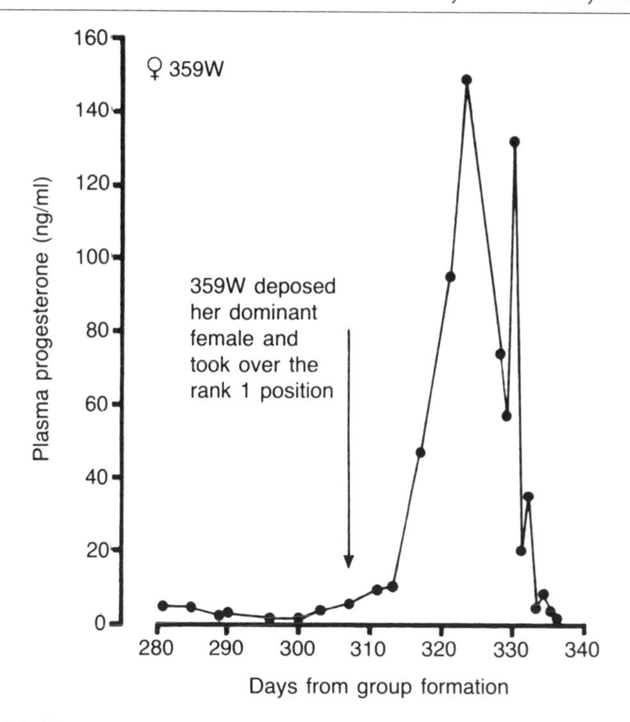

Figure 16.3 Plasma progesterone concentrations in female marmoset 359W before and after she deposed her ovariectomized dominant female to take over as the rank 1 female of the group. Ovulation took place about 5 days later and was followed by an apparently normal luteal phase of 20 days.

16.6 SUBORDINATE FEMALES REMOVED AND HOUSED INDIVIDUALLY

We were particularly interested in finding out whether removing subordinate females from their groups, and housing them singly, was sufficient to initiate their ovulatory cycles. If this was the case, then the only factor preventing normal ovarian cyclicity in mature female marmosets would be the presence of a dominant female imposing subordinate status. This was an important point to ascertain because Ziegler and her colleagues (1987c) had shown that removing anovulatory, subordinate daughter cotton-top tamarins (*Saguinus oedipus*) from their family groups and housing them singly was insufficient to initiate ovulatory cycles. A male had to be introduced to the isolated females for this to occur.

Six subordinate female common marmosets were removed from five separate social groups, after they had experienced 2–18 months of subordinate status. Five out of the six females ovulated for the first

Table 16.1 Time from the removal of subordinate social status until the first ovulation occurred in female marmoset monkeys (sample size is in parentheses)

Social manipulation	*Time to first ovulation (days)*
Removal of the dominant female (a)	9, 12 (2)
Deposing of the dominant female (b)	5 (1)
Removal of subordinate females from their groups and housing them singly (c)	16.8 ± 14.7* (6)
Subordinate female becoming dominant female in a new group (d)	8 (1)
(a)–(d) combined	13.5 ± 3.8* (10)

*Mean ± SEM.

time within 18 days of removal, and the sixth ovulated after 46 days (Table 16.1). However, the latter female had received an intense bout of aggression from her dominant female immediately prior to her removal from the group. This may have contributed to the lengthy time taken to initiate ovulatory cycles in this isolated subordinate.

Figure 16.4 illustrates typical examples from two of the five females which rapidly ovulated in isolation. Once ovulatory cycles began, ovarian activity continued normally in all six females. Thus, in female marmosets, there is no apparent need to pair with a male, either to initiate or to maintain ovarian cycles in previously anovulatory subordinate females. The only factor preventing mature female marmosets from undergoing ovulatory cycles is their subordinate social status.

16.7 DOMINANT FEMALES WITH SUBORDINATE STATUS IN THEIR PREVIOUS GROUP

For various experimental and management requirements, social groups were disbanded from time to time and some of the group members were used to establish new groups. On four such occasions (included in the total of 28 groups), the female that attained the dominant position in the new group had been previously subordinate and acyclic in another group for at least 6 months. On attaining the rank 1 position in their new group, all four females exhibited normal ovulatory cycles, as illustrated by the example of female 255W in Figure 16.5. The remaining three females had been removed from their previous group some time before their inclusion into a new group, and thus had already begun to ovulate.

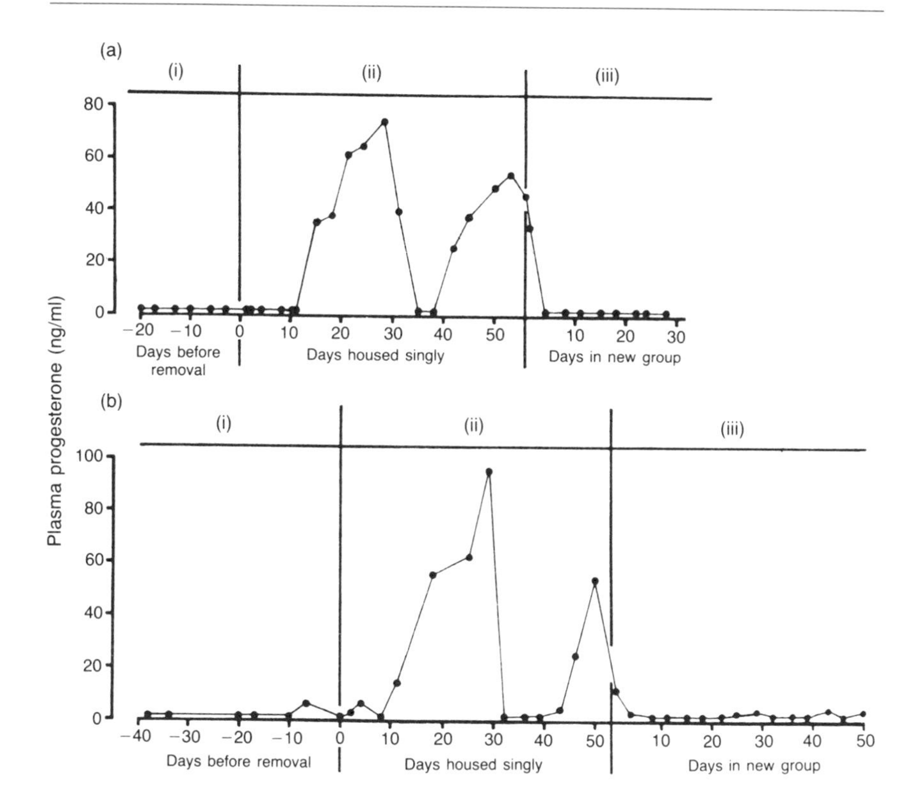

Figure 16.4 Plasma progesterone concentrations in (a) a female marmoset monkey (200W) while (i) subordinate in a social group (established 17 months previously), (ii) housed singly (for 56 days) and (iii) subordinate in a new social group, and in (b) a female marmoset monkey (224W) while (i) subordinate in a social group (established 9 months previously), (ii) housed singly (for 53 days) and (iii) subordinate in a new social group. (Data from Abbott *et al.*, 1988.)

In summary, the above examples illustrate the rapid onset of ovarian cycles in female common marmosets which were removed from subordinate status. The results are combined in Table 16.1, and clearly implicate the dominant females as mediating the suppression of ovulation in subordinates. The results also show that subordinate status does not preclude a female from becoming an effective dominant female in a different social context.

We now corroborate the whole argument by turning to some cases which illustrate the rapid suppression of ovarian activity when previously ovulatory females became socially subordinate.

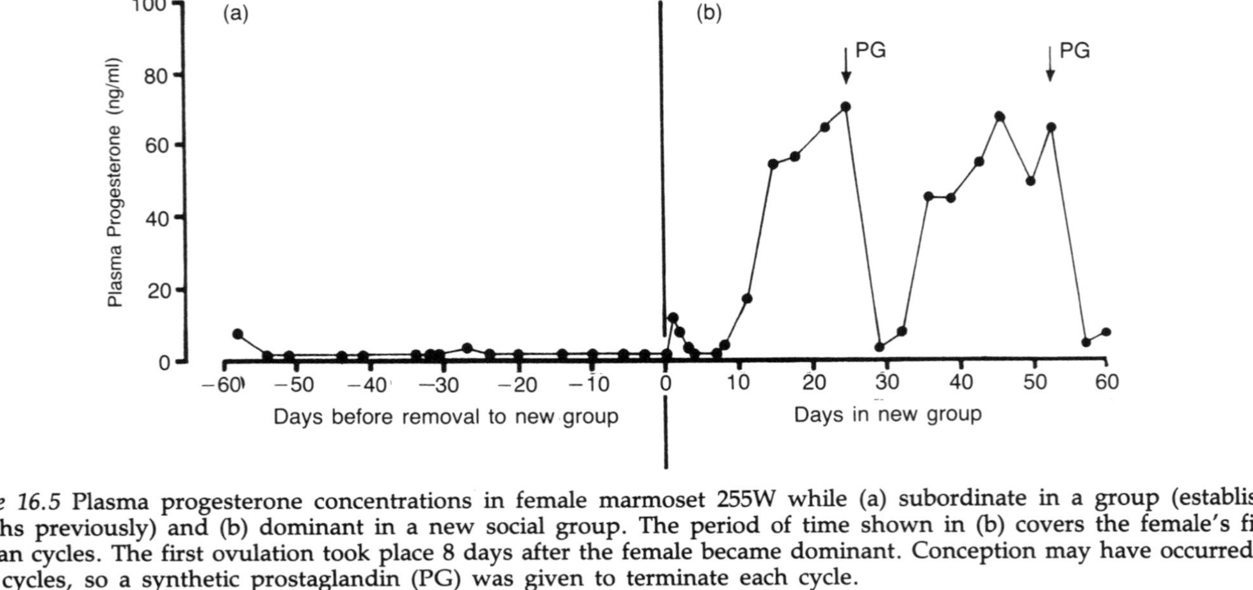

Figure 16.5 Plasma progesterone concentrations in female marmoset 255W while (a) subordinate in a group (established 11 months previously) and (b) dominant in a new social group. The period of time shown in (b) covers the female's first two ovarian cycles. The first ovulation took place 8 days after the female became dominant. Conception may have occurred during both cycles, so a synthetic prostaglandin (PG) was given to terminate each cycle.

16.8 OVULATING FEMALES AS SUBORDINATES IN A NEW GROUP

Five ovulating female marmosets, housed singly, were placed into social groups where they became subordinate. All five females immediately stopped ovulating, as illustrated in the two examples in Figure 16.4. Female 224W in Figure 16.4b also illustrates the premature collapse of the luteal phase, (demonstrated by three of the five females) when she became subordinate. In the remaining two females, the luteal phase was already nearing the end of its normal span before the females became subordinate (as illustrated by female 200W in Figure 16.4a), so the abrupt collapse of the luteal phase in these latter two females could not be definitely associated with becoming subordinate. Nonetheless, the results clearly demonstrate the dramatic consequences for luteal function of becoming a socially subordinate female marmoset. Thus relegation to subordinate status in this monkey is responsible for more physiological disruption than just suppression of ovulation. The terminal effect of subordinate status on the normal luteal phase of the ovarian cycle may well explain the inadequate cycles occasionally demonstrated by a small number of subordinates: luteal insufficiency might well have led to the curtailed cycles.

16.9 A DOMINANT FEMALE AS A SUBORDINATE IN A NEW GROUP

In one example, female 313W had been a dominant, ovulating female in a group for 6 months. She was then removed from this group and was used to form a new group. In that group, she was recorded as a subordinate within 2–3 days. As illustrated in Figure 16.6, she was in the mid-luteal phase of her ovarian cycle when she became subordinate. However, after the apparently normal end of the luteal phase of that cycle, the now subordinate female 313W did not reovulate. She remained anovulatory and subordinate until she was removed after approximately 3 months to form a further group. These observations show that previous dominant status and ovarian activity of a female did not carry over into the new group: social status and reproductive status were established anew. Further social manipulations of this kind are planned to pursue this finding.

16.10 DISCUSSION

One question of particular interest in the whole scenario of reproductive suppression in marmosets relates to whether reproduction is equally

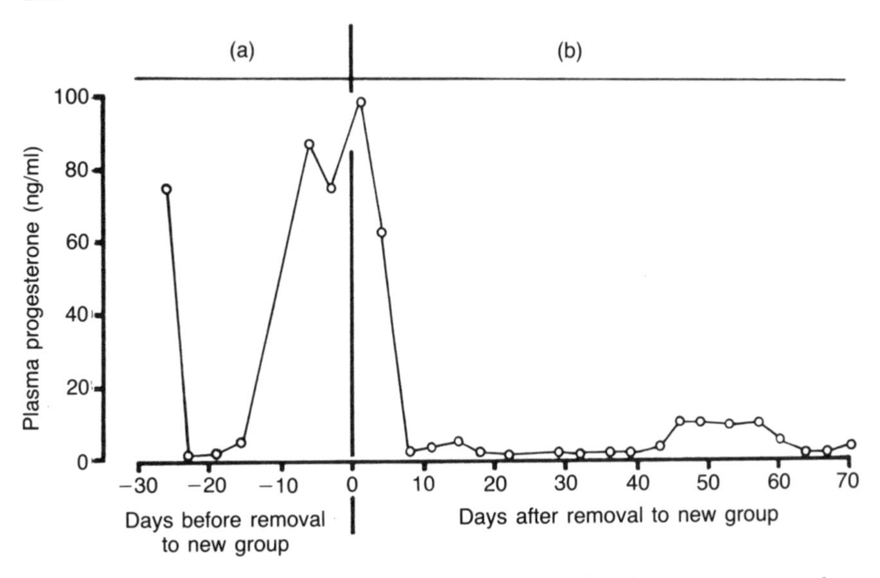

Figure 16.6 Plasma progesterone concentrations in female marmoset monkey 313W while (a) dominant in a group (established 3 months previously) and (b) subordinate in a new social group.

suppressed in all subordinate females. The short answer to this question is 'effectively, yes'. The vast majority of subordinates in our study were completely acyclic, although a minority of them (13 out of 52: 25%) did exhibit an occasional inadequate cycle, but these did not approach the regular pattern of fertile cycles of the dominant females. However, the appearance of these inadequate cycles was significantly related to the rank of subordinate females, with more appearing in rank 2 females (10 out of 28) than in females of ranks 3 and 4 combined (3 out of 24: chi-square (d.f. = 1) = 5.1; $P < 0.05$).

An investigation of the underlying endocrinological cause of the suppressed cycles in subordinates suggested that physiological suppression in rank 2 females was less than in rank 3 females. Rank 4 females were not tested. Reduced stimulation of the ovaries by pituitary gonadotrophins is the apparently immediate cause of the suppressed cycles in subordinates (Abbott, 1987, 1989; Abbott *et al.*, 1988). The suppressed secretion of luteinizing hormone (LH) from the anterior pituitary gland is apparently due to suppressed release of gonadotrophin releasing hormone (GnRH) from the hypothalamus (Abbott, 1987, 1989; Ruiz de Elvira and Abbott, 1986). Pituitary LH responses to exogenous GnRH treatment were poor in subordinates in comparison to dominants, probably because the reduced endogenous

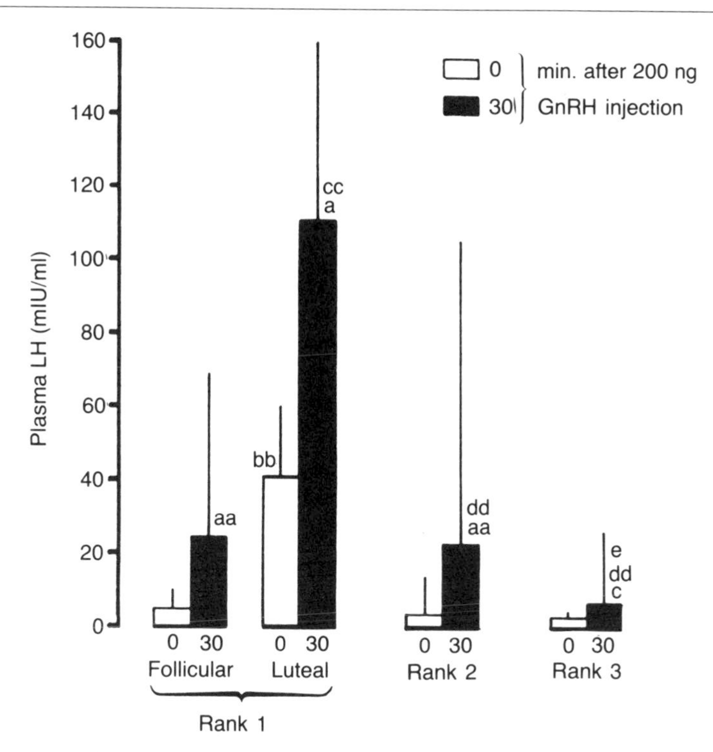

Figure 16.7 Plasma concentrations of luteinizing hormone (LH) (mean + confidence limits) in rank 1 (dominant) and ranks 2 and 3 (subordinate) female marmoset monkeys at 0 and 30 min following administration of 200 ng gonadotrophin-releasing hormone (GnRH). The data are expressed as the antilogue of the transformed means with 95% confidence limits. For clarity, only the upper confidence limits are shown. a, $P < 0.05$ vs 0 min; aa, $P < 0.01$ vs 0 min; bb, $P < 0.01$ vs rank 1 (follicular), rank 2, rank 3 0 min; c, $P < 0.05$ vs rank 1 (follicular) 30 min; cc, $P < 0.01$ vs rank 1 (follicular) 30 min; dd, $P < 0.01$ vs rank 1 (luteal) 30 min; e, $P < 0.05$ vs rank 2 30 min. Duncan's Multiple Range Rest following ANOVA. (Reprinted with kind permission of Wiley Liss, New York.)

GnRH secretion failed to stimulate sufficient synthesis and storage of LH in pituitary gonadotroph cells of subordinate females. Therefore, it was expected that the LH responses of all subordinate females to GnRH treatment would be equally poor. They were not. As illustrated in Figure 16.7, rank 2 subordinate females showed a significant LH response to an injection of 200 ng GnRH, whereas rank 3 subordinates did not. The LH response of the rank 2 subordinates was equivalent to that of rank 1 females in the follicular phase of the ovarian cycle, but was significantly less than that of rank 1 females in the luteal phase. The most probable explanation for this physiological difference in LH responsiveness between rank 2 and rank 3 subordinates is that there is less suppression of endogenous GnRH secretion in rank 2 subordinates and hence, perhaps, the greater preponderance of inadequate cycles in rank 2 females. Such a physiological difference between females of different subordinate ranks may prove useful in investigating the neuroendocrine mechanisms mediating this reproductive suppression. These findings also provide further evidence of the benefits for a social hierarchy amongst subordinate female marmosets. The highest-ranking subordinates (rank 2 females) appear closer, in physiological terms, to full reproductive potential than lower-ranking subordinates. Such a physiological advantage might prove useful in the event of opportunities to attain dominant (rank 1) status in a group, as exemplified by the rank 2 female in Figure 16.2b.

The marmoset is one of several species of primate which has developed the ability to use a controlled amount of behavioural stress, through the female social hierarchy, to limit the reproductive output of a group to one female (Abbott, 1987; Harcourt, 1987a). This may be necessary in free-living groups to ensure the survival of the offspring, in that nonbreeding 'helpers' may be crucial in the successful breeding of the infants (Garber *et al.*, 1984; Abbott, 1987; MacDonald and Carr, 1989; Sussman and Garber, 1987). It is also interesting, however, that the nonbreeding 'helpers' themselves may respond quickly to changes in the social environment, such as when the breeding female is lost through death or emigration to another group, or if they themselves emigrate to another group. When such events occur in the wild, and there is no longer a dominant female to suppress their fertility, the reproductive function of marmosets that were previously subordinate can soon take advantage of the new social situation. As migration between free-living groups appears to be fairly common in marmoset and tamarin monkeys (Terborgh and Goldizen, 1985; Scanlon *et al.*, 1988; Stevenson and Rylands, 1988), the ability to adapt quickly in reproductive status to a new status in a group, might prove beneficial to dominant and subordinate females alike.

16.11 CONCLUDING REMARKS

In female common marmosets, different social ranks have different, but repeatable, consequences for an animal's reproductive physiology. When a female changes her social rank, her reproductive physiology immediately follows suit. Other female primates, such as the talapoin monkey, *Miopithecus talapoin*, and the gelada baboon, *Theropithecus gelada* (Keverne *et al.*, 1984; Dunbar, 1989; see also Chapter 4), show similar but usually less extreme relationships between social rank and reproductive physiology. The effect of social rank should, therefore, always be considered when the reproductive physiology of female primates living in groups is studied.

ACKNOWLEDGEMENTS

We thank our colleagues, including J. Barrett, J.K. Hodges, S.F. Ferrari and B.R. Ferreira, for their collaboration in some of the work reported here; Professor A.P.F. Flint for criticism of the manuscript; M.J. Llovet and the animal technical staff for their care and maintenance of the animals; T.J. Dennett and M.J. Walton for preparation of the illustrations and T. Grose for typing the manuscript. This work was supported by an MRC/AFRC Programme Grant and grants from the Nuffield Foundation and the Association for the Study of Animal Behaviour.

17

Behavioural and physiological indices of social relationships: comparative studies of New World monkeys

SALLY P. MENDOZA

17.1 INTRODUCTION

Prior to embarking on his classic field study of howler monkeys, C.R. Carpenter was impressed with the importance of social interactions in the day-to-day lives of the nonhuman primates (Carpenter, 1934). Largely on the basis of his careful field studies of howler monkeys (*Alouatta*), spider monkeys (*Ateles*), macaques (*Macaca*) and gibbons (*Hylobates*), it was soon recognized that each primate species forms and maintains a characteristic or modal grouping pattern, defined not only by the numbers and kinds of individuals within a group, but also by 'the attractive and repellent forces in the matrix of interactions of group members [which] result in the characteristic local spatial dispersions' (Carpenter, 1952, p. 371). Carpenter considered the central tasks in primatology to be: (1) to describe for each species the characteristic grouping pattern; (2) to discover the extent to which each social system can vary; and (3) to elucidate the processes by which social systems are formed and maintained (Carpenter, 1942).

We have made considerable progress in describing social structures that are typical of primate species, and their range of variation. We are still, however, far from understanding the processes which contribute to species-typical expression of social organization. Contrary to Carpenter's (1942, 1952) expectation, there is no phyletic progression of primate social organization in either form or complexity. As a

Figure 17.1 A pair of titi monkeys (*Callicebus moloch*).

consequence of the seemingly unsystematic diversity of primate social systems, the task of understanding the processes by which social systems are formed and maintained is more difficult than originally conceived (see also Chapter 2. section 2.1).

The most systematic effort to delineate the proximal sources of species-typical social structure has been conducted by W.A. Mason and colleagues, in a comparison of two New World cebid species: *Callicebus moloch* (titi monkey, Figure 17.1) and *Saimiri sciureus* (squirrel monkey, Figure 17.2).

Figure 17.2 Squirrel monkey (*Saimiri sciureus*).

Initial field studies established that the titi monkey lives in small family groups. It is a monogamous and territorial species, and the male participates heavily in the care of the young (Mason, 1966, 1968, 1971, 1974). Squirrel monkeys, on the other hand, live in large mixed-gender social groups, sometimes including 30 or more individuals. They range as a group over large areas and show no evidence of territoriality. Primary associations occur between individuals of the same age and gender, with the exception of very young infants who interact almost exclusively with adult females (Baldwin, 1985; Boinski,

1987a,b). The species are also similar in size and general morphology, eat similar foods, and have the same basic behavioural repertoire. Thus, the differences between them do not appear to be the result of major behavioural or structural adaptations that might limit the ability of members of each species to engage socially (Eisenberg, 1981). Moreover, titi and squirrel monkeys often live in the same forest. Although they utilize their environments in very different ways (Terborgh, 1983; Wright, 1986), this probably reflects their general responsiveness to their physical world, rather than specific adaptations that would pose different ecological constraints for the two species (Fragaszy and Mason, 1983).

The differential social tendencies of titi and squirrel monkeys are maintained when the animals are housed and tested under identical conditions in captivity. When living in heterosexual pairs, titi monkey cage-mates spend more time engaged in mutual assessment, grooming or sitting in passive contact than do squirrel monkey cage-mates, and their locomotor and feeding activities are more highly coordinated (Mason, 1974; Phillips and Mason, 1976). In social preference tests, titi monkeys show a strong and specific attraction to the cage-mate. In contrast, squirrel monkey females prefer strange females to their familiar male cage-mate, and squirrel monkey males do not reliably discriminate between their familiar cage-mate and strangers of either gender (W.A. Mason, 1975). When several established pairs are released into a large enclosure, titi monkeys maintain their original pair relationships, whereas squirrel monkeys quickly organize themselves into sexually segregated subgroups (Mason, 1971; Mason and Epple, 1969).

Titi and squirrel monkeys not only differ in the type of associations they seek and the quality of the relationships they form, but they also differ in the impact these relationships have on the individual's reactions to the nonsocial environment. Observations of feeding behaviour indicate that titi monkeys interact over food more than squirrel monkeys. Even when a single, highly desirable food item is available, titis tend to interact peaceably more often than squirrel monkeys, and passive food-sharing is more likely than in squirrel monkeys (Fragaszy and Mason, 1983). For titi monkeys, the presence of the pair-mate has a substantial facilitative effect on feeding activity, whereas a stranger suppresses feeding; in squirrel monkeys, neither effect is apparent (Mason and Lorenz, 1988). Similarly, the presence of the pair-mate has a powerful and positive influence on the response of titi monkeys to unfamiliar situations or novel objects, whereas squirrel monkeys respond in the same manner, with or without the cage-mate present (M. Andrews, 1986; Fragaszy and Mason, 1978).

A particularly important aspect of this research programme has been to establish that many of the contrasts between titi and squirrel monkeys in their social tendencies reflect more generalized distinctions between individuals of each species in their modes of relating to their environment (Mason, 1978). Squirrel monkeys characteristically display high levels of locomotor output, whether engaged in social or nonsocial activities (Mason, 1974; Cubicciotti and Mason, 1975; Fragaszy and Mason, 1978; Fragaszy, 1980). In nature, they travel long distances each day (Baldwin, 1985), and social interactions are typically boisterous, frequently involving vigorous motoric components (Mason, 1974; Fragaszy and Mason, 1978). In addition, squirrel monkeys readily engage in novel activities, entering previously unexplored areas when given the opportunity to do so (Fragaszy and Mason, 1978; Andrews, 1986). Titi monkeys, on the other hand, have a much more sedentary lifestyle. They spend considerable portions of their time engaged in activities which do not include high motor output, such as prolonged bouts of mutual grooming and sitting in passive contact with tails entwined (Mason, 1974). Their movements tend to be slow and cautious, they generally avoid novel situations or activities, and they travel within their small territories along habitual pathways (W.A. Mason, 1968; Fragaszy, 1978; M. Andrews, 1986; C.R. Menzel, 1986). Although titi and squirrel monkeys do not differ in their cognitive capacities, titi monkeys spend more time engaged in visual exploration of novel objects or problems, and their contact with them is less varied, vigorous or prolonged (Visalberghi and Mason, 1983; see also Chapter 3, section 3.3). In their interactions with each other and with the physical environment, the differences between species form a clear and consistent pattern. Squirrel monkeys are bolder, more active and more impulsive than titi monkeys; squirrel monkeys may be described as rambunctious and titi monkeys as reluctant.

The contrasts between the species in their characteristic modes of responding to the environment have a direct effect on their ability to respond to challenge and change. The clearest evidence for this is provided by their initial response to captive conditions. Squirrel monkeys adjust to captivity readily and prosper under standard laboratory conditions. The situation with titi monkeys is quite different. Most newly imported animals become inactive; they often sit in a hunched-over posture, show a lack of interest in their surroundings, and a loss of appetite, frequently to the point where forced-feeding is required. These symptoms are characteristic of severe clinical depression. The situation is often progressive, and early attempts to establish a titi colony resulted in a mortality rate of nearly 70% during the first 3 months in captivity (Lorenz and Mason, 1971).

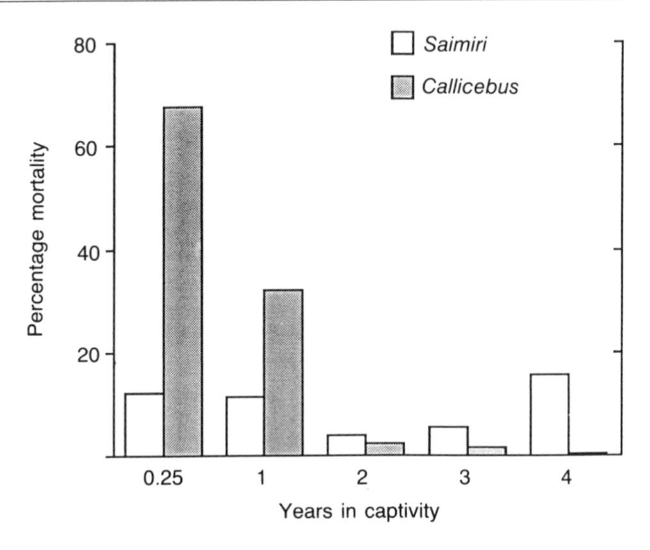

Figure 17.3 Percentage of mortality in *Saimiri* and *Callicebus* at various intervals of time from importation. Time in years represents the upper limit of data included in each bar; the lower limits are determined by the upper limit of the preceding interval. The first interval (0–0.25 years) corresponds to the quarantine period for each species. (Data for *Callicebus* were previously reported by Lorenz and Mason, 1971.)

Once titi monkeys pass through the initial period of adjustment, however, they are not particularly delicate or difficult to keep, and mortality rates drop to a level even below that of squirrel monkeys (see Figure 17.3).

Recently, we have been investigating the possibility that response dispositions affecting social structure are reflected in physiological responsiveness, as well as in behaviour. We have focused on two systems: the pituitary–adrenal and the autonomic nervous system. These physiological systems, traditionally viewed as the major stress-response systems, play an essential part in maintaining important biological functions within normal limits. They participate intimately in the individual's internal adjustments to external events, and they prepare the individual to deal with the particular circumstances in which it happens to find itself. At the very least, since these physiological systems function to provide the individual with the means of altering internal processes in accordance with significant external events, we can use changes in the activity of these systems to further clarify social dispositions. That is, using physiological measurements, it is possible to determine whether a particular situation, context or conspecific presents an individual with a sufficiently salient stimulus to produce alteration in these regulatory systems. The direction and magnitude of change can provide information regarding the potency and valence of the stimulus. Moreover, it is likely that the neuroendocrine products of these systems actually contribute substantially to the generation of social dispositions. In this case, species differences in patterns of responsiveness in these systems become an important proximal source of species-typical social expression. With these possibilities in mind, the physiological responses of titi and squirrel monkeys to a variety of social and nonsocial situations were compared.

17.2 ACTIVITY, REACTIVITY AND REGULATION OF STRESS-RESPONSE SYSTEMS

The clear behavioural contrasts between the two species in general activity levels and reactivity suggested that they may also differ in baseline activity and responsiveness of physiological systems which regulate the availability of energy stores necessary for motoric expression. Although many neuroendocrine changes are involved in regulating an individual's responsiveness to the environment, the pituitary–adrenal and autonomic nervous systems are particularly influential in preparing the individual to engage in activity, and can rapidly alter its capability to do so. We hypothesized that, in comparison to titi monkeys, squirrel monkeys would show higher levels of baseline pituitary–adrenal and sympathetic activity, greater reactivity when presented with challenge or change, and that activity in these systems would be less susceptible to inhibition by negative feedback regulation (pituitary–adrenal) or counter stimulation by the parasympathetic nervous system (sympathetic).

The results of our investigations are generally in accord with our hypotheses. Baseline activity in the pituitary–adrenal system, as measured by plasma levels of cortisol, is nearly five times higher in squirrel monkeys than in titi monkeys (see Figure 17.4). Species differences in basal cortisol levels of this magnitude have been found in several studies (Mendoza and Moberg, 1985; Anzenberger *et al.*, 1986; Cubicciotti *et al.*, 1986; Mendoza and Mason, 1986a). In response to stressful conditions, the already substantial differences between species in adrenocortical activity further increase. For example, in response to 1 h of physical restraint, plasma cortisol levels increase for both species. The magnitude of the increase, however, is substantially higher for squirrel monkeys (Cubicciotti *et al.*, 1986; see Figure 17.5). Similar results have been obtained in the initial reaction to a novel environment; both species respond to novel situations with increased adrenocortical activity, but the magnitude of the response is greater for

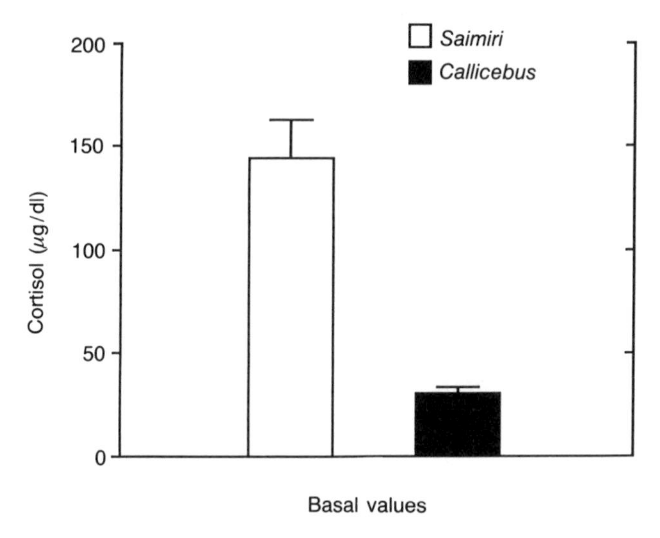

Figure 17.4 Mean basal cortisol levels (± SE) for *Saimiri* (n = 14) and *Callicebus* (n = 13).

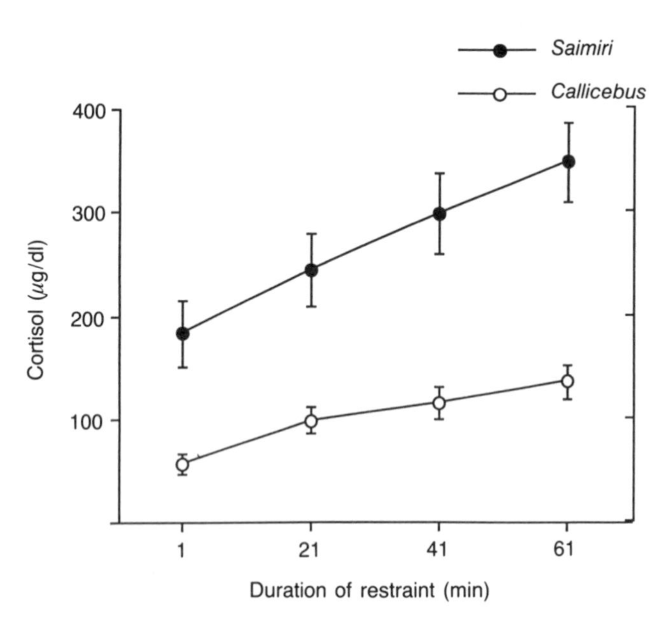

Figure 17.5 Mean cortisol levels (± SE) for *Saimiri* (n = 12) and *Callicebus* (n = 12) in response to 1 h of physical restraint. (From Cubicciotti *et al.*, 1986.)

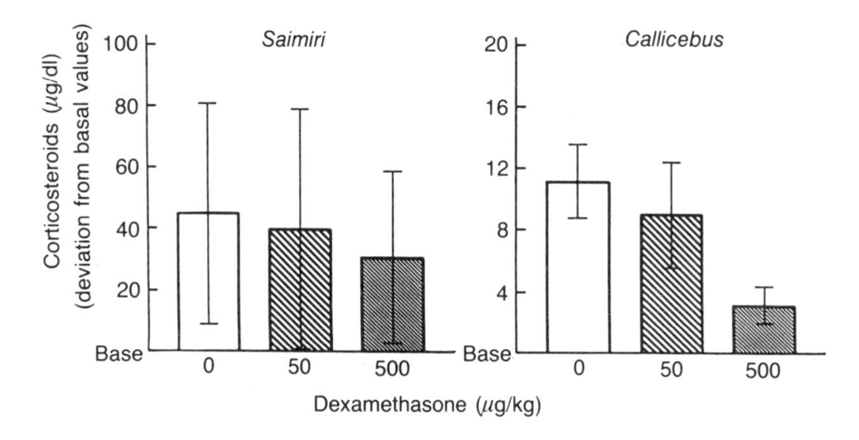

Figure 17.6 Mean cortisol elevation (± SE) following 15 min of physical restraint in *Saimiri* (n = 18) and *Callicebus* (n = 12) that had been pretreated with varying doses of dexamethasone. Note that the data are presented using different scales to facilitate comparison within and between species. (From Mendoza and Moberg, 1985.)

squirrel monkeys than titi monkeys (Cubicciotti *et al.*, 1986).

At least some of the differences between squirrel monkeys and titi monkeys in adrenocortical activity can be attributed to species differences in feedback regulation of the pituitary–adrenal axis. In general, high circulating levels of corticosteroids provide a signal to the hypothalamus and pituitary to shut down release of corticotrophin-releasing hormone (CRH) from the hypothalamus, and, consequently, adrenocorticotrophic hormone (ACTH) from the pituitary, with the net result of reducing the signal to release corticoids from the adrenal cortex. The efficacy of this negative feedback system can be evaluated using a synthetic corticosteroid, dexamethasone, which, when administered in sufficiently large doses, provides the negative feedback signal and, yet, can be distinguished biochemically from endogenous corticosteroids. Administration of dexamethasone to squirrel monkeys and titi monkeys results in suppression of basal levels of corticosteroids for both species, although this effect is greater in titi monkeys than in squirrel monkeys. Following 15 min of physical restraint, titi monkeys show little or no elevation in corticosteroids above basal values when pretreated with high doses of dexamethasone; whereas the adrenocortical response to stress in squirrel monkeys is unaffected by dexamethasone treatment (Mendoza and Moberg, 1985; see Figure 17.6). Thus, both basal adrenocortical activity and the response to stress following dexamethasone treatment indicate that negative

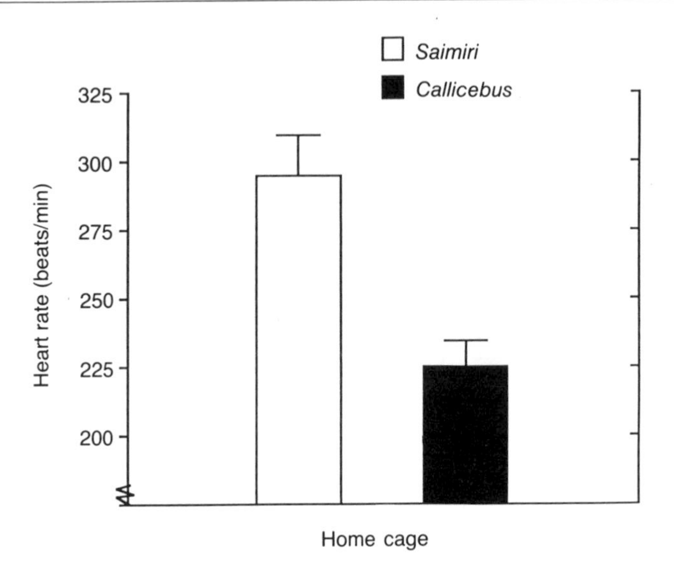

Figure 17.7 Mean heart rate (± SE) for *Saimiri* (n = 12) and *Callicebus* (n = 12) while moving about freely in their home cage. (From Cubicciotti *et al.*, 1986.)

feedback mechanisms regulating pituitary–adrenal activity are more sensitive in titi monkeys than in squirrel monkeys (Mendoza and Moberg, 1985). This can account for elevated basal cortisol levels in squirrel monkeys relative to titi monkeys, and further suggests that in perceptually comparable stressful or arousing situations, the squirrel monkey's adrenocortical response should be greater and more prolonged than that of titi monkeys.

Surprisingly, heightened adrenocortical activity in the squirrel monkey is not complemented by heightened secretion of ACTH from the pituitary (Coe *et al.*, 1978b; Cassorla *et al.*, 1982). The squirrel monkey adrenal cortex is also not particularly sensitive to ACTH (Kittenger and Beamer, 1968; Brown *et al.*, 1970). Reduced ACTH levels, minimal feedback regulation of adrenocortical activity, and the rapidity and magnitude of the adrenal response to perturbation, have led to the speculation that the adrenal cortex of the squirrel monkey is driven by factors other than pituitary stimulation. A particularly likely possibility is sympathetic stimulation of the adrenal cortex (Coe *et al.*, 1978b; Mendoza and Moberg, 1985).

Although direct evaluation of sympathetic contribution to adrenocortical activity has not been evaluated in either squirrel or titi monkeys, the hypothesis of greater sympathetic involvement in squirrel monkeys is consistent with a comparison of cardiac activity in these species. Squirrel monkeys show consistently higher levels of heart rate

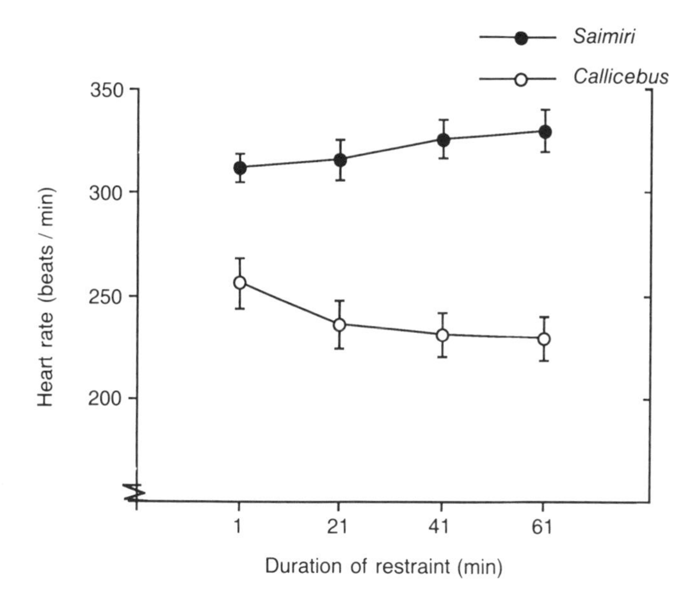

Figure 17.8 Mean heart rate (± SE) for *Saimiri* (n = 12) and *Callicebus* (n = 12) in response to 1 h of physical restraint. (From Cubicciotti *et al.*, 1986.)

than do titis, whether the animals are moving about freely in their home cage (see Figure 17.7) or are responding to stressful events such as physical restraint or exposure to a novel cage (Cubicciotti and Mason, 1975; Cubicciotti *et al.*, 1986). The species also differ in the pattern of cardiac response to stress. During 1 h of physical restraint, for example, the heart rate of squirrel monkeys increases progressively, whereas that of titi monkeys declines (see Figure 17.8). A similar pattern has been obtained in response to an initial reaction to a novel environment.

In order to determine whether heart rate differences could be attributed to species differences in sympathetic and/or parasympathetic activity, heart rate during a 15-min exposure to novelty was monitored following administration of the pharmacological agents propranolol and atropine, which selectively inhibit sympathetic and parasympathetic input to the heart. Using this procedure, it was found that both species respond to novelty with high levels of sympathetic activity, but that this response is substantially greater in squirrel monkeys than in titis. Sympathetic activity in titi monkeys, but not squirrel monkeys, is countered by high levels of parasympathetic activity (Mendoza and Mason, 1984). The species differences in heart rate can largely be attributed, therefore, to a difference in autonomic balance. Squirrel monkeys have unusually high levels of sympathetic activity, which exerts its influence relatively unimpeded by parasympathetic activity; this finding is consistent with previous reports (Kelleher *et al.*, 1972). Titi monkeys also respond to stressful situations with initially high levels of sympathetic activity, but this is quickly counteracted by parasympathetic activity, and thus accounts for the rapid decline in heart rate following the initial response to stressful situations. Autonomic responsiveness can be differentiated in the two species by a lack of inhibitory influence in the squirrel monkeys relative to the titi monkeys, a finding which parallels pituitary–adrenal responsiveness.

Physical restraint and initial exposure to entirely novel environments are potent stressors and reliably elicit heightened sympathetic and adrenocortical activity in most animals (Hennessy and Levine, 1979). We can be relatively certain that both species perceive these situations as stressful and that the different patterns of response to these stressors exhibited by titi and squirrel monkeys reflect fundamental differences in the organization of the stress-response systems.

We have also explored the possibility that some circumstances are perceived differently by the two species. Titi monkeys are generally more neophobic than are squirrel monkeys. Moreover, the high rates of mortality in titi monkeys during the initial period of adaptation to captive conditions suggest that they are slower to adapt to altered circumstances than are squirrel monkeys. In order to determine whether these proclivities were reflected in physiological responses, we compared the adrenocortical activity in squirrel and titi monkeys following prolonged exposure to hospital-like conditions (i.e. alone in a small, novel cage). Both species respond to this situation with increases in cortisol which persist for at least 3 weeks. In this situation, however, the magnitude of the adrenocortical response is greater for titi monkeys than for squirrel monkeys, suggesting that the experimental conditions are considerably more stressful for titi monkeys (S.P. Mendoza and C.L. Coe, unpublished data, see Figure 17.9). Similar results have been obtained in less restrictive environments and in conditions in which the cage-mate was present (Anzenberger *et al.*, 1986). Indeed, titi monkeys require considerable time (several weeks or even months) to adapt to an environment which allows greater freedom than that to which they are accustomed (R.R. Swaisgood, D.D.Cubicciotti, W.A. Mason and S.P. Mendoza, unpublished data).

Titi monkeys are also more likely to respond to minor perturbations in their daily routines than are squirrel monkeys. In our colony, individuals of both species are trained to enter small transport cages

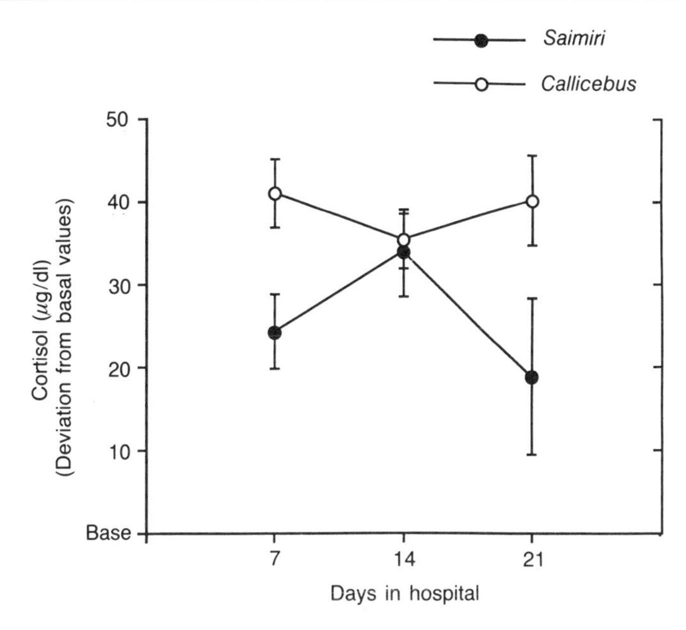

Figure 17.9 Mean cortisol elevations (± SE) for *Saimiri* (n = 6) and *Callicebus* (n = 6) in response to being housed for 3 weeks in hospital-like conditions (i.e. alone in a small, novel cage).

from their larger living cage. This allows capture of animals with minimal disturbance or distress. We compared the adrenocortical response of squirrel and titi monkeys to this relatively innocuous procedure. Animals were caught, held momentarily in the transport cage and released back into their home pen. One hour later, cortisol levels in titi monkeys, but not squirrel monkeys, were elevated above basal levels. Thus, titi monkeys seem to have a lower threshold of response to environmental perturbation than do squirrel monkeys.

The differences between species in physiological activity and responsiveness are in accord with the behavioural differences. Squirrel monkeys are physiologically prepared to sustain high levels of locomotor output, particularly during stressful or arousing circumstances; titi monkeys are not. Titi monkeys, on the other hand, are more likely than squirrel monkeys to respond to minor environmental perturbations, and require considerably longer to adjust to altered circumstances. These contrasts are apparent in the lifestyles of each species.

Individuals bring to any social situation a constellation of characteristics which influence how they will respond to others and to social events (Mason, 1978). These characteristics are not specifically social, but reflect more generalized response dispositions. In the literature on humans, differences of the type we have identified in squirrel and titi monkeys are generally referred to as qualities of temperament (Allport, 1937; Rothbart and Derryberry, 1981). Behaviourally, these qualities are likely to be reflected in the latency to respond, in the vigour of the response, and in the direction of the response, and may be viewed as setting the tone for the kinds of social interactions likely to be engaged in by members of each species.

17.3 SOCIOPHYSIOLOGICAL RESPONSIVENESS

Temperament and associated physiological response profiles determine, to a large extent, how individuals of a given species will interact. Maintenance of a species-typical social structure also requires that interactions will be more likely to occur between certain kinds of individuals. The sexually segregated and age-graded social structure of squirrel monkeys, for example, is dependent on each group member making basic categorical distinctions (i.e. male versus female, infant versus juvenile versus adult) and distributing interactions among potential social partners according to those discriminations. Titi monkeys must make similar distinctions, of course, in order to differentiate between the pair-mate and offspring, but a more important categorical distinction for this species might be familiar versus unfamiliar. Categorical distinctions constitute generic dimensions of the primates' social world and provide an orientation towards the social environment, within which more refined interindividual relationships may be elaborated (Mendoza and Mason, 1989b).

Categorical responses are most clearly demonstrated in responses to strangers, where the factor of established interindividual relationships is ruled out. Comparisons of squirrel and titi monkeys indicate that pituitary–adrenal and autonomic systems are responsive to the presence of strangers representing different social categories. Moreover, the particular responses to strangers can be understood only by taking into account the modal grouping pattern of each species. When presented with an unfamiliar female under conditions in which physical interaction was precluded, male and female squirrel monkeys living in heterosexual pairs showed a reduction in plasma cortisol levels, below basal values. This response did not occur when a male stranger was presented. In contrast, titi monkeys in the same testing situations responded most strongly to a like-gender stranger, as reflected in vigorous behavioural reactions and increases in plasma cortisol levels (Mendoza and Mason, 1986a; see Figure 17.10). These findings indicate that both species differentiate individuals on the basis of gender, and for both species these discriminations are reflected in

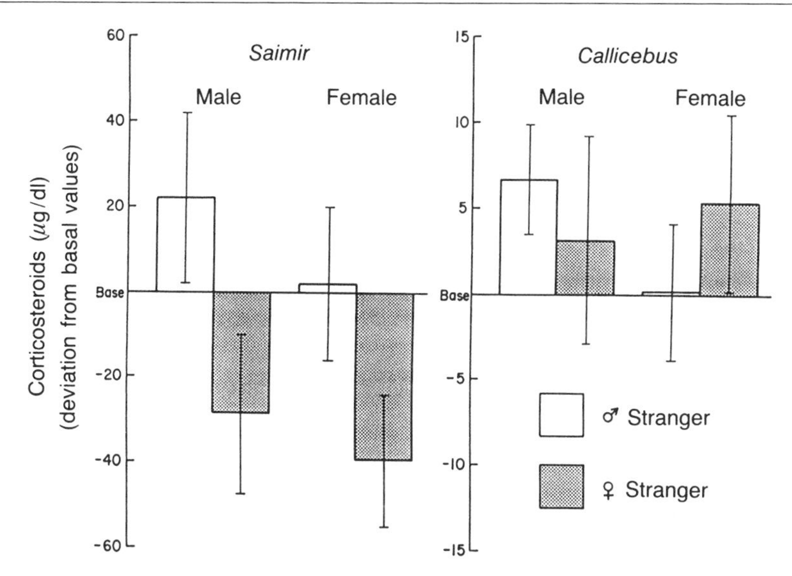

Figure 17.10 Mean cortisol response (± SE) to unfamiliar male and female conspecifics for *Saimiri* (n = 10 male–female pairs) and *Callicebus* (n = 10 male–female pairs). Note that the data are presented using different scales to facilitate comparison within and between species. (From Mendoza and Mason, 1986a.)

their physiological responses. The contrasting reactions of the two species to the stranger's gender probably relate to the fact that squirrel monkeys housed in heterosexual pairs are in an anomalous living arrangement, and the presence of a strange conspecific, particularly a female, offers the potential for establishing a social grouping which more closely approximates the norm for the species. On the other hand, living as a heterosexual pair is the norm for titi monkeys, and a strange, like-gender conspecific constitutes a potential disruption of the species-typical pattern.

The reduction in cortisol below basal levels exhibited by squirrel monkeys in response to some social stimuli is a particularly interesting and surprising finding. Suppression of adrenocortical activity is a relatively rare event and has been linked to the anticipation or induction of pleasurable states (Bunney *et al.*, 1965; Gibbons, 1968; Goldman *et al.*, 1973; Gray *et al.*, 1978). In male squirrel monkeys, the reduction in cortisol in response to females appears to be restricted to situations in which direct interaction is precluded. When one or several female squirrel monkeys are directly introduced to a male, persistent increases in adrenocortical activity ensues (Mendoza *et al.*, 1979; Mendoza and Mason, 1989a). In contrast, for female squirrel monkeys the suppression of adrenocortical activity induced by visual exposure is evident when direct interaction is permitted; furthermore, several days are required for the complete effect to be expressed (see Figure 17.11). Reduction in adrenocortical activity occurs even in entirely novel surroundings, thus overriding the usual tendency for cortisol to be elevated under these circumstances (see Figure 17.9 for comparison). Furthermore, the reduction in basal values persists for several weeks, and perhaps longer. This suggests that regulation of the pituitary–adrenal system may be shifted by the availability of particular types of social companions. The differential influence on male and female squirrel monkeys of opportunities to interact with females results in a reversal of gender differences in adrenocortical activity upon formation of social groups (see Figure 17.12).

Opportunity for direct interaction with like-gender companions can also result in persistent changes in activity of stress-response systems in male squirrel monkeys. Within the first hour of being paired with another male, males showed a substantial reduction in heart rate, as compared to cardiac activity when they were living alone. Following a 2–3-week period of living continuously together, the males showed a further reduction in heart rate (see Figure 17.13). Heart rate variability, a measure of parasympathetic activity, suggests that the reduction in heart rate following pair formation could be attributed to increased parasympathetic activity. This finding is particularly interesting since males living alone, or with a single female, show extremely low levels of parasympathetic activity (Kelleher *et al.*, 1972; Byrd and Gonzalez, 1981; Mendoza and Mason, 1984).

As indicated above, social familiarity may be an important categorical dimension for titi monkeys. When established pairs of titi monkeys are confronted with unfamiliar pairs, heart rate in both male and female members of the established pair increases, and this effect is larger under conditions which preclude visual withdrawal from the strangers. Squirrel monkeys, under identical circumstances, respond physiologically to the novelty of the situation, but not to the rather subtle variations in exposure to strange conspecifics (Anzenberger *et al.*, 1986).

The results of these studies, while far from exhaustive, offer evidence that a substantial portion of each species' modal grouping pattern may be directly related to their categorical responses to potential interactants. The evidence suggests that regulatory control of the stress-response systems are altered by the presence of companions representing different social categories. Such changes in physiological activity can have profound implications for the behavioural tendencies or response capacities of the individuals involved and, hence,

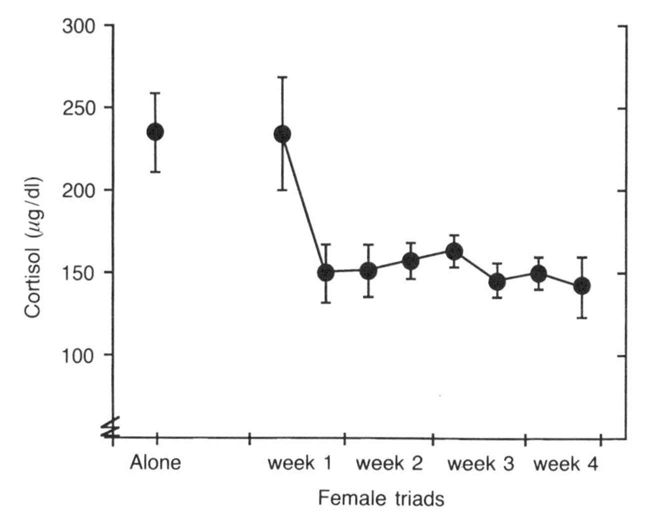

Figure 17.11 Mean basal cortisol levels for female squirrel monkeys (n = 15) when living alone or during the first 4 weeks after the formation of female triads. (From Mendoza and Mason, submitted.)

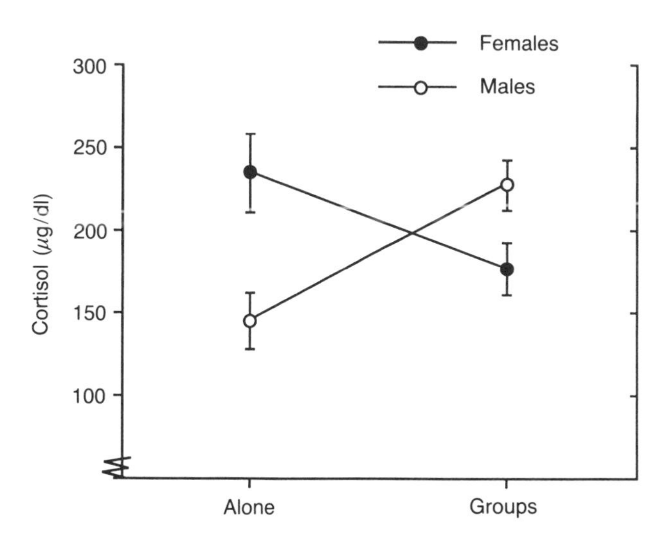

Figure 17.12 Mean basal cortisol levels (± SE) for male (n = 5) and female (n = 15) squirrel monkeys when living alone or in single-male, three-female groups. (From Mendoza and Mason, submitted.)

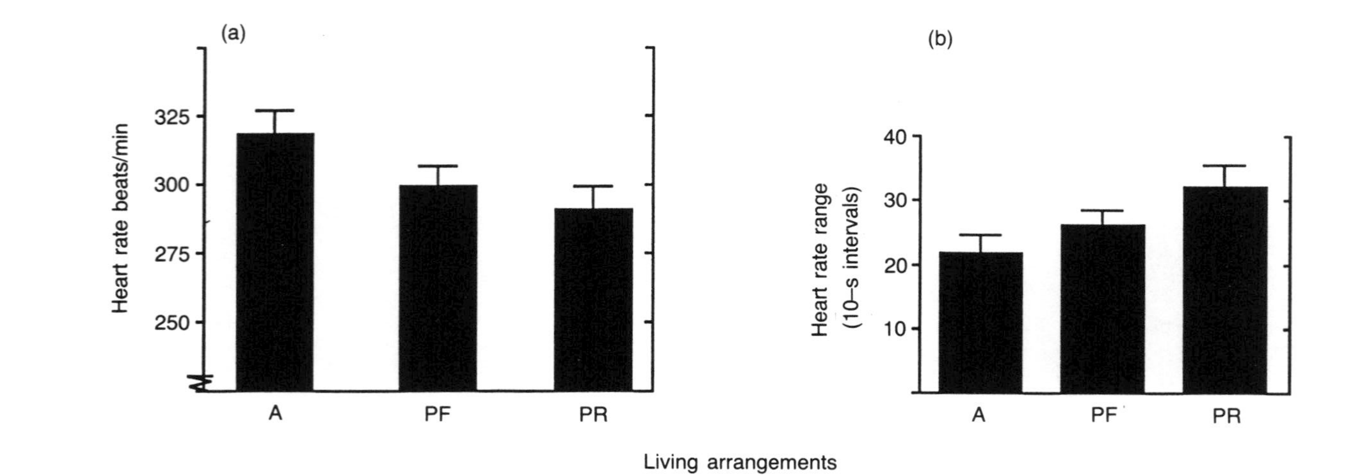

Figure 17.13 (a) Mean heart rate (± SE) and (b) mean heart rate range per 10-s interval (± SE) for male squirrel monkeys (n = 14) when living alone (A), immediately after pair formation (PF) and following 2–3 weeks of continuous cohabitation (PR). All data were collected when the males were placed for 1 h (either alone or with their male partner) in a familiar test cage.

contribute to the kinds of behaviours likely to be expressed in social situations. Categorical distinctions are not immutable, however, and can be modified by prevailing social circumstances. Thus a stranger can represent a potential for forming a new relationship or a threat to the stability of existing ones, depending on the category of the stranger, the context in which the stranger is encountered, and the social norm for the species.

17.4 SOCIOPHYSIOLOGY OF INTERINDIVIDUAL RELATIONSHIPS

Although membership in primate groups is not static, it is generally stable. Individuals typically interact with specific individuals over periods that may cover several years. Long-term associations or relationships, however, are not randomly formed or maintained. One of the species-characteristic features of social organization is the tendency for relationships to form with individuals of a particular category, and not with individuals representing other categories (Mendoza and Mason, 1989b). Relationships also vary, both within and between species, in their intensity, complementarity, symmetry, scope and tone (Hinde, 1983a). It is likely that these qualities are reflected in physiological responses to the formation and disruption of specific relationships. Because the types of relationships formed differ in squirrel monkeys and titi monkeys, discussion of physiological correlates of specific relationships will be presented for each species separately.

In squirrel monkey social groups, adults associate primarily in within-gender associations. Females are generally more active in initiating the agonistic interactions which determine the spatial relations between the genders (Coe and Rosenblum, 1974; Vaitl, 1978; Boinski, 1987a/b). If they do not show this initiative, as is the case for some subspecies of squirrel monkeys, a sexually integrated social structure emerges (Mendoza *et al.*, 1978). Despite the apparent importance of females in establishing and maintaining the structure of the group, recent evidence suggests that males may take a greater role than females in maintaining group cohesion. Male membership in natural groups is stable, whereas females frequently move between groups (Boinski, 1987b). Moreover, male–male associations in captive groups tend to persist for several years, whereas membership in female–female associations, or cliques, tends to be of shorter duration, lasting for a few weeks (Tabor, 1986). Both genders form linear dominance hierarchies, although these are more clearly demarcated among adult males. Thus, within-gender relations are clearly important in squirrel monkeys, but males and females differ in the stability and quality of the relationships they form, and in their contribution to the organization and dynamics of social groups.

Despite the profound physiological effect females have on one another, specific qualities of relationships do not appear to influence their adrenocortical activity. Dominant and subordinate females show equivalent responses to group formation; even subordinate females subjected to considerable harassment by their partners show comparable reductions in cortisol (Mendoza and Mason, submitted). Females living in established groups do not respond to separation from one another with elevations in adrenocortical activity (Hennessy *et al.*, 1982), although prolonged social isolation results in a gradual increase in adrenocortical activity (S.P. Mendoza, J.W. Hennessy and D.M. Lyons, unpublished data). Moreover, cortisol concentrations of females exposed to potentially stressful situations, such as a novel room, are no lower in the presence of a familiar female cage-mate, than when alone, or in the presence of a like-gender stranger (Hennessy, 1986).

In contrast to females, adrenocortical activity in males seems to reflect specific relational attributes. Cortisol levels following formation of male–male relationships reflect the relative status of the males. Whether dominant or subordinate have higher cortisol titres, however, depends on the particular subspecies. For males imported from Bolivia (*Saimiri sciureus boliviensis*), dominant males show higher adrenocortical activity than do subordinate males (Coe *et al.*, 1979; Mendoza *et al.*, 1979), whereas the reverse pattern has been found for the other subspecies (Coe *et al.*, 1983; Mendoza, 1987). Autonomic activity is also differentially altered by the formation of male–male relationships for dominant and subordinate males. As indicated in the previous section, males show a substantial reduction in heart rate following pair formation. This effect is greater for the dominant male (see Figure 17.14). The lower heart rate of the dominant male is not attributable to greater parasympathetic activity, and is most probably due to greater sympathetic activity in the subordinate animals. Changes in physiology produced by male–male relationships occur rapidly (within 24 h for the hormonal measures and within 1 h for heart rate), but take several days or weeks to develop fully. Differences between dominant and subordinate males in autonomic and pituitary–adrenal activity are not evident before they have had an opportunity to interact. It can be concluded, therefore, that physiological differences do not determine the direction of status delineation; rather, physiological differentiation results from relationship formation. The rapidity of the initial response suggests that status determinations occur shortly after males are placed together, and probably well before relationships are fully

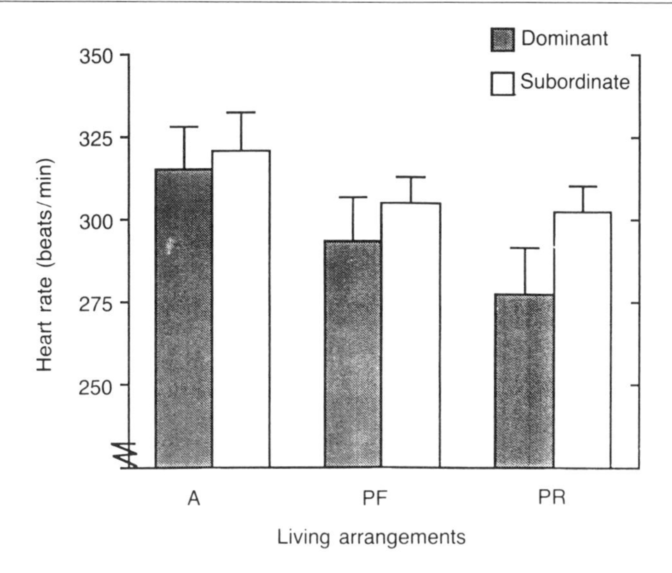

Figure 17.14 Mean heart rate (± SE) for dominant (n = 7) and subordinate (n = 7) male squirrel monkeys when living and tested alone (A), during the first hour of pair formation (PF) and following 2–3 weeks of continuous cohabitation (PR).

developed. Indeed, the physiological differentiation may be apparent only during the early periods of relationship formation, and disappears once males have been living under stable social conditions for several weeks or months (Mendoza, 1984).

Disruption of existing relationships in squirrel monkey groups has relatively little effect on the physiological state of the individuals involved, with the exception of the mother-infant dyad. Involuntary separation induces increased pituitary–adrenal activity in both mother and infant squirrel monkeys, even when the separated individual remains in the home cage surrounded by familiar companions (Mendoza *et al.*, 1980). Furthermore, mother and infant can reduce, and under some circumstances eliminate, for one another the adrenocortical response to potentially stressful events, such as capture and handling procedures (Mendoza *et al.*, 1980). For adult squirrel monkeys, social separation does not activate the pituitary–adrenal system, nor does the presence of familiar adults buffer individuals against experiencing the full range of physiological responses to stressful events (Hennessy, 1986; Hennessy *et al.*, 1982, Mendoza and Mason, 1986a).

For titi monkeys, the relationship between adult pair-mates is based on mutual attraction, vigilance and active exclusion of potential

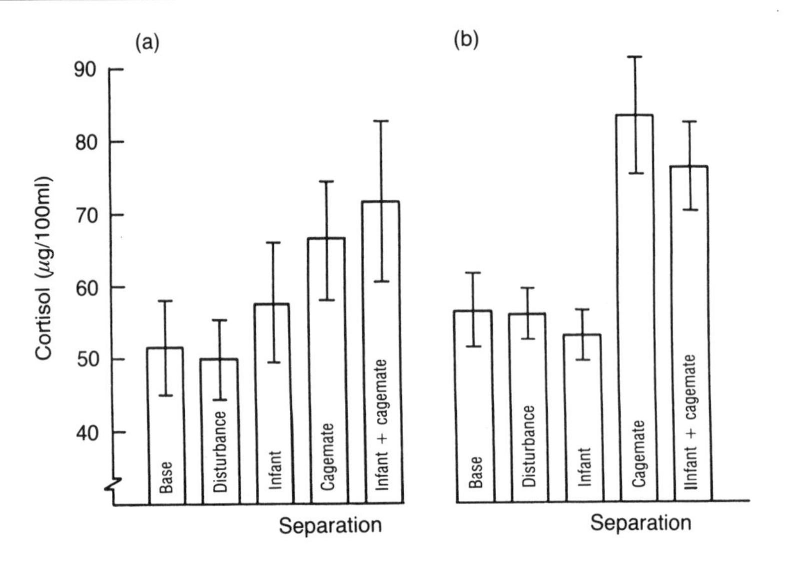

Figure 17.15 Mean plasma cortisol levels (± SE) of (a) father (n = 7) and (b) mother (n = 7) during basal, disturbance control and separation conditions. Separation (1 h) involved capture of parent–infant triads and return of all but the subjects indicated to the home cage. (From Mendoza and Mason, 1986b.)

intruders. Interactions with nongroup members are restricted to long-distance vocalizations and ritualized encounters with neighbouring groups, although occasional instances of extra-group sexual encounters have been observed (W.A. Mason, 1968; Cubicciotti and Mason, 1975, 1978; Anzenberger *et al.*, 1986). We do not know what physiological changes accompany relationship formation in titi monkeys. Titi monkeys do, however, show a vigorous response to disruption of adult heterosexual relationships, in contrast to adult squirrel monkeys. When adults of an established pair of titi monkeys are separated, both male and female show increased heart rate and elevations in plasma cortisol levels (Cubicciotti and Mason, 1975; Mendoza and Mason, 1986a,b). The adult relationship among titi monkeys thus resembles the filial bond found in squirrel monkey mother-infant dyads. Surprisingly, neither titi parent shows a physiological response to separation from their dependent infant, and, in fact, the mother shows a particularly vigorous adrenocortical response when she is separated from her pair-mate and left alone with her infant (Mendoza and Mason, 1986b; see Figure 17.15). This is perhaps the clearest example of physiological responses clarifying social dispositions. The close association of titi father and infant led us to suppose that their relationship could be characterized by an attachment bond. This does not appear to be the case. Fathers are tolerant of their young, but do not appear to form an intense and mutually dependent relationship with them. Mothers are generally intolerant of their infants; prolonged exposure to them, in the absence of the father, is apparently stressful. These and other findings suggest that the close association between titi parents and young is maintained largely by the infant, which spends the majority of time with the more tolerant parent. The bond between the adults ensures the proximity of the nurturant parent, to whom the infant moves periodically for brief feeding bouts. Thus the triadic relationship between parents and young is maintained largely by the infant and is facilitated by the bond between parents (Mendoza and Mason, 1986b; see also Chapter 3, section 3.5).

In considering the proximal sources of social structure in nonhuman primates, an important consideration is whether patterns of interactions between individuals are dependent on the existence of specific and established interindividual relationships, or whether they can be explained by individual tendencies to respond to conspecifics categorically. In both titi and squirrel monkeys, the available evidence suggests that some patterns of association can be described satisfactorily in categorical terms. In squirrel monkeys this appears to be the case for female–female associations and male-female associations. The patterns of physiological responses in both cases do not seem to reflect specific relational attributes that may develop between the individuals involved, but rather can be explained satisfactorily in terms of the category of individuals available. Similarly, the response of titi parents to their offspring does not indicate that they move much beyond interacting in categorical fashion with a small, sometimes irritating, little animal. In contrast, the physiological response of male squirrel monkeys to one another can be understood only by the specific relationship that they form, and each individual's role in that relationship. Similarly, the relationship between adult titi monkeys, or between mother and infant squirrel monkeys, is individual and specific. Responses to disruption of the relationship are not ameliorated by providing another individual of the appropriate age/gender class. Although interpersonal social relationships develop that are capable of altering individual characteristics, including physiological state and responsiveness, we cannot assume that this is the case whenever two or more individuals consistently associate and interact in benign and socially meaningful ways.

17.5 CONCLUDING REMARKS

Social systems are generated and maintained by collections of individuals who bring to any social situation tendencies to respond to events and objects, including conspecifics, in particular ways (Mason, 1976). These tendencies, collectively referred to as temperament, establish the tone of social activities, and hence contribute to the quality of the social system. To the extent that members of a given species share qualities of temperament, and that these individual characteristics vary between species, temperament is an important determinant in the generation of species-typical social structure (Mendoza and Mason, 1989b). Available evidence clearly indicates this is the case in comparisons of squirrel and titi monkeys. Squirrel monkeys are prepared to sustain more vigorous and prolonged motoric output, and they are more reactive to perturbations than titi monkeys. Titi monkeys, on the other hand, adapt to new situations more slowly and are more likely to be disrupted by minor perturbations in their surroundings or daily routines. The qualities that differentiate these species are reflected in baseline activity and the responsiveness of pituitary–adrenal and autonomic nervous systems. Socially, interactions between squirrel monkeys are frequent and active, whereas interactions between titi monkeys tend to be slow and deliberate. Indeed, understanding the qualities of temperament of the individuals of each species is a critical element in describing their respective social systems.

It is also clear, however, that the relationship of temperament to modal grouping patterns is neither simple nor direct. Individual characteristics and response dispositions are altered by the social context, and the physiological systems which orchestrate each individual's internal milieu can be altered by the mere presence of others. The products of these systems affect peripheral metabolic processes, and interact with neural structures to change cellular processes, neural responsiveness and signal transmission. Thus, changes in physiological parameters produced by the social system can alter the individual's perceptions, change its way of behaving and alter its capacity to react. In short, the characteristics of the individual are as much the product of their social system as the individuals are the producers of that social system (Mason, 1978).

Furthermore, the physiological changes produced by social means do not appear to be the simple result of activation or inhibition of the stress-response systems. Instead, it is becoming increasingly clear that the regulatory mechanisms of the systems are themselves changed by social context. This creates a dynamic loop between the social system, the participating individual, and its internal regulatory processes. The interdependencies between the individual and the social processes may be so complete that understanding primate neuroendocrine systems requires consideration of social processes and vice versa (Mendoza, 1984).

More research will obviously be required to elucidate the complex interplay between physiological processes, behavioural dispositions, and social organization among primates. It is already clear, however, from the work with squirrel monkeys and titi monkeys, that there are at least three major areas whereby physiological activity and responsiveness are likely to lead to the establishment of behavioural dispositions which, in turn, contribute to the formation and maintenance of a species-characteristic grouping pattern (see also Mendoza and Mason, 1989b):

1. Temperament, or the individual's characteristic stance toward the world and its customary ways of dealing with challenge and change;
2. Categorical responses, or the tendency to respond to particular types of conspecifics (e.g. male versus female, familiar versus stranger, infant versus juvenile versus adult) in a particular manner (e.g. fearfully, affiliatively, aggressively); and
3. Relational dynamics, including the tendency to enter into particular associations over a relatively long period of time, the means by which these associations are maintained or disbanded, and responses to engagement in and disruption of social bonds.

The value of this approach is in its potential for establishing general social processes which function in all primates, both human and nonhuman, particularly with respect to how individual characteristics, social responses of individuals and the social relationships formed between individuals, interact to produce a dynamic social structure.

ACKNOWLEDGEMENTS

Preparation of this manuscript was supported in part by grant RR00169 from the National Institutes of Health, USA. I thank William A. Mason and David M. Lyons for critical comments on the manuscript.

18

Stress and distress in response to change

DAVID M. WARBURTON

18.1 INTRODUCTION

The aim of this chapter is to consider various changes in the environment, both physical and social, which trigger a set of behavioural, biochemical and physiological responses known as stress responses. It will consider the pattern of physiological changes that result from environmental change in nonhuman primates. The composition of this picture draws on research from the most extensive literature in the fields – that on rodents and humans. Much of the rodent literature dates from the 1960s, prior to the Vietnam War, and the same war gave a much needed impetus to human and nonhuman primate studies.

The biochemical responses consist of changes in the plasma levels of hormones which themselves result in secondary physiological and biochemical changes, such as increased heart rate and elevated levels of plasma free fatty acids (Warburton 1979a). In the brain, there are changes in the processing of information (Warburton, 1979b). Both sets of responses constitute a preparation for action. Action may or may not occur depending on the outcome of the information processing. The elicitation of the stress response *per se* should not be seen only as a response to environmental changes that are adverse and threaten the animal's welfare. However, it can be one index of this sort of threat.

The term 'stressor' is used to describe the environmental agent, and the term 'stress response' refers to the reaction of the individual. 'Distress' is used to refer to the detrimental effects that can occur in the individual, reducing its fitness. Stressors have the potential for causing distress when the stressor is too intense or too prolonged. First, the nature of stress responses is considered.

18.2 STRESS RESPONSES

The stress response must be viewed as a complex pattern of cognitive and physiological changes that prepare an individual for mental and physical action (i.e. a mobilization of resources). For example, there are hormonal changes in response to stressors. The data are most complete for adrenocortical responses. There is also a significant body of literature on urinary catecholamines, adrenalin and noradrenalin. Other responses include plasma growth hormone, prolactin and testosterone (Ursin *et al.*, 1978).

In addition to the hormonal changes, there are electrocortical changes. When an animal is in a state of relaxation, there is a low level of electrocortical arousal, but, if the animal is alerted, there is electrocortical arousal (Lindsley, 1960). In a nonpathological population of animals, an uncertain situation and apprehension increase beta activity (i.e. electrocortical arousal). These changes in electrocortical activity with stressors provide an explanation for the cognitive shifts in information throughput and information storage. For instance, as the intensity of a stressor increases, the selectivity of attention increases, primary memory decreases, work rate increases, but the selectivity of responses and accuracy decrease (Warburton, 1979b).

This pattern of cognitive function is obviously ideal for mental and physical action in a highly stressful situation involving uncertainty and danger. The fuel for this action is mobilized by the stress hormones that are released in such a situation. One set of hormonal changes consists of the release from the adrenal glands of the catecholamines, adrenalin and noradrenalin, and the corticosteroids, including cortisone and corticosterone.

Catecholamines are released from the adrenal medulla via the pathways of the sympathetic nervous system, which is controlled from the hypothalamus. The control of corticosteroid secretion from the adrenal cortex is more complex; the median eminence of the hypothalamus secretes corticotrophin releasing factor (CRF) into a specialized blood vessel system to the adenohypophysis of the pituitary gland, where it triggers the release of adrenocorticotrophic hormone (ACTH). ACTH is carried, via the general systemic circulation, to the adrenal cortex, from where corticosteroids are released into the bloodstream and act on many tissues throughout the body, including the brain. Catecholamines do not pass through the blood-brain barrier easily and their major action is on tissues outside the brain.

Very early in the study of the stress response, investigators were interested in the concomitant changes in the immune response. Marsh and Rasmussen (1960) found that both avoidance learning and confinement in mice are accompanied by adrenal hypertrophy (indicative of adrenal hypersecretion), lymphocytopenia (a sign of immune deficiency), as well as a slowly developing involution of the thymus (reducing lymphocyte production) which occurs in association with an increased susceptibility to viral infections. Studies in vitro confirmed that cortisol can be immunosuppressive and reduces lymphocyte production (Hirshhorn *et al.*, 1963). Receptors for catecholamine have been found on lymphocytes (Melmon *et al.*, 1976). Both adrenalin and noradrenalin also decrease some types of immune responses (Kram *et al.*, 1975; Schmutzler and Freundt, 1975), including the suppression of lymphokinine release from lymphocytes (Melmon *et al.*, 1976).

In summary, the stress response is an adaptive response that enhances the likelihood of survival in the short term. However, the immunosuppressive effects of the hormones constituting the stress response may have deleterious effects if exposure to the stressor is prolonged.

18.3 FEEDBACK CONTROL OF THE HORMONAL RESPONSE TO STRESS

It is known that the concentration of corticosteroids in the blood is controlled by a feedback monitor in the central nervous system, a 'hormonostat' (Yates and Urquhart, 1962). As was mentioned earlier, the corticosteroids pass through the blood-brain barrier and act on the brain. Consequently, if the level of plasma steroids varies from a specific value, the hormonostat in the hypothalamus senses the disparity, and secretes corticotrophin releasing factor into the adenohypophyseal vein to the pituitary, which in turn releases ACTH to stimulate the adrenal cortex to adjust the levels appropriately.

The responsiveness of the hormonostat is determined to some extent by the early experience of the infant, at least in rodents. In an extensive research programme, Levine and his colleagues (see review in Levine and Mullins, 1968) studied the effects of infantile stimulation on the corticosteroid response in adults. In these studies, the stimulation ranged from handling newborn rats to shaking, and mild electric shocks. It was found that daily handling of the rat pups from birth speeded up the maturational rate of the stress response. Thus, the release of plasma corticosteroids following test electric shocks occurred at 3 days, with comparable increases not occurring until 9 days in the nonhandled animals (Levine, 1965).

Animals stimulated by these methods also showed differences in their physiological responses to stressors when tested as adults. The acute response to stressors was much more intense in stimulated animals than those not stimulated in infancy. For example, adult rats handled in infancy were given a brief electric shock, and then killed at 0.25, 0.50, 1.0, 5.0 and 15.0 min after the shock. The free plasma corticosteroids were measured in each group and compared with similarly treated groups which were not handled in infancy. The time-response curve of stimulated groups shows a much more rapid rise in steroids from a lower baseline, and a quicker return to baseline (Levine, 1962).

Levine and Mullins (1968) suggested that infant stimulation increases the levels of plasma adrenal steroids, which feed back to the hypothalamus and modify the set value of the hormonostat during a critical period in development. Consequently, the hormonostat varies in a graded manner in the adult, with several possible values between the minimum and the maximum. The sensitivity of the hormonostat is determined by the number of values the set point can have between the minimum resting level and the maximum achieved under extreme levels of a stressor. In the nonstimulated infant, the set point develops fewer values because there is less variation in steroid concentration at the critical period.

A second explanation is that, instead of the circulating steroids prenatally and postnatally modifying the range of the steroid response, they increase the sensitivity of the feedback receptors in the hypothalamus (Warburton, 1975). This hypothesis would explain the slightly lower adult baseline levels in handled animals. It would explain both the smaller response to weak stressors and the more rapid return to resting level after stressful stimuli. Whatever the explanation, the implication of these rodent studies is that the early experience of the infant is crucial for the magnitude of the stress response and immunity to disease.

18.4 NEURAL MEDIATION OF THE STRESS RESPONSE

The changes in hormone secretion that are produced by stressors are controlled via neural pathways that converge on the hypothalamus from other parts of the brain. The control of corticosteroid secretion has been investigated most frequently and will be considered first. It seems that there is a noradrenergic system that inhibits corticosteroid secretion while a cholinergic pathway is initiating release. This pathway appears to be a branch of the cholinergic pathways controlling cortical arousal and the information-processing changes which are part of the stress response (Warburton, 1979a).

The amount of the arousal is a function of size of the mismatch between expectancy and input (see Pribram, 1967). The animal establishes simple and complex expectancies about the world around it, but novel information represents an increase in uncertainty and produces an increase in electrocortical arousal. The hippocampus is responsible for identifying low information stimuli and for preventing an electrocortical arousal feedback pathway to the midbrain region. This implies that there must be storage of information about simple, predictable stimuli in the hippocampus (Warburton, 1979a).

One neural output of the hippocampus is the medial corticohypothalamic tract from the presubiculum to the hypothalamus, which could provide a direct pathway for hippocampal influence over the pituitary–adrenocortical system (M.M. Wilson, 1985), as well as the indirect input from the hippocampus to the hypothalamus, by way of the septum to the origins of the cholinergic pathways (Warburton, 1979a). This notion links neatly with the proposal of Gray (1982) that the hippocampus is a comparator which processes information. It describes the current state of the world and expectancies about the future state from a generator of predictions (i.e. information gained from the past experience of the animal). These regularities are of two principal kinds, each mapping on to classical and instrumental conditioning: namely, stimulus–stimulus associations and response–stimulus associations. The comparator then decides whether there is a match or mismatch between the description of the world and the prediction.

Some information, such as a loud noise, will require little neural analysis for eliciting the stress response. However, information about social status requires extensive analysis and this complex sort of experience must be mediated to a great extent by cortical systems (Warburton, 1979a). In the various sensory regions of the neocortex, sensory information is transformed by such factors as experience and environmental context to give meaning. In this way the psychological meaning of the information for the individual is derived, and then the cortex will excite the hippocampus. There is a cascading of inputs from several cortical areas through all adjacent regions leading to the entorhinal cortex, suggesting that the hippocampus receives highly analysed, abstracted information from all modalities, rather than information about any specific modality (M.M. Wilson, 1985).

There is also evidence for the involvement of parts of the amygdala in control of the stress steroid response (J.W. Mason *et al.*, 1961). Parts of the amygdala receive direct or indirect inputs from the secondary sensory areas of the neocortex where complex elaboration of information occurs. It is beginning to look as if the amygdala is an entrance to the limbic system for this kind of input, which can then act upon those other areas (e.g. septum, hypothalamus and brainstem) which excite the relevant motivational response, in the light of the animal's prediction of functional 'costs' and 'benefits' of a particular course of action (Herbert, 1984).

In the following sections, the kinds of information which can change limbic activity and initiate the release of stress hormones are considered.

18.5 NONSOCIAL STRESSORS

A wide range of physical stimuli, such as pain and heat, act as stressors and elicit stress responses, which are simple monotonic functions of the intensity of the physical stimulus. However, they are not necessarily associated with activation of the adrenocortical system, when the animal does not readily detect the change (J.W. Mason, 1975). When physical manipulations induce a stress response, it is due to the suddenness of change (i.e. a mismatch with expectancy), rather than to the particular physical stressors to which they are exposed.

The stress response to psychological stimuli can be equally as large as that produced by physical stressors. The response also increases as a simple monotonic function of the intensity of the stimulus and, as with physical stimuli, some psychological stimuli act as stressors, not merely because of their intensity but because of their distribution in time. Information produces a much greater stress response if it is presented at unpredictable intervals, even though the density of these stimuli over time is the same (Warburton, 1979b). Thus, stressors must be considered in terms of their uncertainty.

This point can be illustrated by considering the pattern of excretion in a unit of soldiers in Vietnam who were expecting an attack (Bourne, 1969). Intuitively, one might predict that all soldiers would show dramatic increases in corticosteroid production under the threat of combat. In fact, the ordinary soldiers showed a 40% decrease in urinary corticosteroids on the day of an anticipated attack, and only the officer and radio operator showed 25% increases. These unexpected changes can be related to the levels of uncertainty under which the two sets of soldiers were operating. On the day of the attack, the ordinary soldiers performed routine tasks for which they had been highly trained (i.e. tasks involving little uncertainty). In contrast, the officer was required to stay alert for new instructions which might call for novel decisions and patterns of behaviour. The radio operator was in a similar position in the sense that he was transmitting and receiving new messages, and so was dealing with high information inputs (uncertainty).

Bourne (1969) argued that corticosteroid release was the result of the uncertainty in the situation. However, it could also be argued that the situation did involve some aversive component. The importance of uncertainty is given credence by an animal study (Levine *et al.*, 1972). Rats were trained on either a variable-interval or a fixed-interval schedule with the same density of reinforcement, and then switched to the other schedule. Animals that were changed from an unpredictable variable-interval schedule to the predictable fixed interval schedule, showed no change in steroid levels, but rats switched from the predictable fixed interval to the unpredictable variable interval showed an elevation of plasma corticosterone concentration. Thus, uncertainty seems to be the common denominator in the initiation of the corticosteroid response.

Uncertainty is also a factor in the catecholamine response, because the behavioural situations, in which catecholamine excretion occurs, are similar to those in which corticosteroid excretion occurs (J.W. Mason, 1968). For example, Patkai (1971) found increased catecholamine secretion in uncertain situations in which the subjects were enjoying themselves, such as gambling. Although the neural systems remain to be established, it is a fact that the release of corticosteroids and catecholamines, especially adrenalin, are highly correlated, especially in situations that are unpredictable and may require action. Release in these circumstances makes sense in terms of their metabolic actions which mobilize the energy resources that would be required for action.

One aspect of uncertainty is exposure to an unfamiliar environment, such as a change in location. Changing cages produced significantly elevated levels of corticosteroids in squirrel monkeys (*Saimiri sciureus*) each time it was done, but not if other members of the group were moved with the individual monkey being tested (Vogt *et al.*, 1981). A similar phenomenon has been reported in Old World monkeys. For example, a marked pituitary adrenal response occurred when an adult female rhesus monkey (*Macaca mulatta*) was moved to a new cage, but this response was attenuated by the presence of her infant (Gonzalez *et al.*, 1981a,b). Similarly, 1–year old rhesus monkeys moved to a novel environment showed a smaller pituitary–adrenal response when their peers were present (Gunnar *et al.*, 1980).

An extreme example of novelty is a completely unpredictable environment. Exposure to a constantly changing environment has dramatic consequences. Individual vervet monkeys (*Cercopithecus aethiops*) were exposed to repeated stimulus change in terms of bright lights, loud noises, bells, buzzers and cage shaking (Hill *et al.*, 1967). In comparison with controls which were not so treated, Hill *et al.* found that corticosteroids were elevated. There was also an immunosuppression with its consequent threat to the health of the individuals.

However, one must be cautious about generalizing from one species to another. In a comparative study of titi (*Callicebus moloch*) and squirrel monkeys, both species responded to novel environments with increased adrenocortical activity, but the magnitude of the response was greater for squirrel monkeys than for titi monkeys (Cubicciotti *et al.*, 1986). It was known that the baseline plasma levels of cortisol are nearly five times higher in squirrel monkeys than in titi monkeys (Mendoza and Moberg, 1985), and this difference was found in this study as well. Thus, the substantial differences in adrenocortical activity shown by the two species are increased further in response to novelty (see Chapter 17 for an extensive discussion).

Habituation with repeated exposure to stimuli

In one of his seminal papers in the stress field, Selye proposed that after prolonged exposure to stressors there is a stage of pituitary or adrenal insufficiency, referred to as 'stage of exhaustion' (Selye, 1937). At this stage, there is no more resistance and death may occur. There are important implications for responsiveness to change.

In an important study, J.W. Mason *et a*l. (1968) assessed the effects of repeated exposure to a stressful experience. After 5 days of training on a Sidman avoidance schedule, rhesus monkeys were given alternate cycles of 6 h of shock avoidance and 6 h of rest for up to 30 consecutive days. Adrenocortical excretion was assessed from urinary 17-hydroxycorticosteroid (17-OHCS).

During the first 24 h of avoidance, the typical pattern of the plasma 17-OHCS levels indicated that, during the first 2 h, there was an initial, acute elevation of about 20%. The size of the response appears to be very large but it represented about half the increase that would be produced by a large dose of ACTH. Then the levels declined, and reached extremely low values by 6 and 12 h after avoidance onset in some individuals. These individual differences are discussed in section 18.6. Given the submaximal values, it seems that these declines in plasma 17-OHCS levels on the first day suggest the operation of suppression mechanisms, rather than pituitary or adrenal 'exhaustion'.

Additional support comes from a study of rats following prolonged exposure to stressors. Pollard *et al.* (1976) found that, after 20 days of exposure to shock avoidance, animals failed to show any rise in corticosterone. The intense secretory activity observed in the cells of the adenohypophysis that secrete ACTH was no longer observed after 20 days, and corticosteroid activity returned to control levels. The lack of pituitary activation following prolonged exposure again strongly supports the argument that animals adapt to repeated stressful stimuli by some suppression mechanism.

This suppression mechanism could be either humoral or neural, or a combination of both. The fact that in some cases profound decreases in corticosteroids at 6 or 12 h were observed, following rather mild initial rates of rise, while in other animals no comparable secondary decreases are seen following much higher initial elevations, argues against a humoral negative feedback mechanism as a sole explanation for the observed changes.

However, there is evidence for a neural suppression mechanism. A similar temporal pattern of plasma 17-OHCS response occurred during sustained stimulation of the amygdaloid complex in the rhesus monkey (J.W. Mason *et al.*, 1960, 1961). Partial amygdalectomy consistently reduced, but did not necessarily abolish, the acute rise in plasma 17-OHCS during the first 2 h of avoidance. Interruption of hippocampal outflow lessened the usual drop in plasma 17-OHCS levels 12 h after onset of a 6-h avoidance session in animals subjected to partial amygdalectomy. These findings strongly suggest that amygdala–hippocampal interactions may participate in the integration of the 17-OHCS response during sustained avoidance.

The question of the existence of mechanisms suppressing the pituitary–adrenal cortical activity is also raised by the progressive decline in 'basal' urinary 17-OHCS levels as the avoidance sessions were repeated. The 17-OHCS response to avoidance was also markedly diminished in the second and third sessions and the response peak was shifted to the right (J.W. Mason *et al.*, 1968). This change can be interpreted as indicating that some suppression mechanism was decreasing the rise time for the hormonal response, and reducing the magnitude of response (i.e. the threshold for the 'hormonostat' was lower). It is important that large increases in 17-OHCS could be obtained in these subjects with ACTH infusion in the presence of chronically low 17-OHCS levels. This argues strongly that the hypothalamic-pituitary–adrenal cortical axis still has a substantial response capacity, and suggests that the underlying mechanism is a neural one which is actively suppressing the response.

The shift of the 17-OHCS response peak with repeated experience is interesting in terms of a possible neural mechanism. It is significant that a very similar pattern of delayed, as well as diminished, 17-OHCS response to avoidance has been observed in monkeys subjected to amygdalectomy (J.W. Mason *et al.*, 1960, 1961). Again, altered amygdala function is implicated in the neural suppression mechanism.

In summary, these studies demonstrate that stressful psychological stimuli do not elicit responses of the same size after repeated presentations, as they did initially. This phenomenon is analogous to habituation and it is known that the hippocampal-amygdala system is involved in this process. However, it is significant that there are marked individual differences in the original response, and rates of adaptation, which argue for differences in limbic function.

18.6 INDIVIDUAL DIFFERENCES IN THE STRESS RESPONSE

Early studies of the stress response emphasized the breadth of neuroendocrine responses to a wide variety of stressors, but neglected variation in an individual animal's response. The following chronic avoidance study examined this question (Levine *et al.*, 1972). Rhesus monkeys were given 72–h shock avoidance sessions. After the first few sessions, they had no increase in adrenocortical activation, and had levels even lower than baseline when chronically exposed. However, there were marked individual differences in the response, depending on social status. When dominant and aggressive rhesus monkeys were put in shock avoidance they showed a very rapid increase in adrenocortical activity. However, submissive and less-aggressive monkeys failed to show any increase of adrenocortical activity when subjected to shock avoidance.

When the monkeys were subdivided, in terms of aggressiveness and speed of learning, the aggressive ones, which were also slow to learn, showed evidence of continued emotionality (17-OHCS elevation) during the whole experimental period, while the nonaggressive, which were quick to learn, showed little effect after the first few sessions. Since the task was the same for both groups, and their avoidance performance was similar, it was concluded that the factor producing this difference was some neural mechanism that is related to aggressiveness.

A monkey's level of aggressiveness is a function of both genetic factors and experience. No indications of genetic evidence was available but Levine *et al.* (1972) speculated on the possible influence of experience on responsiveness. They argued that, for subordinate and submissive animals, avoidance of punishment is not as novel or as potentially stressful an event as it is for a dominant male. They noted that the two monkeys with the highest 17-OHCS avoidance response in a previous study were 'laboratory-naive', and the monkeys which showed the smallest reaction had extensive laboratory experience. It is possible that this previous history made it easier for some monkeys to adapt to prolonged exposure to the repeated avoidance. Clearly, an animal's past history is a relevant factor in the present response to a stressor.

18.7 CONTROL OF ENVIRONMENTAL CHANGE

It has already been noted that environmental change does not inevitably produce a stress response, and the factor of predictability was used to explain the findings. In this section, the hypothesis will be considered that a completely predictable environment is not ideal either. Another crucial factor in the response to change is whether the individual can exert control over the change. A good example of the effect of control on the stress response has been seen in avoidance-conditioning studies in rats. During acquisition of the avoidance response, high levels of corticosteroid are released but the stress steroid levels decline markedly as the avoidance response is learned (Coover *et al.*, 1973). Thus, it is not the threat of shock which is important in activating the stress response but the animal's lack of control of the situation. Competence at performing the response represents acquisition of control.

Loss of previously established control by the individual has also been shown to be a stressful experience. For example, if rats which have been trained in shuttle-box avoidance are put in the box with a barrier which prevents the response, then there is a large increase in corticosteroids in comparison with control rats just placed in the box without the barrier, even though no shock is given to the group with the barrier (Coover *et al.*, 1973). Removal of the lever in a lever-pressing avoidance has the same effect. Delivery of unavoidable shock produces a much greater increase in plasma corticosterone in rats which had learned an avoidance response than in those which had never learned a response, and so had never experienced control over their environment (Overmier *et al.*, 1980). When animals perform a response and are not rewarded, there is also a marked corticosteroid elevation (Levine *et al.*, 1972), which can also be interpreted as loss of control (Weinberg and Levine, 1980).

From these studies, it can be argued that a situation is a stressor because it is imposed on the animal and is not chosen. In fact, animals may choose to separate themselves from the group and no stress response is seen, but enforced separation is stressful (Katz, 1983). Similarly, there is evidence that novelty is not always a stressor. When animals are placed in a novel environment, there is a marked corticosteroid elevation. In contrast, when animals can control their exposure to the same environment by exploring it from their home cage and becoming gradually familiar with it, then there is no elevation of steroids (Misslin and Cigrang, 1986).

Moreover, homeostatic models of the stress phenomenon (Miller, 1980) have assumed that all change is aversive, but this need not be so when the change is initiated by the individual. In fact, as Toates (1987) pointed out, an unchanging environment may be less suitable for animal welfare than one in which there is an opportunity for self-determined change. Control is especially important if by exploiting it an animal can escape from a negative event. Toates (1987) goes on to point out that, if the animal has a control strategy which has a response-outcome expectancy that is not too demanding, then we need not be too concerned for the animal's welfare (see also Chapter 19). Sophisticated organisms, like nonhuman primates, have a need for control over their environment. While some environmental changes result in a stress response, an important aspect is not environmental change *per se* but the amount of control that the animal can exert over the change.

18.8 SOCIAL STRESSORS

Simian primates are characterized by the variety and complexity of their social lives (Smuts *et al.*, 1987). Hence, changes in the social environment may be potent stressors. These may influence health, fitness and reproduction (see Chapter 4 for a detailed discussion of social stress and reproductive fitness).

Recent research on humans has established both a theoretical basis and strong empirical evidence for a causal impact of social relationships on health. Prospective studies, which control for baseline health status, consistently show increased risk of death among people with a low quantity, and sometimes low quality, of social relationships. Experimental and quasi-experimental studies of humans also suggest that social isolation is a major risk factor for mortality from widely varying causes (see review by House *et al.*, 1988).

The prospective mortality data are made more convincing by their agreement with the evidence from experimental and clinical research, which shows that reduced social contact produces physiological effects that may produce illness and even death, if prolonged. For nonprimate species, Cassel (1976) reviewed evidence that the presence of a familiar member of the same species can buffer the impact of experimental stressors on ulcers and hypertension in rats, mice and goats. In this chapter, already, evidence has been presented that social buffering reduces the release of steroid hormones in response to stress. In people, clinical and laboratory data indicate that the presence of, or physical contact with, another person can modulate human cardiovascular activity and reactivity in general, and in stressful contexts such as intensive care units (Lynch, 1979).

Bovard (1985) argues that such influences are mediated via the amygdala, which inhibits the posterior hypothalamus. In this way the secretion of adrenocorticotrophic hormone, cortisol, catecholamines and associated sympathetic autonomic activity are lowered and the immune response will function unimpaired, with a consequent lowering of mortality.

Social changes that act as stressors include, for example, the introduction of a stranger into an established group, the formation of new groups, and separation from important conspecifics. The majority of studies has been conducted on the rhesus monkey in all these areas, although there is some work on other species and this will be referred to here. Those readers who are interested in particular species should consult the encyclopaedic comparative review of Kaplan (1986), which proved an invaluable reference source for this chapter.

When an unfamiliar conspecific is introduced into an established group, for example, group members may demonstrate xenophobia – aggressive intolerance towards the stranger. Strangers may elicit behavioural reactions as well as a marked stress response in terms of a rapid rise in cortisol (Scallet *et al.*, 1981). In that study, strange rhesus monkeys were placed behind a Plexiglass screen so that they were in visual contact with the group but no injury could be inflicted. In free-living groups, of course, there is the possibility of escape, but some laboratory situations may involve such stressful conditions. The magnitude of the cortisol elevation was correlated with the amount of threat that had been directed toward the animal, and females that were exposed to threat showed larger increases than males.

Formation of new groups

The formation of new social groups in captivity may also result in aggressive behaviour, as in the case for the introduction of a stranger (Bernstein *et al.*, 1974), the outcome of which in many cases is the establishment of dominance relations. The problems of defining social dominance for individuals and groups have been discussed at length elsewhere (e.g. Bernstein, 1981). However, it is of interest in the context of stress responses that Chamove and Bowman (1976, 1978), for example, found that in groups of young rhesus monkeys, there was an association between social subordination and increased secretion of cortisol, or its indicator 17–hydroxycorticosteroid, during group formation, although it has also been found that all juveniles showed elevated corticosteroids in the first phases (Golub *et al.*, 1979). Elevations in corticosteroid and prolactin release have also been reported for adult talapoin (*Miopithecus talapoin*) males and females

(Bowman *et al.*, 1978; Eberhart and Keverne, 1979) in group-formation studies.

Heart rate has been used as a measure of autonomic arousal, and has been examined during group formation in squirrel monkeys (Candland *et al.*, 1970; Leshner and Candland, 1972). It was found that during social disruption the relationship of heart rate with dominance status was U-shaped, with higher rates for dominant and subordinate group members compared with mid-ranking individuals.

In studies with two other species, rhesus monkeys and hamadryas baboons (*Papio hamadryas*), very similar effects were observed (Cherkovich and Tatoyan, 1973). When two males and two females were formed into a new group, the dominant male had the lowest heart rate and least autonomic arousal. However, when the dominant male was removed from the group, the subordinate male's heart rate fell below that of the females. Since heart rate is an index of catecholamine release, then the fitness of the previously subordinate male will have increased due to a higher immune response.

With reference to Candland's work, and a U-shaped relation with dominance during new group formation, it was also found that this relationship may be completely reversed when the groups have stabilized. Candland *et al.* (1973) argue that prior to group formation, the enhanced baseline levels of stress steroids of dominant animals increase aggression and help to maintain dominance, while for the subordinate animals the increased corticosteroids and autonomic activity are the consequence of defeat.

An alternative explanation can be given in terms of uncertainty of social status. During group formation, the dominant animals are uncertain about maintaining their status, while the subordinate cannot predict their next attack, so that both show elevated adrenal function. After group formation, both dominant and subordinate animals are certain of their social status and the social uncertainty is in the middle ranks. As a consequence, adrenal activity will be lower for both groups in comparison with middle-order individuals. The implications of these data and the interpretation given here is that studies in this area should focus on social dynamics, the changes that occur over time in an individual's social relations with others of that particular species.

The importance of considering social dynamics comes from studies of atherosclerosis. This is the deposition of lipids in the arteries and is believed to be due to adrenal activation by stressors. In three studies of cynomolgus monkeys (*Macaca fascicularis*), serum cholesterol levels were studied in single-gender groups of four individuals over a period of 16 months. In stable groups, the highest serum cholesterol levels were seen in the subordinate animals (Hamm *et al.*, 1983). Support for this work has come from a study of vervet monkeys. In this work on stable groups, higher serum cholesterol was found in the more submissive animals (Bramblett *et al.*, 1981). Consistent with this picture, the dominant cynomolgus monkeys had the lowest amount of atherosclerosis of the coronary artery for both males and females (Kaplan *et al.*, 1982; Hamm *et al.*, 1983). In related work, the groups of four monkeys were rearranged at intervals to produce new groups. In these conditions of repeated social disruption, it was the dominant monkeys which had the most atherosclerosis of the coronary artery (Kaplan *et al.*, 1982, 1983). Clearly, it was not dominance or subordination *per se* which was important but the social dynamics.

It was mentioned at the beginning of this chapter that, under many circumstances, the stress response is a normal part of natural adaptive responsiveness and is advantageous in the short term. By contrast, prolonged environmental changes may lead to distress responsiveness, which is not advantageous to health, fitness and reproduction.

18.9 DISTRESS

In this final section, the phenomenon of distress will be considered. Distress has an everyday meaning and was the origin of the word 'stress'. Modern emotion theory (Ortony *et al.*, 1988) has given it clearer specification.

1. Distress is an emotion type (i.e. a distinct kind of emotion that can be manifested in varying states and with varying degrees of intensity). For example, fear is an emotion type which could be manifested by behaviour which might be described as 'concern', 'fright', 'panic', and so on.
2. Distress is an emotion type whose eliciting condition is an event, rather than an agent or an object, where an event is the organism's interpretation about something that happened in terms of consequences for the individual, independent of causality. Thus, the distress of separation from a conspecific is a reaction to that event and not to the experimenter who caused it.
3. Distress refers to an immediate response and can be differentiated from fear in terms of future consequences; fear implies that there will be future consequences of the event, while distress refers to the immediate consequences of the event for that individual.
4. Distress is a negatively valenced reaction. In common language, the event is undesirable in terms of its immediate consequences.
5. As was mentioned earlier, distress can have several manifestations which differ in terms of the specific events that the individual interprets as undesirable. For example, some theorists have interpreted

the response of a nonhuman primate infant separated from its mother as bereavement or grief (Bowlby, 1958). The bereavement response is not just a negative reaction to the separation, but is specifically a reaction to the loss of a significant individual in the infant's life, such as the mother. Reactions to this separation are indices of the individual's distress. Virtually all infants separated from their mother show a 'protest' phase, which is followed by a 'despair' phase in most individuals in most species (Mineka and Suomi, 1978).

Protest phase

Protest seems to be a universal response to separation in primates although it varies in degree (Plimpton and Rosenblum, 1983). All individuals display hyperactivity and increased vocalization. This protest phase can be interpreted as the infant's active behavioural attempts to cope with separation, but the physiological data indicate that maternal loss is an extremely stressful event for the young monkey. Squirrel monkeys show a significant increase in plasma cortisol when mother and infant are placed alone without each other in separate cages (Vogt and Levine, 1980), or separately but in a social group (Coe *et al.*, 1978a). In these monkeys, at least, separation of the infant from its mother evokes pituitary–adrenal activation in the infant equivalent to, or greater than, that induced by physical trauma. During sustained separations, adrenal activation persisted for over 2 weeks. In addition, no habituation was observed in the cortisol response to repeated acute separations. For example, no change in the cortisol elevation was observed after six 4-h separations, or after 20 30–min separations (Coe *et al.*, 1985). Similarly, there is a large increase in plasma cortisol levels in rhesus monkeys at 3 h after separation, in the same manner as for the squirrel monkeys (Smotherman *et al.*, 1979; Golub *et al.*, 1981). However, the adrenal activation subsided 24 h after separation (Gonzalez *et al.*, 1981; Coe *et al.*, 1985). Clearly, there are qualitative differences in the duration of the effect in different species.

There is also massive activation of the sympathetic nervous system. There was increased production of catecholamines in rhesus infants (Breese *et al.*, 1973), and increases in heart and body temperature in pig-tailed macaques (*Macaca nemestrina*) which correlated with the protest phase (Reite *et al.*, 1978). In the latter study, return to baseline occurred when the infants were returned to their mothers. Given the massive activation of the sympathetic nervous system after separation, it is not surprising that the immune system is also impaired. In addition to the hormonal responses, Coe *et al.* (1985) found alterations in the immune competence of separated squirrel monkey infants, in terms of immunoglobulin, and antibody production. Consonant with the relatively weak adrenal response of the separated rhesus infant however, separation has little effect on the immune system in rhesus macaques. Coe *et al.* (1985) found no effect of a week's separation on immunoglobulin or complement levels in the rhesus infant at 1, 3, 4 or 7 days after separation. Similarly, examination of the effect of a week's separation on the immune responses to a bacteriophage in the rhesus macaque also revealed only a small difference between four control and four separated infants.

Coe *et al.* attribute the absence of marked immune changes in the rhesus macaque to the absence of sustained adrenal activation, in contrast to its presence in the squirrel monkey. They believe that this activation mediates the immunological changes. However, they also point out that the immune system of the rhesus macaque proved to be ten times more resilient than the squirrel monkey's following separation.

Once again, there are important species differences in the stress response to change. Surprisingly, heightened adrenocortical activity in the squirrel monkey is not complemented by heightened secretion of ACTH from the pituitary (Coe *et al.*, 1985). Reduced ACTH levels, minimal feedback regulation of adrenocortical activity, and the rapidity and magnitude of the adrenal response to disturbance, have led to the speculation that the squirrel monkey's adrenal cortex is driven by sympathetic stimulation (Coe *et al.*, 1985; Mendoza and Moberg, 1985).

Despair phase

After the vigorous protest phase, which may last for a few days, there is a hypoactive 'despair' phase in rhesus monkeys and pig-tailed monkeys to total social isolation (Seay *et al.*, 1962; Hinde *et al.*, 1966; Kaufman and Rosenblum, 1967). There was total lack of interest in the social and nonsocial environment. Exploration was decreased and the infants just sat in a huddled posture. These symptoms have been described as analogous to human depression and the degree to which it is homologous will be discussed later. The symptoms disappeared rapidly when the infant was reunited with its mother, and social contact was re-established immediately in every study with these two species of monkey. However, the clear-cut picture which is seen with rhesus and pig-tailed macaques is not seen with all species, and the marked species differences in severity of responsiveness have defied simple explanation (Mineka and Suomi, 1978).

If separated infants are put into cages with peers the cortisol response to separation is attenuated (Golub *et al.*, 1981), and these separated infants develop strong, mutual attachments. Moreover, within the same species, there are individual differences in the nature of the response and its magnitude. Some infant rhesus appear to be more prone to despair than others (Lewis *et al.*, 1976). For example, 5 out of 14 pig-tailed infants showed only a brief, mild despair phase (Rosenblum, 1971). Variation in autonomic changes have been reported, especially in the rate of return to baseline (Reite and Short 1980). It has been argued that these differences have a constitutional component (Suomi, 1981). The best predictor of whether an adult will show a despair response is whether it shows such a response as an infant (Mineka *et al.*, 1981), which, once again, points to a genetic predisposition for the despair phenomenon.

The despair response has also been described as a 'depressive' reaction because of the analogy with retarded depression with its hypoactivity, slouched postures and lack of interest in the social and nonsocial environment. Pharmacological tests have been given in order to discover if the responses are homologous – that is, have the same underlying mechanism (e.g. Lewis and McKinney, 1976; Suomi *et al.*, 1978; Suomi, 1982), and the reversal of the despair symptoms by treatments for human depression suggest the same underlying biochemical pathology for the two disorders, which, in the case of human depression, is thought to be lowered functional catecholamines (Warburton, 1975).

A test of the hypothesis that lowered functional catecholamines cause separation-induced despair was performed by Kraemer and McKinney (1979). They used alpha-methylparatyrosine, a drug which interferes with catecholamine synthesis and so reduces functional catecholamines in the brain. They gave small doses of the drug to juvenile rhesus monkeys during the social separation and reunion parts of the repeated separation paradigm of Suomi *et al.* (1970). Alpha-methylparatyrosine potentiated the intensity of the despair symptoms, transforming them from mild to weak. There was no effect of the drug on behaviour during the reunion part of the paradigm.

It is of significance that the dose required to precipitate and exaggerate the despair symptoms differed for individuals with different social histories. Juveniles who had been separated from their mother and reared with their peers required doses of alpha-methylparatyrosine a quarter to an eighth the size of those required by juveniles who had remained with their mother during infancy. A comparison of peer-reared monkeys who had only experienced a few separations with those who had had many separations, showed that only half the dose of alpha-methylparatyrosine was required for the much-separated group.

These results show that distress in responsiveness to change, in this case separation from peers, depends on the neurochemical state of the individual. In addition, the importance of genetic background must be emphasized; there are clear species differences and there are individuals within one species who are congenitally more prone to disturbance after separation than others. Thus, in considering change as a stressor, nonhuman primates must be considered as individuals whose genetic and social history determine their behaviour, depending on the physical and social environment with which the individual is interacting (Suomi, 1981, 1982). The mediation of this interaction is via the neurochemical state of the individual.

The commonalities of this despair state with human depression have implications for the fitness of the individual. It has long been believed that separation and depression are related to susceptibility to disease in people (Schmale, 1958). Patients showing reactive depression following an emotional loss showed significant increases in corticosteroids when they confronted their loss and experienced distress (Sachar *et al.*, 1967). An investigation of immune function in the recently bereaved found the lymphocyte response to two mitogens was significantly suppressed for up to 8 weeks (Bartrop *et al.*, 1977).

The work of Coe *et al.*, on elevation of plasma cortisol levels in separated squirrel monkeys, has already been described. In a follow-up to this work, measures of the complement protein levels were made. Complement protein levels are an index of the function of the complement system, which is the principal effector in the humoral immune system and is involved, among other things, in the immune response to bacteria and the formation of antigen–antibody complexes. Complement proteins were decreased in the infants on separation from their mother and the decline was greatest and sustained longest in the isolated, in comparison with the group housed (Coe *et al.*, 1985). Separation also decreased the serum levels of immunoglobulins (Coe *et al.*, 1985).

These workers also examined whether these changes had any functional significance by assessing the antibody response to an antigen bacteriophage after 7 days of separation. There was a 10–fold reduction in antibody production by the separated infants in comparison with infants staying with their mothers. When the infants who had been separated were returned to their mothers after challenge, the immunosuppression influence of separation was halved (Coe *et al.*, 1985).

These marked effects in squirrel monkeys contrasted with an absence of marked immune changes in the rhesus monkey (Coe *et al.*, 1985). Clearly, there are qualitative differences in the effect in different species. These differences correlate well with the differences in the cortisol response of the two species, with a much greater cortisol response in squirrel monkeys (Coe *et al.*, 1985) in comparison with rhesus macaques (Gonzalez *et al.*, 1981). Coe *et al.* believe that the rhesus monkey and squirrel monkey represent two ends of a continuum of physiological responsiveness to separation.

18.10 CONCLUDING REMARKS

The stress response is an adaptive response enhancing the likelihood of survival in the short term. However, the immunosuppression effects of the hormones constituting the stress response may have deleterious effects if exposure to the stressor is prolonged. Stressors are a consequence of an interaction between the individual animal and the situation in which early experience changes the sensitivity of the neural mechanism, which controls the stress response, and the responses may change in magnitude in ways analogous to habituation. Uncertainty seems to be a common factor of many situations in which there are such responses and they can be seen as a preparation for action. One aspect of uncertainty is situational uncertainty, which is the result of the many alternative meanings that the input has for the individual and the available responses. Consequently, uncertainty must be defined subjectively, and it will depend upon the individual's past experience.

Recent primate research has indicated that it is not environmental change *per se*, but the amount of control that the animal can exert over the change. While the magnitude of the stress response correlates with the amount of environmental unpredictability, studies indicate that a predictable one is not ideal, and the optimum situation is one in which environmental change is initiated by the individual.

More than any other mammal, primates reference their behaviour to conspecifics more than to location. Studies of different sorts of psychosocial changes that act as stressors, result in increased release of stress steroids with the magnitude of the response, depending on the social dynamics.

Distress is used to refer to the detrimental effects that can occur in the individual, which can reduce its fitness. Stressors have the potential for causing distress when the stressor is too intense or too prolonged. While the stress response is a normal part of natural adaptive responsiveness and advantageous in the short term, distress is disadvantageous.

19
Criteria for the provision of captive environments

TREVOR B. POOLE

19.1 INTRODUCTION

The previous chapters of this part of the book have been concerned with primates in captivity. The topics considered have been numerous and varied. It is the aim of this final chapter to discuss aspects of captive environments from the perspective of the needs of different primate species in captivity.

A captive environment is one in which an animal is confined within an area with well-defined boundaries, and is wholly dependent on feeding by human agency. This is in contrast to the wild state where, even though they may be restricted by geographical barriers, animals acquire food entirely by their own efforts. Islands, such as Cayo Santiago off Puerto Rico (Rawlins and Kessler, 1986), represent a half-way house, because the animals are constrained by a natural boundary but depend on provisioning by people to sustain their numbers. Safari park zoos create a similar situation, except that the boundary is artificial. Thus, captive environments for primates range from large areas containing vegetation and topographical features (Mager and Griede, 1986) to laboratory situations, where monkeys may be singly housed in very small, but legally approved, metal cages (Boot *et al.*, 1985), often arranged in two tiers (see Chapter 14, section 14.2).

The aim of this chapter is to discuss captivity as such; to consider criteria for the adequate provision of captive environments for different species, with special reference to the relatively restricted conditions of zoos and laboratories, because these are the commonest conditions of captivity for primates, and also represent the greatest departure from the natural state.

19.2 NEEDS OF THE KEEPERS

In examining captive conditions, a most important consideration is to be aware of the aims of the people keeping the animals. In zoos, for example, primates are kept for a variety of reasons which include conservation, education and, of course, entertainment of the public. In laboratories, primates may be used to investigate aspects of their biology, or for biomedical research as models for human physiology, for product safety testing, or to assess the effectiveness of remedies for disease. The aims of the keepers frequently place constraints on the system of housing and husbandry employed.

The reason for keeping the animals may determine whether or not they are bred in captivity. For example, primates used in long-term studies in the laboratory are often bred in self-sustaining colonies. These are usually in the country of use, but there are some organizations which breed the animals in the country of origin (Hobbs *et al.*, 1987). Breeding may be achieved either by the creation of a breeding group, or through timed matings of singly or same-gender housed individuals. However, for short-term studies, monkeys may be imported directly from the wild, and are usually housed singly or, occasionally, in same-gender groups.

19.3 NEEDS OF THE ANIMAL

Captive animals should, as Webster (1984) pointed out, be able to enjoy what he termed 'the five freedoms'. These are: freedom (1) from malnutrition, (2) from thermal and physical discomfort, (3) from injury and disease, (4) to express most normal patterns of behaviour, and (5) from fear and stress. Awareness of freedoms (1)–(3) led laboratories and zoos to place animals in bare hygienic cages with ample supplies of food. Hediger (1955) was the first to point out that this was not enough to satisfy the captive animal's needs, and that behavioural requirements were not being met by these conditions. There was a considerable lapse of time, however, before Markowitz (1979, 1982) and Maple (1979) described some practical improvements to captive environments aimed at meeting the animals' behavioural needs.

Nowadays, the first three freedoms are generally taken for granted, and the emphasis in this chapter will be on those of (4) and (5). Wild primates live in a complex and constantly varying environment, to which they have adapted over millions of years. They may exist over a wide range of habitats and have evolved an appropriate repertoire of behaviour, which promotes their survival and reproduction. Often, the individual lives in a social group where membership entails complex personal interrelations (Smuts, *et al.*, 1987; Dunbar, 1988), and recognition of kinship ties (Walters, 1987a). Even relatively solitary species such as the orang-utan, which have large interindividual distances, may communicate with, and have a detailed knowledge of, other individuals overlapping their home range (Horr, 1975).

Primates in captivity often lack opportunities to carry out much of the behaviour indicated above and, even when the animal is housed in groups, the sociological structure of the group often bears little resemblance in complexity to that seen in the wild (Stevenson, 1983). It is clear that, in considering 'the five freedoms', providing primates with the freedom to express most normal behaviour patterns creates the greatest difficulties. The aim should be to provide an artificial habitat which lies within the adaptive range of the species, so that individuals have the appropriate behavioural responses to thrive in it. To this end, the two key considerations when considering the captive environment are: (1) The degree to which it departs from the natural habitat; the greater this is, the more critical each element of this situation becomes in terms of meeting the animal's needs. (2) The development of practical techniques for satisfying not only physical and physiological requirements, but also the animal's behavioural or psychological needs.

In providing a captive environment for primates it is clear that, like many other wild mammals, they appear to experience a need to carry out complex behaviour, and require the facilities to do so. We can readily empathize with the ape sitting by itself in a featureless cage, separated from the real world by bars. It does not really surprise us that such animals appear apathetic and abnormal, because we ourselves would be likely to react similarly in an equivalent situation. This would appear to be the equivalent to human boredom (Wemelsfelder, 1984). Moreover, in the absence of opportunities for complex behaviour, a variety of abnormal behaviours may develop.

In restricted captive environments, opportunities to carry out certain types of behaviour seem to be of particular benefit to primates. The two most important appear to be opportunities for increased foraging (Chamove *et al.*, 1982; see also Chapter 14, section 14.4) and for physical contact with conspecifics (Harris, 1988; Chapter 14, section 14.3). Complex devices which can be manipulated may also relieve boredom and help to maintain activity and alertness (Beaver, 1989).

19.4 OBJECTIVE ASSESSMENTS OF WELFARE

There are basically two scientific approaches to the assessment of the welfare of primates; namely, physiological and behavioural. Of these, behavioural methods may be preferable because they may be more sensitive, do not require the use of elaborate techniques, and are invariably noninvasive. Physiological parameters may show little perturbation because the animal works to achieve homeostasis. On the other hand, a combined behavioural and physiological approach is ideal if it can be achieved, because behavioural indicators are sometimes difficult to interpret (see Chapter 3, section 3.4).

Natural behaviour may be defined as that occurring under wild conditions. However, some such behaviour may be harmful in captivity. Fights which result in serious injury or death, infanticide by a new dominant male (see Hrdy, 1976) and fear responses to predators are good examples. Moreover, the differences between the natural habitat and the confined situation encourage behaviour that is not expected under natural conditions, as in learning artificial foraging tasks (see section 19.7). If such behaviour is adaptive to the situation and increases the well-being of the animal or its conspecifics, it can be regarded as beneficial. Some behaviour not expected in nature, however, is clearly regarded as abnormal. It does not promote the survival of the animal, and usually appears to be 'goalless', as with stereotyped head-shaking, self-clutching, or self-mutilation. Significant levels of abnormal behaviour are regarded by most animal keepers and stockmen as indicative of poor husbandry (Cronin, 1985). There are a variety of forms which abnormal behaviour may take, and a few which are commonly seen will be briefly discussed. For a more detailed coverage of abnormal behaviour, see Erwin and Deni (1979) and Poole (1988).

The gross level of activity may be abnormal. The individual may be underactive and may not make full use of the environment; it may not interact with conspecifics, and show little curiosity towards environmental change. Alternatively, the animal may be hyperactive and overreact to even minimal changes in the environment.

Abnormal behaviour commonly takes the form of behavioural stereotypies, which are repetitive, goalless actions involving the whole body or parts of it. There is some evidence that this behaviour induces the secretion of beta-endorphins in both pigs (Cronin, 1985) and horses (Bridneach, 1988; Universities Federation for Animal Welfare Annual Reports). This suggests that the animal creates a sensation of euphoria through carrying out the stereotypy. Cronin regards this as evidence of poor welfare, as it appears to be an attempt to cope with an intolerably unstimulating environment over which the animal can exert little control. In singly housed animals, abnormal behaviour may also include self-directed social behaviour, juvenile behaviour, and self-directed aggression. Even when animals are housed socially they may be hyperaggressive, fail to mate, or be infanticidal or neglectful of their young.

There are considerable differences between species and among individuals in the ways in which primates react to captivity. Hence, it is important that behaviour should be regularly monitored to ensure that the needs of the animals are being accommodated. In determining whether a captive environment is satisfactory, several behavioural indicators can be considered. A restricted repertoire of behaviour is an indicator of an inadequate environment, and a comparison with the wild situation is particularly helpful, where appropriate studies have been made. If quantitative data are available, then the time which individuals of different species spend in different activities in captivity can be compared with that in the wild. It is encouraging to learn that Fragaszy and Adams-Curtis (see Chapter 13), for instance, have devised housing conditions for capuchins (*Cebus apella*) in which behaviour is frequently monitored and which induces similar activity levels to those shown by wild cebus. In the absence of field data, however, it is generally true that enriching the range of behaviour not associated with stressful environmental stimuli represents an improvement for the animals.

19.5 ANIMAL WELFARE GUIDELINES

The ethical obligation to consider the animal's needs is a welfare consideration which is now incorporated in statutes in both Europe and the United States. A variety of guidelines have been drawn up in different countries but their legal status varies. In both the USA and Europe, there is a legal requirement for laboratories to take into account the animal's behavioural and psychological needs (Novak and Suomi, 1989). However, guidelines set minimum standards, which are often far from ideal. They tend to be based on current practice, and economic considerations play an important role. For instance, there is little doubt that a British zoo would be prosecuted if it kept a primate in a cage recommended for laboratories by the Council of Europe (1986). Examples of approved cages are shown in Figure 19.1. These must be viewed in the light of the Council's recommendation that: 'Any restriction on the extent to which an animal can satisfy its physiological or ethological needs shall be limited as far as possible.' Readers are invited to draw their own conclusions as to the adequacy of the cages illustrated.

We may now consider specifically some characteristics of the captive environment.

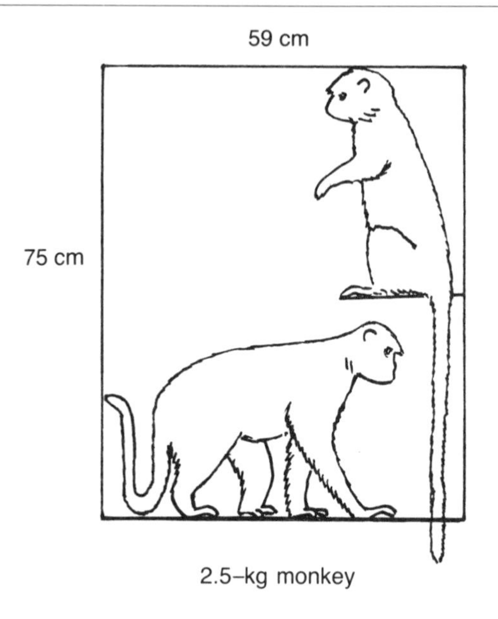

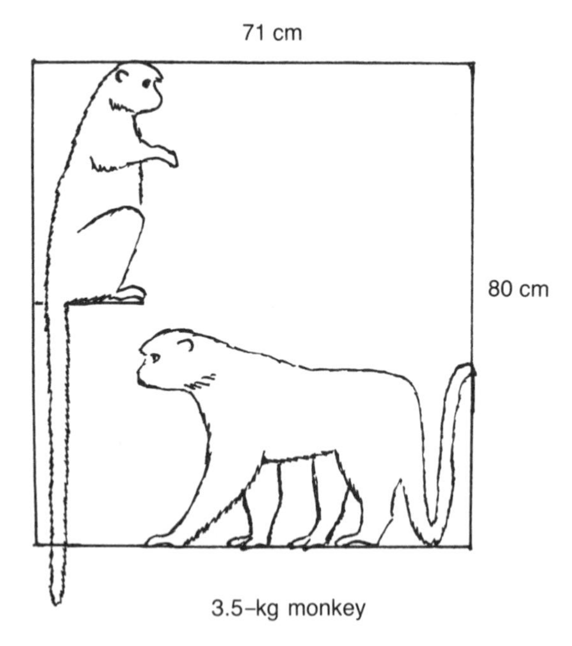

Figure 19.1 Council of Europe recommended cage with long-tailed macaque or cynomolgus monkey weighing 2.5 or 3.5 kg drawn to scale. The measurement of the cage base is the square root of the recommended floor area. Even if a perch is supplied, the animal cannot sit with its tail freely suspended.

19.6 THE PHYSICAL ENVIRONMENT

Space and complexity

The natural environment provides complexity and also ensures a degree of environmental change. These features are highly recommended for captive situations, but they may be difficult to achieve in practice. In considering the provision of a suitable physical environment, the factors of major importance are: the size of enclosure, the materials with which it is constructed, its complexity, the methods of maintaining hygiene and the financial costs incurred. What is clear, however, as we have noted, is that primates kept in a restricted, unvarying environment may survive, but they pay a high cost, often becoming apathetic or showing severely disturbed behaviour. It would appear that higher primates experience a need for spontaneous activity and stimulation, and that they suffer when kept in an environment which is over predictable and where they have little control. In some cases, reproduction may be suppressed or the animals suffer ill-health as an indirect result of stress.

First, then, there is the question of size. The view that the larger the enclosure the better it will be for the animal must be qualified at the outset. Space alone may be of little value because the primate may not be able to use it, or have any reason for doing so. A large room with bare walls, for example, provides only a floor as usable space. For primates, space must be considered as three dimensional, so that it allows the individual to display its normal repertoire of locomotor behaviour; namely, to walk, climb, jump and run. This means that in an open situation, such as a compound, vertical climbing frames or trees should be provided, while, in a cage, vertical climbing surfaces and perches are essential. Where it is necessary for an animal to be confined in a restricted space and housed singly for periods of time, the provision of a large complex exercise area with regular, but limited, access may be practicable (see Jaeckel, 1989). An exercise cage may also provide a solution to the problem of improving existing accommodation. To take the cynomolgus monkey (*Macaca fascicularis*) as an example. It has been found that in the wild the leap length averaged 2.2 m, and that the distance covered by leaping, in relation to the total journey length, ranged from 2.7% during foraging and up to 10.8% when travelling (Cant, 1988). In fact, this species is commonly kept in the laboratory, but it is seldom provided with the facility to leap.

A problem which is often raised in relation to the provision of more generous accommodation in the laboratory is that the animal may become relatively inaccessible. This may be difficult in cases where it may have to be dosed several times a day, or be weighed regularly. However, training a monkey to accept an injection, or to enter a smaller crush cage is not difficult, provided that a suitable reward is given (Jaeckel, 1989).

In most laboratories, primates are kept indoors within a restricted range of temperature and humidity. For example, those recommended by The Royal Society/UFAW (1986) and The Home Office (1989) suggest 20–28°C for New World primates, 15–24°C for Old World primates, and relative humidities of 45–65%. There is much to be said, however, for a degree of climatic variability, provided that tropical species have access to a warm humid indoor area. Redshaw and Mallinson (Chapter 12, section 12.2) describe the outdoor living accommodation for golden lion tamarins (*Leontopithecus rosalia*), which thrive outdoors in Jersey; variations in climate induce the animals to show a wider repertoire of behaviour, and they show adaptation by growing thicker coats. This species is also kept successfully outdoors in Apenheul Zoo in The Netherlands (Mager and Griede, 1986). Further, *Leontopithecus* species are kept at the Rio de Janeiro Primate Centre, within the habitat area of the species, and experience winter temperatures as low as 4°C without access to heated quarters (A. Pissinatti, personal communication). Such natural temperatures would not be acceptable in a UK laboratory!

The degree of complexity in the physical environment that can be provided not only depends upon the size of enclosure but is directly related to the size of animal. Smaller primates can be provided with greater space at less cost, and they are much less destructive. Chimpanzees, gorillas and orang-utans make nests out of trees, so that any unprotected trees in their enclosures are rapidly destroyed, while common marmosets (Callitrichidae) only damage vegetation by gouging holes in branches.

Any primate soon becomes accustomed to the complexity of its captive environment. This is evidenced by the regular use of walkways and climbing routes. The situation is also true in nature but on a larger scale. However, a total lack of change in the captive situation may result in boredom (Wemelsfelder, 1984), and even the presence of behavioural stereotypies. A good example of well-meaning efforts to provide a complex 'naturalistic' environment of this kind is seen in some zoos where the animals live in an artificial, virtually indestructible plastic jungle, with metal leaves and artistically painted lichens on the plastic trunks of the simulated vegetation. This creates an environment which deceives the public into thinking that the animal is living under conditions approaching the natural, but only provides a stage setting for the monkeys. It cannot be denied that this type of artificial environment provides a complex climbing frame and increases usable space in the cage. However, against this must be set the extremely high financial cost, which might be spent more effectively to improve the welfare of the animals in other ways.

19.7 PROVIDING VARIABILITY IN THE PHYSICAL ENVIRONMENT

Two principal forms of environmental change can be incorporated into the captive situation. Firstly, unpredictable environmental perturbations, which require an adaptive response from the animal, and, secondly, a facility for the animal itself to generate changes.

To provide changes in the physical environment, which are of value to the animal, several options are open. Firstly, foraging time can be increased by providing some of the animal's food in such a way as to make its delivery or discovery unpredictable. Food can be concealed in the substrate or scattered about the enclosure. Chamove *et al.* (1982) found that increased opportunities for foraging provided by food items concealed in the substrate increases the animal's range of behaviour, while reducing aggression within the group. In addition, where the floor is covered with wood chips and food is concealed, the usable area of the cage increases. The animals spend more time on the ground and this applies even to the more arboreal species.

For the small, more insectivorous primates, cricket or mealworm dispensers can be provided. These dispensers need not be expensive or elaborate; the cricket dispenser used at London Zoo is simply a hollow log with holes in it, from which the crickets emerge spontaneously. The mealworm dispenser (Shepherdson, 1989) is a corked plastic tube containing oats and mealworms; the tube has holes in it through which every now and then the mealworms will fall (Figure 19.2).

Another example is described by Scott (Chapter 14, section 12.5). Food dispensers are devices which can give the animal increased control over its environment. Marmosets can be provided with artificial gum trees (McGrew *et al.*, 1986), and artificial 'termite mounds', from which sticky food must be fished with a stick, have been effective not only for chimpanzees, which have been observed to show similar behaviour in the wild, but also for orang-utans (McEwen, 1986). The results of providing an artificial termite mound or hollow log for an orang-utan are shown in Figure 19.3; it can be seen that the animal spent more time in probing, manipulating, sitting and eating than in other activities.

Electronic games and devices which dispense food have been used

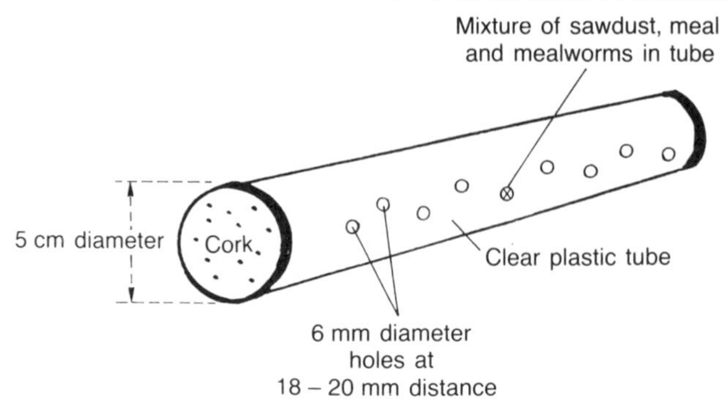

Figure 19.2 A simple mealworm dispenser. The mealworms fall through the holes in the tube at unpredictable intervals. (After Shepherdson, 1989.)

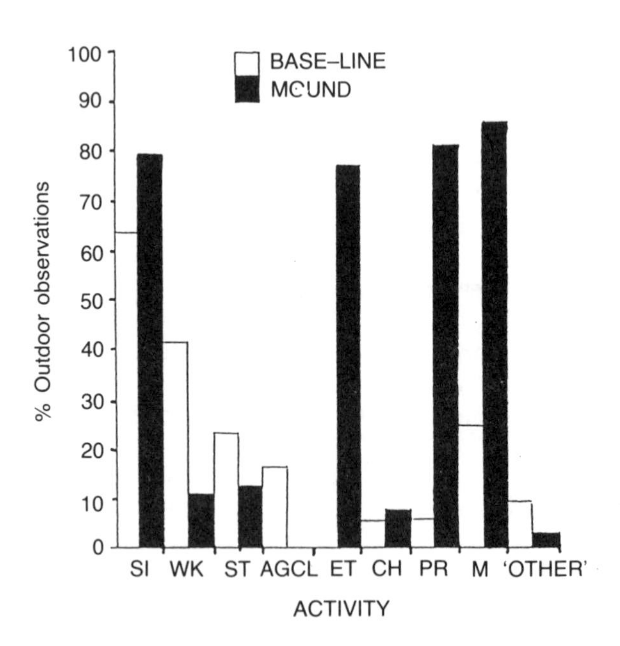

Figure 19.3 The results of providing an orang-utan at London Zoo with a 'termite mound' in the form of a hollow log provisioned with baby food flavoured with raspberry jam or chocolate powder. Significant changes were recorded in sitting (SI, $P < 0.05$), walking (WK, $P < 0.001$), eating (ET, $P < 0.001$), probing (PR, $P < 0.001$) and manipulating (M, $P < 0.001$). Other categories were standing (ST), above ground/climbing (AGCL), chewing (CH) and 'other'. (After McEwen, 1986.)

in zoos and laboratories (Markowitz, 1982; Markowitz and Spinelli, 1986). These may be of value for socially deprived or closely confined animals, but they are costly and may prove addictive if their use is not restricted. Their value in zoos has been questioned (for a review, see Forthman-Quick, 1984; see also Chapter 12, section 12.1) because they may restrict, rather than enrich, the behavioural repertoire, give a false impression to the public, and the activity which they induce is unnatural. In some circumstances, however, they can lead to an increase in well-being for the animal, but they are not generally available commercially and are expensive to produce on a limited scale.

The provision of destructible materials can be used to provide environmental change. Natural materials, such as logs and branches, are often provided and can occupy the animal for long periods of time. However, artefacts such as cardboard boxes, telephone directories and magazines have also proved effective (Rupniak and Iverson, 1989). In assessing the extra labour involved in cage cleaning, and the aesthetic aspects of the mess produced by the destruction of such artefacts, the gain in well-being of the primates has to be considered.

Finally, the use of mobile artefacts can provide environmental enrichment. A simple swing is always valuable, and most primates will incorporate it into play sessions. It encourages jumping and balancing and, when more than one individual uses it, the position of the swing in space becomes difficult to predict (personal observations). Barrels, balls, baskets, simple puzzles and other toys have their place, although their novelty may wear off so that they need to be changed regularly. For some species, particularly macaques, the provision of a paddling pool can be a very effective way of improving the environment (K. Ryz, unpublished report).

Fragaszy and Adams-Curtis (Chapter 13) show that capuchin monkeys can be readily trained to manipulate objects with varying properties of edibility, portability, manipulability and novelty. While their aim was to increase our understanding of the behaviour of capuchins, this approach is also valuable for enrichment of the environment. An important finding of their work is that, for the animal, the nature of environmental change and the individual's ability to exercise control over that change are more significant than the degree of novelty. It is also clear that the monkeys appreciate a challenge, and may be capable of quite complex manipulative tasks. Objects presented to animals can promote the performance of a wide array of behaviour.

An inhibitory factor in providing a complex variable environment for primates and other animals has been the veterinary concern for health and hygiene. Ease of cleaning cages has often led to sterile metal cages whose size is usually determined, not by the requirements of the animal, but by whether the cage will fit into a cage-washing machine. In zoos, tiled concreted cages have frequently been provided, and many still exist. Fortunately, the situation is now changing, and it is realized in most zoos, and in an increasing number of laboratories, that a 'soft' environment may not be any less hygienic. Further, that an animal whose behavioural needs are catered for is much less likely to be a sick animal. There is need for more research in this area, but Chamove *et al.* (1982) found that a woodchip substrate, for example, was actually bactericidal and that the provision of natural materials for primates does not necessarily lead to a reduction in standards of hygiene. An additional factor that has to be considered is that some primates, such as prosimians and marmosets and tamarins, scent-mark their cage so that cleaning may remove their chemical signals and create an unfamiliar and, therefore, more stressful environment.

19.8 THE SOCIAL ENVIRONMENT

Structure of the social group

Most of the nonhuman primates kept in captivity are highly sociable, but there are great variations in their social structure, both between and within species. As was mentioned earlier, social relationships between animals in nature provide a source of environmental complexity and a framework for the individual's behaviour. This social organization will change with the death of individuals and the birth or immigration of new animals into the group. Group members are both competitive and cooperative; they form alliances and initiate changes which result in realignments in the structure of relationships. Hausfater (1975), for example, found that the dominance hierarchy in wild yellow baboons (*Papio cynocephalus*) changed, on average, for males every 13 days and for females every 57 days. It is apparent that primates are adapted to living in structured societies which are subject to unpredictable but frequent modifications.

Furthermore, in the wild, the individual is often a member of a genetically related group, and its behaviour towards other individuals is dependent upon perceived degrees of relatedness to other members. By contrast, individuals in captivity are often unrelated, and a female biased gender ratio is frequently maintained, to avoid fighting between males (Erwin, 1986). For example, in most laboratories breeding groups of macaques constitute one male and several females. This harem structure does not resemble that of a wild group, which is normally made up of numerous adult males and females. Likewise the

Figure 19.4 Orang-utans housed socially at Singapore Zoological Gardens. (Photo: Helen M. Poole.)

females, which would be closely related in a wild group, may, in captivity, be derived from different social groups or even from geographically remote areas. However, they may adjust to this situation satisfactorily.

Given the discrepancies between the natural and the captive state, the question has to be asked as to the criteria which must be considered to provide a satisfactory social environment in captivity. The obvious answer at first sight appears to be to simulate as exactly as possible the natural social grouping. The problems likely to arise, however, are not merely the probability that individuals will be unrelated, but also the effects of confinement which make it difficult or impossible for individuals to avoid aggressors or to emigrate to another social group. However, a most impressive example of a captive group simulating that of the wild state is that of the chimpanzees of Arnhem Zoo, whose social interrelations were so aptly described by de Waal (1982).

Difficulties in simulating features of a natural social organization do not preclude the formation of well-integrated social units in captivity. However, this may differ widely from the natural state. For example, it is well known that orang-utans adapt well to social life in captivity (Figure 19.4) and show play and affiliative behaviour whereas, under natural conditions, they are largely solitary (Edwards, 1982; Poole, 1987).

In creating a captive social group, the main criterion should be that the animals appear to be contented and well adjusted. One useful indicator of a good social situation is the incidence of social play, because this is behaviour which only occurs in a relaxed situation (Fagen, 1981). Moreover, a conspecific represents a force for change in the environment and enriches the experience of its fellows; it provides a very complex set of stimuli, far greater than any inanimate artefact (see also Chapter 15, section 15.1). In addition, the presence of a conspecific enables an individual to utilize and develop its repertoire of social behaviour. Thus, the most important message to the keeper of primates is to ensure that they are kept with companions. Harris (1988) found that captive female cynomolgus monkeys spent 55–60% of their time in social behaviour. Of the social behaviour observed, only 0.001% was agonistic, and that did not involve physical assault.

Species do differ considerably, however, in the extents to which they are likely to show aggressive behaviour under confined conditions. For example, French and Inglett (Chapter 15) have shown that hostility to competitors may differ considerably, even between species within the same family of primates (Callitrichidae). For example, the breeding female cotton-top tamarin (*Saguinus oedipus*) physiologically suppresses reproductive cycling in other females within or entering the social group. In this species, there is little need for female–female aggression, so they show little hostility to one another. In contrast, the breeding female in the related golden lion tamarin, who does not suppress reproductive cycling in other females, is highly aggressive to female strangers. Mendoza (see Chapter 17) has also observed marked degrees of difference in both physiological and behavioural responsiveness to environmental change between titi (*Callicebus moloch*) and squirrel monkeys (*Saimiri sciureus*). These differences correlate with the relative ease of adaptation to the captive environment; titi monkeys are known to be difficult to keep in captivity, while squirrel monkeys adapt well, even to laboratory housing. In keeping primates in captivity, it is therefore important to take into account the endocrinological responsiveness and associated temperament of the species. This can differ within the same genus, as Clarke *et al.*, (1988a) have shown for three species of macaque (*Macaca mulatta*, *M. radiata* and *M. fascicularis*). They have suggested that these represent distinct methods of coping with environmental change (see also Chapter 3, section 3.3).

While it is important that facilities should be provided to ensure that animals can avoid one another and be out of visual contact, in situations where group behaviour is controlled by a dominant individual, the ability to be out of sight of the controlling monkey may lead to more, not less, aggression. This was demonstrated by Erwin (1986), who kept a pig-tailed macaque (*Macaca nemestrina*) harem in an indoor room. When he doubled the space, by providing a second interconnecting room, aggression between females increased because they were able to fight unhindered by the dominant male, when he was out of sight. Hence, the general lesson in creating compatible groups of unrelated individuals is that the individuals' reactions to one another should always be carefully monitored before placing them in physical contact (Reinhardt *et al.*, 1988).

In some cases, it may be difficult to provide the animal with a sufficiently rich social environment and, in this case, relations with human keepers will be an important aspect of environmental enrichment. In many instances, it is possible to train primates to collaborate and to carry out particular tasks. This not only provides variation and gives the animal some control over its environment, but it also may be of great value if the animal has to be given medical treatment, weighed or moved from one place to another (Heath, 1989). Trained monkeys are of particular value in long-term studies where monkey and experimenter may spend many years working together (Jaeckel, 1989).

19.9 REPRODUCTION IN CAPTIVITY

One of the greatest problems for the keepers of captive monkeys and apes is that of providing a social environment which is suitable for the rearing of offspring. Primiparous females that have no experience of observing other females caring for their infants, or that have not had the opportunity of helping to rear other females' offspring, may be at a loss when faced with an infant of their own, and kill it through unfamiliarity or neglect (see, for example, Gardin *et al.*, 1989). Sometimes human caretakers panic and label a female a poor mother when this occurs, but it has to be remembered that, even in the wild situation, primiparous females may have problems with their first infant. Even in captivity an inexperienced female may gain some experience from its initial failure and subsequently prove to be a good mother.

In some laboratory colonies, several factors may mitigate against the production of normal offspring. It is common practice in the laboratory for mother and infant to be kept in a separate cage from other monkeys so that the young has no experience of living in a group. In many breeding colonies of macaques, the young are separated from their mothers at 3–6 months, to encourage the mother to return to reproductive cycling (for example, see Hobbs *et al.*, 1987). Such young, reared without a proper social background, appear to be less adaptable in the sense that they may have problems in coping with a restricted environment as adults. Goosen (1989) found higher levels of abnormal stereotyped behaviour in singly housed rhesus monkeys, which had been weaned early, as compared with individuals which had been reared in the wild, or captive animals which had remained in a social group with their mother. Harris (1988) obtained similar results for cynomolgus monkeys. There is need for more data on the influence of early social environment on adult behaviour for a variety of species.

From the evidence available, it does seem clear that in order to rear normal infant primates successfully in captivity, they should be allowed to live in a suitable social group of mixed ages and genders so that the individual can develop its social skills. An isolated mother and infant is rarely observed in nature in primates, other than in the orang-utan (MacKinnon, 1974; Horr, 1975; Rijksen, 1978). The young of many primates reared in isolation, with peers, or even alone with their mother, may show deficiencies in both their social behaviour and in the flexibility and adaptability of their responses to environmental change.

Separation of the infant from its mother has long been used as a model for human depression, so that the behavioural implications are well known (Reite *et al.*, 1981), but recent evidence shows that abnormalities may extend to the endocrine and even to the immune systems. Reite (1987) found that when infant pig-tailed macaques were separated from their mothers at 6 months old for a period of 14 days, they differed from unseparated controls in showing suppression of lymphocyte proliferation in response to three mitogens when they were adults (see Chapter 18).

Premature separation of the young from its mother clearly affects its well-being and therefore represents a welfare problem where deprived or abnormal animals result. It is also in the interests of both zoos and laboratories that their captive primates should be as normal as possible. The validity of most experimental procedures in laboratories depends on the assumption that the subjects of research are normal in their physiological and immune systems. Clearly, science based on abnormal animals can only be poor science (unless abnormality is the subject of study); thus in most cases, good animal welfare and good science are inseparable (see also Chapter 3).

Where a species is being bred for conservation, with the ultimate aim of introducing it to the wild, even though every effort may be made to provide as normal a rearing environment as possible for the individual, there may be serious problems for higher primates because of a lack of traditional knowledge in the group. Survival of the social group in species such as gorillas and chimpanzees may depend on the knowledge and experience of older adults who determine the group's responses to adverse environmental changes. The work of Fossey (1983) on gorillas and that of Moss (1988) on elephants have shown how large-brained, long-lived animals build up a lifetime of knowledge which enables them to respond adaptively to environmental changes and crises (see Chapter 2, section 2.2 and Chapter 3, section 3.2). Without a matriarch or patriarch, a group may be unable to survive. Adaptive responses of this kind may depend not only on the 30–40 years' experience of older animals but may have been handed down for many generations from adult to offspring.

If captive-bred primates are to be released into a wild habitat, and they are to survive and breed, it is important that they should not simply be abandoned, but that the adaptiveness of their behaviour should be monitored. In the event of their being unable to cope with environmental change, it may even be necessary to take practical action to assist them. Redshaw and Mallinson (see Chapter 12) recognize a need for training, even for shorter-lived monkeys such as the golden lion tamarin, and the environment at the Jersey Wildlife Preservation Trust takes this need into account.

19.10 CONCLUDING REMARKS

The captive environment should incorporate sufficient space and environmental complexity to allow primates to show a wide repertoire of behaviour which is appropriate to the species. An enclosure with natural vegetation and topographical features is ideal. The greater the departure from the wild environment the greater the need to install forms of environmental enrichment. If an outdoor enclosure can be provided, natural climatic variation will also help to enrich the lives of the animals.

As a large proportion of the behavioural repertoire of nonhuman primates relates to social behaviour, providing compatible companions greatly extends the range of activities possible to the individual. Primates should, unless there are compelling reasons for not doing so, be housed socially.

Where single caging is unavoidable, the primate's environment can be improved by behavioural therapy in the form of manipulanda, training or learning to use devices for delivering rewards. Such forms of environmental enrichment encourage a varied daily time budget, exercise, both physical and mental, and the development of motor skills.

Assessing environmental quality can best be achieved by monitoring behaviour to identify indicators of poor welfare, such as restricted behavioural repertoire, low activity levels, or abnormal behaviours. Additional, noninvasive assessments of physiological parameters, such as urinary corticosteroids and heart rates, may also prove valuable. The animal's behaviour can be compared with its natural repertoire and, if they are available, with field studies to obtain comparative data on natural time budgets.

If the behaviour of the animal is found to indicate poor welfare, then improvements to the environment should be made along the lines suggested. The individual's behaviour should be reassessed using the earlier behavioural data as a control to ensure that improvements are not ephemeral. Samples of behaviour should be taken at intervals over an extended period of time.

Finally, while nonhuman primates should be provided with a stable home environment, there should be sufficient variability in the form of temporally or spatially unpredictable events to provide adequate levels of stimulation. It is also very important that the animal should be able to exert some control over its environment.

Postscript

HILARY O. BOX

In putting this book together we aimed to address a wide range of theoretical and practical issues from the perspective of responsiveness to environmental change. This general framework has stimulated the authors to put new perspectives on familiar problems, and it has created new lines of enquiry. Many of the problem areas with which we have been concerned refer to environmental changes brought about by human activity because these have critical importance in a contemporary context. Various studies described in this book show how differences in responsiveness among species to such environmental changes have implications for their fitness and chances of survival.

It is also the case that, despite the relatively stable tropical habitats in which most simian primates live, there are many natural sources of social and physical variation. A variety of recent studies has also stimulated new areas of enquiry. Hence, in addition to traditional demographic and social behavioural approaches, questions about the influences of ecological parameters are extended in various ways. For example, although the effects of regular change (such as seasonal variations within particular habitats) have been studied in some detail for many years, there is interest in changes which occur less frequently.

Consideration is given to assessments of the rarity of events, and the use that people make of them in interpreting their observations (Weatherhead, 1986; see also Chapter 2). The influences of conditions of severe drought are good examples. Cases include those of Hamilton III (1986) on the shift of dietary items by chacma baboons (*Papio ursinus*) during a period of such change in the Namib Desert in south-west Africa, and that of Berenstain (1986) during a period of severe food and water shortage on cynomolgus monkeys (*Macaca fascicularis*) in Borneo. The ability of these monkeys to switch their sources of food was also a major contributory factor in their survival, although it could be that some monkeys at least survived due to the superabundance of insect larvae at that time (see Chivers, Chapter 1 for a detailed discussion).

Another interesting case is given by Sapolsky's (1986) long-term studies of behavioural endocrinology among troops of olive baboons (*Papio anubis*) in East Africa. By comparison with basal concentrations of testosterone for a number of years previously, concentrations during the severe drought of 1984 showed a very marked significant decline in males of all social status. Evidence from the influence of various other environmental stressors upon steroid output in the animals suggested that the drought presented a considerable stress. However, individuals were able to maintain their body weight by increasing their foraging time.

Studies such as this latter one also bear reference to the value of field techniques for physiological measurement, and these extend still further the range of questions that we can ask about responsiveness to environmental change. The collection of urine and faecal samples, from which the reproductive condition of individuals may be monitored, for instance, is an important methodological advance (see Chapter 4).

In more general terms, a key methodological development is the extent to which we can present specific environmental events under controlled experimental conditions. In nature, examples include playback experiments of vocalizations (Cheney and Seyfarth, 1980) and the provision of novel foods, as with comparative studies of the social transmission of new information (Cambefort, 1981). In many cases, such experiments have the advantage that responsiveness may be studied from the beginning, over particular periods of time in defined social and physical contexts.

Although there is much more methodological sophistication about recent field research, it is also the case that captive conditions offer the environmental controls necessary to answer particular kinds of questions. Early studies by Alison Jolly (1964) on the responsiveness of prosimians to various inanimate problem-solving tasks, and by Glickman and Sroges (1966) using a variety of vertebrate species with simple objects, pointed the way to delineating and interpreting differences in responsiveness to unfamiliar objects in terms of the different behavioural ecology of the species.

The Glickman and Sroges study, for example, showed, first, that the patterns of responsiveness of primates were more varied than those of members of all the other taxa. Second, that differences among species in terms of curiosity about the physical environment were positively associated with such factors as natural freedom from predation and strategies of feeding. The studies added new perspectives in an area in which numerous experiments had been concerned with theoretical concepts of curiosity (Wünschman, 1963), exploratory behaviour (e.g. Fowler, 1965), and the characteristics of the central nervous system implicated in behavioural responsiveness (e.g. Sokolov, 1963), but in which there had been relatively few on functional interpretations of behavioural responses to change in terms of natural strategies of life.

In this context there is now more interest in what we often describe as 'ecological validity'. Hence, various contributions in this book emphasize the use of tasks and situations that occur naturally in the lives of the species concerned, as with manipulative tasks for capuchin monkeys (see Chapter 13), and the influence of changes in the social status of adult female marmosets and tamarins upon their reproductive condition (see Chapters 4 and 16).

Simulated foraging tasks have also proved to be very useful. For example, Caine and Marra (1988) presented groups of red-chested tamarins (*Saguinus labiatus*) and squirrel monkeys (*Saimiri sciureus*) with a variety of food items covered in wood chips in troughs attached to their home cages. The aim was to consider the potential influence of social organizational differences between the two species in their relative vigilance to the social and nonsocial environment during the task. As predicted, when the tamarins looked up from foraging they were more attentive to the physical environment than to their conspecifics. By contrast, the squirrel monkeys paid much more attention to other squirrel monkeys. Caine and Marra suggested that the naturally cooperative and relatively placid patterns of social interactions within established social units of tamarin monkeys contrast sharply with the social hierarchy and tension common within groups of squirrel monkeys and, at least potentially, allows for greater vigilance of the physical environment.

In principle, environmental challenges offer all kinds of experimental possibilities, but it is as well to preserve a healthy cautiousness about the advantages of studying environmental change under what may appear to be very stringent conditions.

One important consideration, for example, is that most primate colonies are small. Compared with the numbers of captive animals from many other groups that are available for study, it is rare for there to be large numbers of simians with which to work. This fact is often compounded by the potential variability in behaviour within a species in response to different captive conditions (Box, 1982). Simian primates are, after all, renowned for their behavioural flexibility. These difficulties, however obvious upon reflection, are all too often ignored.

How serious these problems become depends, of course, upon the kinds of question that one wants to ask. The robustness of experimental results is obviously much improved by carrying out identical studies in different colonies. For this reason, we really need to sample responsiveness among different captive populations, as we do for behavioural indices of different populations of species in nature (Kummer and Kurt, 1965).

The following remarks also relate to captive studies but in no way detract from the studies with captive animals which are reported in this volume. Those studies were chosen not only because they address particular kinds of issues but for the robustness of the techniques which were used, given the following considerations. For example, squirrel monkeys and titi monkeys are mentioned on numerous occasions in the book, and with good reason. There is much detailed comparative data from captive studies on various aspects of their physiological and behavioural responsiveness to change. There are also field studies on their natural patterns of behaviour, including feeding ecology and social organization. Difficulties may arise, however, where we know little about species in nature.

For example, the longstanding paucity of field data on species of marmoset and tamarin monkeys (Callitrichidae), until recently persistently misled observers into extrapolating that a monogamous mating system was the natural condition. Thus, an important aspect of callitrichid husbandry, as, for instance, of the familiar common marmoset (*Callithrix jacchus*), has involved respecting the observed social and reproductive exclusivity of an adult breeding pair, and maintaining 'monogamous' pairs and their offspring in at least physical and visual isolation from other family groups. However, we now know that the mating system of callitrichids in nature may vary among and within species (Goldizen and Terborgh, 1986), a finding which is in line with evidence for a wide variety of primate taxa (Dunbar, 1988). Thus, we should ask whether the social, reproductive exclusivity seen in captive callitrichids is a response to the ecological conditions of captivity, in which the quality and quantity of various resources are limited compared with the natural habitat. In such cases, we do not know how much a behavioural tendency in captivity is a *reflection* of a natural propensity, and how much it is a *response* to the ecological constraints of captivity as such.

Another interesting example refers to the question of gender differences in responsiveness which, as it happens, represents a considerable gap in our knowledge in this field. There is a variety of observations in nature (aside from reproductive activities) which deserve much closer attention. To give one example, Hannah and

McGrew (Chapter 9) found that female chimpanzees were more adaptive in some senses than the males during rehabilitation from captivity to an island habitat. It was the females, for instance, that initiated palm-nut cracking. In other similar observations with palm-nut foods, Hannah and McGrew (1987) found that male chimpanzees were much more easily distractible than females. However, until more work is specifically addressed to this area, there is little to say about the nature, extent and function of such gender differences.

Many of the points of uncertainty revolve around the question of whether we are dealing with differences that delineate males from females, in ways which have functional significance. Differences between the genders may be absolute in the sense that a characteristic occurs in one of them only or, as is so, for example, with many behavioural characteristics, there are qualitative differences in the frequency with which they occur.

One thing to appreciate at this point is that emphasized in an excellent discussion by Goldfoot and Neff (1985). Their concern is that many workers, with few data from limited social environmental conditions in captivity, assume that differences in behaviour between males and females are due to biological differences. The point is that both environmental conditions, as well as the past experiences of individuals, mediate gender differences. Moreover, a biological propensity may require particular social conditions to develop and/or be observable. In a very familiar example, adult male rhesus monkeys do not usually show paternal behaviour, either in nature or under conditions of captivity with adult females in the group, but they may do so in captivity when no females are present. In fact, they may show persistent and adequate care of infants (Redican and Mitchell, 1973).

There are also cases in which differences in behaviour between males and females may be wrongly attributed to a source of environmental change in captivity. We may take a deceptively simple case from our own studies as an example. Hence, various people have suggested that adult male common marmosets are significantly more responsive by scent-marking than their females when they are paired with an unfamiliar partner; that this response is one consequence of the social change (Ingram, 1975; Evans, 1981).

Olfactory systems are highly developed among the Callitrichidae (Dawson, 1978; see also Chapter 15), and one might expect that scent-marking would be important in behavioural profiles of responsiveness. On the other hand, A.P. Bates (unpublished data) found that the males of established pairs of marmosets marked significantly more than their females for several days when the pairs were introduced into a new living area. Hence, as Woodcock (1982) pointed out, differences in marking between males and females may not necessarily be a consequence of a change in social situation but to one in the physical environment. The establishment of new pairs of animals often takes place in an unfamiliar cage.

A central area of growing interest in studies of environmental change is the influence of other animals of the same species. Hence, whereas earlier studies of exploratory behaviour, for example, tended to use individual animals as a data base, there is now more interest in the social context of responsiveness. This is realistic because the natural context of individual responsiveness is a social group (Hall, 1963). Most simian primates naturally spend a good deal of their lives in complex social interactions. Moreover, it has been experimentally demonstrated, for instance, that the responses made by primates to changes in their physical and social environments reflect different social organizational predispositions (Fragaszy and Mason, 1978), as well as quantitatively different relationships among particular individuals (Ainsworth *et al.*, 1971; Stevenson-Hinde *et al.*, 1980; see Chapters 3 and 17).

An interesting development to studies of the social context of responsiveness is to delineate the environmental parameters which provide destructive as well as constructive change. There is a growing concern, for example, with the fact that in some western countries many adult humans, as with many captive nonhuman primates, spend a great deal of their time alone, whereas it is known that social contact is necessary in the regulation of a variety of physiological systems (Gruter and Masters, 1986). We also have good reason to believe that the 'highly developed' cognitive abilities of simian species are used predominantly in social contexts which are potentially complex in extent and form compared with other mammals. Hence, the social conditions of captive primates should especially concern us, particularly when they affect health and fitness. In fact, mental and physical health *per se* in the context of changing social and physical environments is a major scientific challenge of our time. For example, the stress response (see Chapter 18 for a general discussion) and its influence upon the immune system is an important line of developing enquiry. Moberg (1985), for instance, has advocated that in the long run the competence of an animal's immune system, together with its reproduction, growth and development, are better indices of the influence of environmental stressors than individual physiology or behaviour. Prepathological states are especially useful in this respect. Moreover, from many points of view, studies of the interface between physiological and behavioural parameters of responsiveness extend our knowledge of responsiveness to environmental change. Differences in 'temperament' and 'response style' are especially useful (see Chapters 3 and 17) in guiding our appreciation of the organization of behaviour and the varying influences which similar environmental events may have among different species.

In more practical terms, we need to make careful behavioural, physiological and veterinary checks on captive populations, and especially on endangered species. As habitats are increasingly destroyed we shall need to keep more species in captivity, where questions about responsiveness to change will be of paramount importance to maintain viable populations for conservation, research and education, and to prepare at least some individuals for reintroduction to natural habitats. Natural habitats, however, also present us with many uncharted influences of environmental change, which stem primarily from rapid and extensive changes in human land use. In the last analysis, all aspects of the biology of individuals of varying age, gender and developmental history as members of different species influence their responsiveness to environmental events in nature and in captivity. The major task for the future is to understand the interrelationships among these various characteristics and their potential flexibility.

References

Abbott, D.H. (1978) The physical, hormonal and behavioral development of the common marmoset, *Callithrix jacchus jacchus*. In: *Biology and Behaviour of Marmosets* (H. Rothe, H.-J. Wolters and J.P. Hearn, eds). *Proceedings of the Marmoset Workshop* Eigenverlag-H. Rothe, Göttingen, 99–106.

Abbott, D.H. (1979) The sexual development of the common marmoset monkey, *Callithrix jacchus jacchus*. Unpublished PhD thesis, University of Edinburgh.

Abbott, D.H. (1984) Behavioural and physiological suppression of fertility in subordinate marmoset monkeys. *Amer. J. Primatol.* **6**, 169–86.

Abbott, D.H. (1987) Behaviourally mediated suppression of reproduction in female primates. *J. Zool. Lond.* **213**, 455–70.

Abbott, D.H. (1988) Natural suppression of fertility. In: *Reproduction and Disease in Captive and Wild Animals* (J.P. Hearn and G.R. Smith, eds). Symposia of the Zoological Society of London, Vol. 60. Oxford University Press, Oxford, 7–28.

Abbott, D.H. (1989) Social suppression of reproduction in primates. In: *Comparative Socioecology: The Behavioural Ecology of Humans and Other Mammals* (V. Standen and R.A. Foley, eds). Blackwell Scientific, Oxford, 285–304.

Abbott, D.H. and J.P. Hearn (1978) Physical, hormonal and behavioural aspects of sexual development in the marmoset monkey (*Callithrix jacchus jacchus*). *J. Reprod. Fertil.* **53**, 155–66.

Abbott, D.H., A.S. McNeilly, S.F. Lunn, M.J Hulme and E.J. Burden (1981) Inhibition of ovarian function in subordinate female marmoset monkeys (*Callithrix jacchus jacchus*). *J. Reprod. Fertil.* **63**, 335–45.

Abbott, D.H., E.B. Keverne, G.F. Moore and U. Yodyinguad (1986) Social suppression of reproduction in subordinate talapoin monkeys (*Miopithecus talapoin*). In: *Primate Ontogeny, Cognition and Social Behaviour* (J.G. Else and P.C. Lee, eds). Proceedings from the 10th Congress of the International Primatological Society, Vol. 3. Cambridge University Press, Cambridge, 329–41.

Abbott, D.H., J.K. Hodges and L.M. George (1988) Social status controls LH secretion and ovulation in female marmoset monkeys (*Callithrix jacchus*). *J. Endocrinol.* **117**, 329–39.

Abbott, D.H., O.L. Wilson, L.M. George, S.E. Aldridge and E. Wilkinson (1989a) Ultrasound examination of ovarian activity in dominant and subordinate female marmoset monkeys. *J. Reprod. Fertil. Abstr. Ser.* No. 3, 34.

Abbott, D.H., K.T. O'Byrne, J.W. Sheffield, S.F. Lunn and L.M. George (1989b) Neuroendocrine suppression of LH secretion in subordinate female marmoset monkeys (*Callithrix jacchus*). In: *Comparative Reproduction in Mammals and Man. Proceedings of the NCRR Conference*, Nairobi, November 1987 (R.M. Eley, ed.). National Museums of Kenya, Nairobi, 63–7.

Abbott, D.H., J. Barrett, C.G. Faulkes and L.M. George (1989c) Social contraception in naked mole-rats and marmoset monkeys. *J. Zool. Lond.* **219**, 703–10.

Adams, D.B., A.R. Gold and A.D. Burt (1978) Rise in female-initiated sexual activity at ovulation and its suppression by oral contraceptives. *N. Engl. J. Med.* **229**, 1145–9.

Adams, M.R., J.R. Kaplan and D.R. Koritnik (1985) Psychosocial influences on ovarian endocrine and ovulatory function in *Macaca fascicularis*. *Physiol. Behav.* **35**, 935–40.

Ainsworth, M.D. and S.M. Bell (1970) Attachment, exploration and separation: illustrated by the behaviour of one year olds in a strange situation. *Child Develop.* **41**, 49–67.

Ainsworth, M.D.S., S.M. Bell and D.J. Stayton (1971) Individual differences in strange situation behaviour of one-year-olds. In: *The Origins of Human Social Relations* (H.R. Schaffer, ed.). Academic Press, London, 17–52.

Albrecht, H. and S.C. Dunnet (1971) *Chimpanzees in Western Africa*. Piper-Verlag, Munich.

Allchurch, A.F. (1986) *The Nutritional Handbook of the Jersey Wildlife Preservation Trust*. JWPT, Jersey.

Allport, G.W. (1937) *Personality: A Psychological Interpretation*. Henry Holt, New York.

Almeida, O.F.X., K.E. Nikolarakis and A. Herz (1988) Evidence for the involvement of endogenous opioid peptides in the inhibition of luteinizing hormone by corticotrophin-releasing factor. *Endocrinology* **122**, 1034–41.

Altmann, J. (1979) Age cohorts as paternal sibship. *Behav. Ecol. Sociobiol.* **6**, 161–9.

Altmann, J. (1980) *Baboon Mothers and Infants*. Harvard University Press, Cambridge, MA.

Altmann, J. (1983) Costs of reproduction in baboons. In: *Behavioral Energetics: Vertebrate Costs of Survival* (W.P. Aspey and S.I. Lustick, eds). Ohio State University Press, Columbus OH, 67–88.

Altmann, J. and S. Alberts (1987) Body mass and growth rate in a wild primate population. *Oecologia* **72**, 15–20.

Altmann, J. and P. Muruthi (1988) Differences in daily life between semi-provisioned and wild-feeding baboons. *Amer. J. Primatol.*

Altmann, J., S.A. Altmann, G. Hausfater and S.A. McCuskey (1977) Life history of yellow baboons: physical development, reproductive parameters and infant mortality. *Primates* **18**, 315–30.

Altmann, J., G. Hausfater and S.A. Altmann (1985) Demography of Amboseli baboons 1963–1985. *Amer. J. Primatol.* **8**, 113–25.

Altmann, J., S.A. Altmann and G. Hausfater (1988) Determinants of reproductive success in savannah-baboons (*Papio cynocephalus*). In: *Reproductive Success: Studies of Individual Variation in Contrasting Breeding Systems* (T.H. Clutton-Brock, ed.). University of Chicago Press, Chicago, 403–18.

Altmann, S. (1974) Baboons, space, time and energy. *Amer. Zool.* **14**, 221–48.

Altmann, S.A. and J. Altmann (1970) *Baboon Ecology*. University of Chicago Press, Chicago.

Altmann, S.A. and J. Altmann (1979) Demographic constraints on behaviour and social organization. In: *Primate Ecology and Human Origins* (I.S. Bernstein and E.O. Smith, eds). Garland, New York, 47–64.

Amerasinghe, F.P., B.W.B. Van Cuylenberg and C.M. Hladik (1971) Comparative histology of the alimentary tract of Ceylon primates in correlation with diet. *Ceylon J. Sci. Biol.Sci.* **9**, 75–87.

Andrews, M. (1986) Contrasting approaches to spatially distributed resources by *Saimiri* and *Callicebus*. In: *Primate Ontogeny, Cognition and Social Behaviour* (J.G. Else and P. Lee, eds). Cambridge University Press, Cambridge, 79–86.

Andrews, P. (1981) Species diversity and diet in monkeys and apes during the Miocene. In: *Aspects of Human Evolution* (C.B. Stringer, ed.). Taylor and Francis, London, 25–61.

Andrews, P. and L. Aiello (1984) An evolutionary model for feeding and positional behaviour. In: *Food Acquisition and Processing in Primates* (D.J. Chivers, B.A. Wood and A. Bilsborough, eds). Plenum Press, New York, 429–66.

Andrews, P. and C.P. Groves (1976) Gibbons and brachiation. In: *Gibbon and Siamang* (D.M. Rumbaugh, ed.), Vol. 4. Karger, Basel, 167–218.

Anzenberger, G.A., S.P. Mendoza and W.A. Mason (1986) Comparative studies of social behavior in *Callicebus* and *Saimiri*: behavioral and physiological responses of established pairs to unfamiliar pairs. *Amer. J. Primatol.* **11**, 37–51.

Archer, J. (1975) Rodent sex differences in emotional and related behaviour. *Behav. Biol.* **14**, 451–79.

Archer, J. and L. Birke (1983) *Exploration in Animals and Humans.* Van Nostrand Reinhold, Wokingham, Berkshire.

Armitage, K.B. (1986) Individuality, social behaviour and reproductive success in yellow-bellied marmots. *Ecology* **65**, 1186–93.

Ashmore-Declue, P. (1988) The socio-ecological complex of the genus *Macaca. Amer. J. Primatol.* **14**, 408.

Ashton, E.H. and C.E Oxnard (1963) The musculature of the primate shoulder. *Trans. Zool. Soc. Lond.* **29**, 553–650.

Ashton, E.H., M.J.R. Healy, C.E. Oxnard and J.P. Spence (1965) The combination of locomotor features of the primate shoulder girdle by canonical analysis. *J. Zool. Lond.* **147**, 406–29.

Aveling R.J. and C. Aveling (1987) Report from the Zaire Gorilla Conservation Project. *Primate Conserv.* **8**, 162–4.

Ayres, J.M. (1981) *Observacoes sobre a Ecologia e o Comportamento dos Cuxius* (Chiropotes albinasus *e* Chiropotes satanas, *Cebidae: Primates*). CNPq, INPA, FUE Manous, Brazil.

Ayres, J.M.C. (1986) Uakaris and Amazonian flooded forest. Unpublished PhD thesis, University of Cambridge.

Bailey, R.C., R.S. Baker, D.S. Brown, F. von Hildebraud, R.A. Mittermeier, L.E. Sponsel and K.E. Wolf (1974) Progress of a breeding project for non-human primates in Colombia. *Nature, Lond.* **248**, 453–5.

Baker, A.J. (1987) Emigration in wild groups of golden lion tamarins (*Leontopithecus rosalia*). *Amer. J. Primatol.* **8**, 500.

Baldwin, J.D. (1985) The behavior of squirrel monkeys (*Saimiri*) in natural environments. In: *Handbook of Squirrel Monkey Research* (L.A. Rosenblum and C.L. Coe, eds). Plenum Press, New York, 35–53.

Baldwin, J.D. (1986) Behavior in infancy: exploration and play. In: *Comparative*

Primate Biology, Vol. 2A: *Behavior, Conservation, and Ecology* (J. Erwin, ed.). Alan R. Liss, New York, 295–326.
Baldwin, J.D. and J.I. Baldwin (1974) Exploration and social play in squirrel monkeys (*Saimiri*). *Amer. Zool.* **14**, 303–14.
Barrett, J. and D.H. Abbott (1989) Pheromonal cues suppress ovulation in subordinate female marmoset monkeys, *Callithrix jacchus*. *J. Reprod. Fertil. Abstract Ser.* No. 3, 33.
Bartrop, R.W., E. Luckhurst, L. Lazarus, L.C. Killoh and R. Penny (1977) Depressed lymphocyte function after bereavement. *Lancet* **i**, 834–6.
Bateson, G. (1963) The role of somatic change in evolution. *Evolution* **17**, 529–39.
Baulu, J. and D.E. Redmond (1980) Some sampling considerations in the quantitation of monkey behavior under field and captive conditions. *Primates* **19**, 391–9.
Beaver, B.V. (1989) Environmental enrichment for laboratory animals. *ILAR News* **31**, 2.
Beck, B.B. (1980) *Animal Tool Use*. Garland, New York.
Beck, B. (1987) Reintroduction of golden lion tamarin: Round Three – A draft proposal.
Becker, J.D.Y. and G. Berkson (1979) Response to neighbors and strangers by capuchin monkeys (*Cebus apella*). *Primates* **20**, 547–51.
Beer, C.G. (1986) The evolution of intelligence: costs and benefits. In: *Animal Intelligence: Insights into the Animal Mind* (R.J. Hoage and L. Goldman, eds). Smithsonian Institution Press, Washington, DC, 115–34.
Beischer, D.E. and D.E. Furry (1964) *Saimiri sciureus* as an experimental animal. *Anat. Rec.* **148**, 615–24.
Bekoff, M. (1977) Mammalian dispersal and the ontogeny of individual mammalian phenotypes. *Amer. Nat.* **111**, 715–32.
Bekoff, M. and J.A. Byers (1985) The development of behaviour from evolutionary and ecological perspectives in birds and mammals. In: *Evolutionary Biology* (M.K. Hecht, B. Wallace and G.T. Prance, eds). Plenum Press, New York, 215–86.
Bennett, E.L. (1984) The banded langur: ecology of a colobine in West Malaysian rain-forest. Unpublished PhD thesis, University of Cambridge.
Berenstain, L. (1986) Responses of long-tailed macaques to drought and fire in Eastern Borneo: a preliminary report. *Biotropica* **18**, 257–62.
Bercovitch, F.B. (1986) Male rank and reproductive activity in savanna baboons. *Int. J. Primatol.* **7**, 533–50.
Bercovitch, F.B. (1988) Coalitions, cooperation and reproductive tactics among adult male baboons. *Anim. Behav.* **36**, 1198–209.
Bernor, R.L. (1983) Geochronology and zoogeographic relationships of Miocene Hominoidea. In: *New Interpretations of Ape and Human Ancestry* (R.L. Ciochon and R.S. Corruccini, eds). Plenum Press, New York, 149-64.
Bernstein, I.S. (1967) A field study of the pig-tailed monkey (*Macaca nemestrina*). *Primates* **8**, 217–28.
Bernstein, I.S. (1970) Primate status hierarchies. In: *Primate Behaviour: Developments in Field and Laboratory Research*. (L. Rosenblum, ed.). Academic Press, New York, 71–109.
Bernstein, I.S. (1981) Dominance: the baby and the bathwater. *Behav. Brain Sci.* **4**, 419–57.
Bernstein, I.S., T.P. Gordon and R.M. Rose (1974) Aggression and social

controls in rhesus monkey (*Macaca mulatta*) groups revealed in group formation studies. *Folia Primatol.* **21**, 81–107.
Bernstein, I.S., P. Balcean, L. Dresdale, H. Gouzoules, M. Kavanagh, T. Patterson and P. Neyman-Warner (1976) Differential effects of forest degradation on primate populations. *Primates* **17**, 401–11.
Bishop, R.C. (1980) Endangered species: an economic perspective. *Transactions of North American Wildlife and Natural Resources Conference* **45**, 208–18.
Boesch, C. (1978) Nouvelles observations sur les chimpanzés de la fôret de Tai (Côte d'Ivoire). *Terre et Vie* **32**, 195–201.
Boesch, C. and H. Boesch (1981) Possible causes of sex differences in the use of natural hammers by chimpanzees. *J. Hum. Evol.* **13**, 415–40.
Boesch, C. and H. Boesch (1983) Optimization of nut-cracking with natural hammers by wild chimpanzees. *Behaviour* **83**, 265–86.
Boesch, C. and H. Boesch (1984) Possible causes of sex differences in the use of natural hammers by wild chimpanzees. *J. Hum. Evol.* **13**, 415–40.
Boinski, S.H. (1986) The ecology of squirrel monkeys in Costa Rica. Dissertation, University of Texas, Austin.
Boinski, S. (1987a) Birth synchrony in squirrel monkeys (*Saimiri oerstedi*): a strategy to reduce neo-natal predation. *Behav. Ecol. Sociobiol.* **21**, 383–400.
Boinski, S. (1987b) Mating patterns in squirrel monkeys (*Saimiri oerstedi*). *Behav. Ecol. Sociobiol.* **21**, 13–21.
Boinski, S.H. (1988) An attack by *Cebus capucinus* on a poisonous snake (*Bothrops asper*) using a branch as a club. *Amer. J. Primatol.* **14**, 177–9.
Bookstaber, R. and J. Langsam (1985) On the optimality of coarse behaviour rules. *J. Theor. Biol.* **116**, 162–93.
Boot, R., A.B. Leussink and R.F. Vlug (1985) Influence of housing conditions on pregnancy outcome in cynomolgus monkeys (*Macaca fascicularis*). *Lab. Anim.* **19**, 42–7.
Borner, M. (1979) *Orang Utan: Orphans of the Forest*. Book Club Associates: London.
Borner, M. (1985) The rehabilitated chimpanzees of Rubondo Island. *Oryx* **19**, 151–4.
Bourne, P.G. (1969) Urinary 17–OHCS levels in two combat situations. In: *The Psychology and Physiology of Stress* (P.G. Bourne, ed.). Academic Press, New York and London, 95–116.
Bovard, E.W. (1985) *Perspectives in Behavioral Medicine*. Academic Press: New York.
Bowen, R.A. (1980) The behaviour of three hand-reared lowland gorillas, with emphasis on the response to a change in accommodation. *Dodo, J. Jersey Wildl. Preserv. Trust* **17**, 63–79.
Bowen, R.A. (1981) Social integration in lowland gorillas at the Jersey Wildlife Preservation Trust. *Dodo, J. Jersey Wildl. Preserv. Trust* **18**, 51–9.
Bower, T.G.P. (1982) *Development in Infancy*, 2nd edn. W.H. Freeman, San Francisco.
Bowlby, J. (1958) The nature of the child's tie to his mother. *Int. J. Psycho-Anal.* **39** (5), 1–24.
Bowman, L.A., S.R. Dilley and E.B. Keverne (1978) Suppression of oestrogen-induced LH surges by social subordination in talapoin monkeys. *Nature, Lond.* **275**, 56–8.
Box, H.O. (1982) Individual and intergroup differences in social behaviour

among captive marmosets (*Callithrix jacchus*) and tamarins (*Saguinus mystax*). *Soc. Biol. Hum. Affairs* **47**, 49–68.

Box, H.O. (1984) *Primate Behaviour and Social Ecology*. Chapman and Hall, London and New York.

Box, H.O. (1988) Behavioural responses to environmental change. Observations on captive marmosets and tamarins (Callitrichidae). *Anim. Technol.* **39** (1), 9–16.

Box, H.O. and D.M. Fragaszy (1986) The development of social behaviour and cognitive abilities. In: *Primate Ontogeny, Cognition and Social Behaviour* (J.G. Else and P.C. Lee, eds). Cambridge University Press, Cambridge, 119–28.

Box, H.O. and B. Rohrhuber (1987) Species differences in behaviour among tamarins (Callitrichidae) in response to unfamiliar tasks. *Int. J. Primatol.* **8**, 462.

Boyd, C.E. and C.P. Goodyear (1971) Nutritive quality of food and ecological systems. *Arch. Hydrobiol.* **69**, 256–70.

Boyd, R. and J. Silk (1983) A method of assigning cardinal dominance ranks. *Anim. Behav.* **31**, 45–58.

Bramblett, C.A., A.M. Coelho and G.E. Mott (1981) Behavior and serum cholesterol in a social group of *Cercopithecus aethiops*. *Primates* **22**, 96–102.

Breese, G.R., R.D. Smith, R.A. Mueller, J.L. Howard, A.J. Prange, Jr, N.A. Lipton, L.D. Young, W.T. McKinney and J.K. Lewis (1973) Induction of adrenal catecholamine synthesizing enzymes following mother–infant separation. *Nature New Biol.* **246**, 94–6.

Brennan, J., J.G. Else and J. Altmann (1985) Ecology and behaviour of a pest primate: vervet monkeys in a tourist lodge habitat. *Afr. J. Ecol.* **23**, 35–44.

Brewer, S. (1978) *The Chimps of Mt. Asserik*. Alfred A. Knopf, New York.

Brewer, S. (1982) Essai de rehabilitation de chimpanzés au Niokola-Koba de chimpanzés auparavant en captivité. *Mém. Inst. Fond. Afr. Noire,* Dakar **92** (XXVI) 341–62.

Bridneach, B. (1988) Alleviation of vices in stabled horses. *Universities Federation of Animal Welfare Annual Report*, **9**.

Brown, G.M., L.J. Grota, D.P. Penney and S. Reichlin (1970) Pituitary-adrenal function in the squirrel monkey. *Endocrinology* **86**, 519–29.

Bruner, J.S. (1974) The organization of early skilled action. In: *The Integration of a Child into a Social World* (M.P.M. Richards ed.). Cambridge University Press, London, 167–84.

Bryant, C.E., N.M.J. Rupniak and S.D. Iverson (1988) Effects of different environmental enrichment devices on cage stereotyping and autoaggression in captive cynomolgus monkeys. *J. Med. Primatol.* **17**, 257–70.

Bunnell, B.H., W.T. Gore and M.N. Perkins (1980) Performance correlates of social behaviour and organisation: social rank and complex problem solving in crab-eating macaques (*M. fascicularis*). *Primates* **21**, 515–23.

Bunney, W.E. Jr, E.L. Hartmann and J.W. Mason (1965) Study of a patient with 48 hr manic-depressive cycles: II. Strong positive correlation between endocrine factors and manic defense patterns. *Arch. Gen. Psychiat.* **12**, 619–23.

Burnham, K.P., D.R. Anderson and J.L. Laake (1980) Estimation of density from line transect sampling of biological populations. *Wildl. Monogr.* **72**, 1–202.

Burt, D.A. and M. Plant (1983) Observations on marmoset breeding at Fisons. *Anim. Technol.* **34**, 29–36.

Butynski, T.M. (1985) Primates and their conservation in the impenetrable (Bwindi) Forest, Uganda. *Primate Conserv.* **5**, 68–72.
Bygott, J.D. (1979) Agonistic behaviour, dominance and social structure in wild chimpanzees of the Gombe National Park. In: *The Great Apes* (D.A. Hamburg and E.R. McCown, eds). Benjamin/Cummings: Palo Alto, CA, 405–28.
Byrd, L.D. and F.A. Gonzalez (1981) Time-course effects of adrenergic and cholinergic antagonists on systemic arterial blood pressure, heart rate and temperature in conscious squirrel monkeys. *J. Med. Primatol.* **10**, 81–92.
Byrne, R. and A. Whitten, eds (1988) *Machiavellian Intelligence.* Clarendon Press, Oxford.
Cade, W.H. (1981) Alternative male strategies: genetic differences in field crickets. *Science* **188**, 563–4.
Caine, N.G. (1984) Visual scanning by tamarins: a description of the behavior and tests of two derived hypotheses. *Folia Primatol.* **43**, 59–67.
Caine, N.G. and S.L. Marra (1988) Vigilance and social organisation in two species of primates. *Anim. Behav.* **36**, 897–904.
Caldecott, J.O. (1986) Sexual behaviour, societies and ecogeography of macaques. *Anim. Behav.* **34**, 208–20.
Caldecott, J.O. and M. Kavanagh (1983) Can translocation help wild primates? *Oryx* **17**, 135–9.
Cambefort J.P. (1981) A comparative study of culturally transmitted patterns of feeding habits in the chacma baboon *Papio ursinus* and the vervet monkey *Cercopithecus aethiops. Folia Primatol.* **36**, 243–63.
Campbell, B.G. (1986) *Human Evolution.* Aldine-Atherton, Chicago.
Candland, D.K., D.C. Bryan, B.L. Nazar, K.J. Kopf and N. Sendor (1970) Squirrel monkey heart rates during formation of status orders. *J. Comp. Physiol. Psychol.* **70**, 417–27.
Candland, D.K., L. Dresdale, J. Leiphart, D. Bryan, C. Johnson and B. Nazar (1973) Social structure of the squirrel monkey (*Saimiri sciureus iquitos*): relationships among behavior, heart rate and physical distance. *Folia Primatol.* **20**, 211–40.
Cant, J.G.H. (1988) Positional behaviour of long-tailed macaques (*Macaca fascicularis*) in Northern Sumatra. *Amer. J. Phys. Anthropol.* **76**, 29–37.
Carpenter, C.R. (1934) A field study of the behavior and social relations of howling monkeys. *Comp. Psychol. Monogr.* **10**, 1–168.
Carpenter, C.R. (1942) Societies of monkeys and apes. *Biol. Symp.* **8**, 177–204.
Carpenter, C.R. (1952) Social behavior of non-human primates. *Colloques Internationaux du Centre National de la Recherche Scientifique* **34**, 227–46. Reprinted in: *Naturalistic Behavior of Nonhuman Primates* (C.R. Carpenter, ed.). Pennsylvania State University Press, University Park, 365–85.
Carpenter, C.R. (1972) Breeding colonies of macaques and gibbons on Cayo Santiago Island, Puerto Rico. In: *Breeding Primates* (W. Beveridge, ed.). Karger, Basel, 76–87.
Carroll, J.B. (1982) Breeding the golden lion tamarin *Leontopithecus rosalia. Dodo, J. Jersey Wildl. Preserv. Trust* **19**, 42–6.
Carroll, J.B., D.H. Abbott, L.M. George, J.E. Hindle and R.D. Martin (1990) Urinary endocrine monitoring of the ovarian cycle and pregnancy in Goeldi's monkey. *J. Reprod. Fertil.* **89**, 149–61.
Carter, J. (1981) A journey to freedom. *Smithsonian* **12**(1), 91–101.
Cartmill, M. (1974) Rethinking primate origins. *Science* **184**, 436–43.

Cassel, J. (1976) Social support and health. *Amer. J. Epidemiol.* **104**, 107–15.
Cassorla, F.G., B.D. Albertson, G.P. Chrousos, J.D. Booth, D. Renquist, M.B. Lipsett and D.L. Loriaux (1982) The mechanism of hypercortisolemia in the squirrel monkey. *Endocrinology* **111**, 448–51.
Caswell, H. (1983) Phenotypic plasticity in life history traits: demographic effects and evolutionary consequences. *Amer. Zool.* **32**, 35–46.
Caulfield, C. (1985) *In the Rain Forest*. Heinemann, London.
Chalmers, N.R. (1968) Group composition, ecology and daily activities of free-living mangabeys in Uganda. *Folia Primatol.* **8**, 247–62.
Chamove, A.S. (1983) Role or dominance in macaque response to novel objects. *Motivation and Emotion* **7**, 213–28.
Chamove, A.S. (1989) Assessing the welfare of captive primates: a critique. In: *Laboratory Animal Welfare Research: Primates*. UFAW Symposium. Universities Federation for Animal Welfare, Potters Bar, Hertfordshire, 39–49.
Chamove, A.S. and R.E. Bowman (1976) Rank, rhesus social behavior and stress. *Folia Primatol.* **26**, 57–66.
Chamove, A.S. and R.E. Bowman (1978) Rhesus plasma cortisol responses at four dominance positions. *Aggress. Behav.* **4**, 43–55.
Chamove, A.S., J.R. Anderson, S.C. Morgan-Jones and S.P. Jones (1982) Deep woodchip litter: hygiene, feeding and behavioural enhancement in eight primate species. *Int. J. Stud. Anim. Prob.* **3**, 308–18.
Chance, M.R.A. and A.P. Mead (1953) Social behaviour and primate evolution. *Symp. Soc. Exp. Biol.* **7**, 395–439.
Cheney, D.L. (1978) The play partners of immature baboons during intergroup encounters. *Anim. Behav.* **26**, 1038–50.
Cheney, D.L. (1987) Interactions and relations between groups. In: *Primate Societies* (B.B. Smuts, D.L. Cheney, R.M. Seyfarth, R.W. Wrangham and T.T. Struhsaker, eds). University of Chicago Press, Chicago, 267–81.
Cheney, D.L. and R.M. Seyfarth (1980) Vocal recognition in free-ranging vervet monkeys. *Anim. Behav.* **28**, 362–7.
Cheney, D.L., P.C. Lee and R.M. Seyfarth (1981) Behavioural correlates of non-random mortality among free-ranging female vervet monkeys. *Behav. Ecol. Sociobiol.* **9**, 153–61.
Cheney, D., R.M. Seyfarth and B. Smuts (1986) Social relationships and social cognition in nonhuman primates. *Science* **234**, 1361–6.
Cheney, D.L., R.M. Seyfarth, S.A. Andelman and P.C. Lee (1988) Reproductive success in vervet monkeys. In: *Reproductive Success* (T.H. Clutton-Brock, ed.). University of Chicago Press, Chicago, 384–402.
Chepko-Sade, B.D. and T.J. Olivier (1979) Coefficients of genetic relationships and the probability of inter-genealogical fission in *Macaca mulatta*. *Behav. Ecol. Sociobiol.* **5**, 263–78.
Chepko-Sade, B.D. and D.S. Sade (1979) Patterns of group splitting within matrilineal kinship groups. *Behav. Ecol. Sociobiol.* **5**, 67–86.
Cherkovitch, G.M. and S.K. Tatoyan (1973) Heart rate (radio telemetrical registration) in macaques and baboons according to dominant-submissive rank in a group. *Folia Primatol.* **20**, 265–73.
Chivers, D.J. (1974) The siamang in Malaya: a field study of a primate in tropical rain forest. *Contr. Primatol.* **4**, 1–335. Karger, Basel.
Chivers, D.J., ed. (1980) *Malayan Forest Primates*. Plenum Press, New York.
Chivers, D.J. and C.M. Hladik (1980) Morphology of the gastrointestinal tract

in primates: comparison with other mammals in relation to diet. *J. Morph.* **166**, 337–86.

Chivers, D.J., B.A. Wood and A. Bilsborough, eds (1984) *Food Acquisition and Processing in Primates*. Plenum Press, New York.

C.I.T.E.S. (1977) U.S. Department of Interior convention on international trade in endangered species of wild fauna and flora. *Fed. Register* **42**(35), 10462–88.

Clark, A.B. (1982) Sibling relationships and social development in the great galago *Galago crassicaudatus*. Paper presented at the Midwest Animal Behavior Conference, Urbana, IL.

Clark, A.B. and T.J. Ehlinger (1987) Pattern and adaptation in individual behavioural differences. In: *Perspectives in Ethology* (P.P.G. Bateson and P.H. Klopfer, eds). Plenum Press, New York, 1–45.

Clarke, A.S. (1985) Behavioural, cardiac and adrenocortical responses to stress in three macaque species. PhD dissertation, University of California, Davis.

Clarke, A.S. and J.W. Mason (1988) Differences among three macaque species in responsiveness to an observer. *Int. J. Primatol.* **9**, 347–64.

Clarke, A.S., W.A. Mason and G.P. Moberg (1988a) Differential behavioural and adrenocortical responses to stress among three Macaque species. *Amer. J. Primatol.* **14**, 37–52.

Clarke, A.S., W.A. Mason and G.P. Moberg (1988b) Interspecific contrasts in responses of macaques to transport cage training. *Lab. Anim. Sci.* **38**, 305–9.

Clutton-Brock, T.H. (1977) Activity patterns of red colobus. *Folia Primatol.* **21**, 161–88.

Clutton-Brock, T.H. and P.H. Harvey (1977) Primate ecology and social organization. *J. Zool. Lond.* **183**, 1–39.

Coe, C.L. and L.A. Rosenblum (1974) Sexual segregation and its ontogeny in squirrel monkey social structure. *J. Hum. Evol.* **3**, 1–11.

Coe, C.L., S.P. Mendoza, W.P. Smotherman and S. Levine (1978a) Mother–infant attachment in the squirrel monkey: adrenal response to separation. *Behav. Biol.* **22**, 256–63.

Coe, C.L., S.P. Mendoza, J.W. Davidson, E.R. Smith, M.F. Dallman and S. Levine (1978b) Hormonal response to stress in the squirrel monkey (*Saimiri sciureus*). *Neuroendocrinology* **26**, 367–77.

Coe, C.L., S.P. Mendoza and S. Levine (1979) Social status constrains the stress response in the squirrel monkey. *Physiol. Behav.* **23**, 633–8.

Coe, C.L., E.R. Smith, S.P. Mendoza and S. Levine (1983) Varying influence of social status on hormone levels in male squirrel monkeys. In: *Hormones, Drugs and Social Behavior in Primates* (H.D. Steklis and A.S. Kling, eds). Spectrum, New York, 7–32.

Coe, C.L., S.G. Weiner, L.T. Rosenberg and S. Levine (1985) Endocrine and immune responses to separation and maternal loss in nonhuman primates. In: *The Psychobiology of Attachment and Separation* (M. Reite and T. Field, eds). Academic Press, New York, 163–99.

Cohler, E.A. and F. Grunbau (1977) Disturbance of attention among schizophrenic, depressed and well mothers and children. *J. Child Psychol. Psychiat.* **18**, 115–35.

Coimbra-Filho, A.F. and R.A. Mittermeier (1976) Exudate-eating and tree-gouging in marmosets. *Nature, Lond.* **262**, 630.

Coimbra-Filho, A.F. and R.A. Mittermeier (1977) Reintroduction and translocation of lion tamarins: a realistic appraisal. In: *Biology and Behaviour of*

Marmosets: Proceedings of the Marmoset Workshop (H. Rothe, H.-J. Wolters and J.P. Hearn eds). Eigenverlag-H. Rothe, Göttingen, 41–6.

Cooper, L. and H. Harlow (1961) Note on a cebus monkey's use of a stick as a weapon. *Psychol. Rep.* **8**, 418.

Coover, G.D., H. Ursin and S. Levine (1973) Plasma-corticosterone levels during active-avoidance learning in rats. *J. Comp. Physiol. Psychol.* **82**, 170–4.

Costello, M. (1987) Tool use and manufacture in manipulanda-deprived capuchins (*Cebus apella*): ontogenetic and phylogenetic issues. *Amer. J. Primatol.* **12**, 337.

Costello, M. and D.M. Fragaszy (1988) Prehension in *Cebus* and *Saimiri*. I: grip type and hand preference. *Amer. J. Primatol.* **15**, 235–45.

Council of Europe (1986) European convention for the protection of vertebrate animals used for experimental and other scientific purposes. Council of Europe, Strasbourg.

Cousins, D. (1983) Man's affiliation with the anthropoid apes in Africa. *Acta Zool. Pathol. Antverp.* **77**, 19–40.

Cowan, P.E. (1977) Neophobia and neophilia: new object and new place reactions of three *Rattus* species. *J. Comp. Physiol. Psychol.* **91**, 63–71.

Cowan, P.E. (1983) Exploration in small mammals: ethology and ecology. In: *Exploration in Animals and Man* (J. Archer and L. Birke, eds). Van Nostrand Reinhold, Wokingham, Berkshire, 147–75.

Crockett, C.M. (1984) Emigration by female red howler monkeys and the case for female competition. In: *Female Primates: Studies by Female Primatologists* (M.F. Small, ed.). Alan R. Liss, New York, 159–73.

Cronin, G.M. (1985) The development and significance of abnormal stereotyped behaviours in tethered sows. Thesis, Agricultural University of Wageningen.

Crook, J.H. and J.S. Gartlan (1966) Evolution of primate societies. *Nature, Lond.* **210**, 1200–3.

Cubicciotti, D.D., III and W.A. Mason (1975) Comparative studies of social behavior in *Callicebus* and *Saimiri*: male–female emotional attachments. *Behav. Biol.* **16**, 185–97.

Cubicciotti, D.D., III and W.A. Mason (1978) Comparative studies of social behavior in *Callicebus* and *Saimiri*: heterosexual jealousy behavior. *Behav. Ecol. Sociobiol.* **3**, 311–22.

Cubicciotti, D.D., III, S.P. Mendoza, W.A. Mason and E.N. Sassenrath (1986) Differences between *Saimiri sciureus* and *Callicebus moloch* in physiological responsiveness: implications for behavior. *J Comp. Psychol.* **100**, 385–91.

Dasser, V. (1985) Cognitive complexity in primate social relationships. In: *Social Relationships and Cognitive Development* (R.A. Hinde, A. Perret-Clermont and J. Stevenson Hinde, eds). Oxford University Press, Oxford, 9–22.

Datta, S.B. (1983) Relative power and the acquisition of rank. In: *Primate Social Relationships: An Integrated Approach* (R.A. Hinde, ed.). Blackwell Scientific, Oxford, 93–102.

Datta, S.B. (1986) The role of alliances in the acquisition of rank. In: *Primate Ontogeny, Cognition and Social Behaviour* (J.G. Else and P.C. Lee, eds). Cambridge University Press, Cambridge, 219–27.

Datta, S.B. (1989) Demographic influences in dominance structure among female primates. In: *Comparative Socioecology* (V. Standen and R. Foley eds). Blackwell, Oxford, 265–84.

Davies, A.G. (1984) An ecological study of the red leaf monkey (*Presbytis rubicunda*) in the dipterocarp forests of northern Borneo. Unpublished PhD thesis, University of Cambridge.
Davies, A.G. and J.F. Oates, eds (in press) *Evolutionary Ecology of the Colobine Monkeys*. Cambridge University Press, Cambridge.
Davies, A.G., J.O. Caldecott and D.J. Chivers (1984) Natural foods as a guide to nutrition of Old World Primates. In: *Standards in Laboratory Animal Management*. Universities Federation for Animal Welfare, Potters Bar, Hertfordshire, 225–44.
Davis, R.T., P.H. Settlege and H.F. Harlow (1950) Performance of normal and brain-operated monkeys on mechanical puzzles with and without food incentive. *J. Genet. Psychol.* **77**, 305–11.
Dawson, G.A. (1978) Composition and stability of social groups of the tamarin (*Saguinus oedipus geoffroyi*) in Panama: ecological and behavioral implications. In: *The Biology and Conservation of the Callitrichidae* (D.G. Kleiman, ed.). Smithsonian Institution Press, Washington, DC, 23–37.
Dawson, G.A. (1979) The use of time and space by the Panamanian tamarin (*Saguinus oedipus*). *Folia Primatol.* **31**, 253–84.
Deag, J.M. (1974) A study of the social behaviour and ecology of the wild Barbary macaque (*Macaca sylvanus* L.). Unpublished PhD thesis, University of Bristol.
Deag, J.M. (1980) Interactions between males and unweaned Barbary macaques: testing the agonistic buffering hypothesis. *Behaviour* **75**, 54–81.
Deag, J.M. and J.H. Crook (1971) Social behaviour and 'Agonistic Buffering' in the wild Barbary macaque (*Macaca sylvanus L.*). *Folia Primatol.* **15**, 183–200.
Delson, E. (1980) Fossil macaques, phyletic relationships and a scenario of deployment. In: *The Macaques: Studies in Ecology, Behavior and Evolution* (D.G. Lindburg, ed.). Van Nostrand Reinhold, New York, 10–30.
Demment, M.W. (1983) Feeding ecology and the evolution of body size of baboons. *Afr. J. Ecol.* **21**, 219–33.
Demment, M.W. and P.J. Van Soest (1983) A nutritional explanation for body size patterns of ruminant and nonruminant herbivores. *Amer. Nat.* **125**, 641–72.
de Silva, G.S. (1971) Notes on the orang utan rehabilitation project in Sabah. *Malay. Nature J.* **24**, 50–77.
de Waal, F.B.M. (1982) *Chimpanzee Politics*. Harper and Row, New York.
de Waal, F.B.M. and J. Hoekstra (1980) Contexts and predictability of aggression in chimpanzees. *Anim. Behav.* **28**, 929–37.
Dietz, J.M. and D.G. Kleiman (1986) Reproductive parameters in groups of free-living golden lion tamarins. *Primate Rep.* **14**, 77.
Dietz, J.M. and D.G. Kleiman (1987) Sexual dimorphism and attractive reproduction tactics in the golden lion tamarin. *Int. J. Primatol.* **8**, 506.
Dittus, W.P.J. (1975) Population dynamics of the toque monkey, *Macaca sinica*. In: *Socioecology and Psychology of Primates* (R.H. Tuttle, ed.). Mouton, The Hague, 125–51.
Dittus, W.P.J. (1977) The social regulation of population density and age-sex distribution in the toque monkey. *Behaviour* **63**, 281–322.
Dittus, W.P.J. (1979) The evolution of behaviors regulating density and age-specific sex ratios in a primate population. *Behaviour* **69**, 265–302.
Dittus, W.P.J. (1980) The social regulation of primate populations: a synthesis.

In: *The Macaques: Studies in Ecology, Behavior and Evolution* (D. Lindburg, ed.). Van Nostrand Reinhold, New York, 263–86.

Dittus, W.P.J. (1986) Sex differences in fitness following a group take-over among toque macaques: testing models of social evolution. *Behav. Ecol. Sociobiol.* **19**, 257–66.

Dixson, A.F. (1983) The hormonal control of sexual behaviour in primates. In: *Oxford Reviews of Reproductive Biology*, Vol. 5 (C.A. Finn, ed.). Oxford University Press, Oxford, 131–219.

D'Mello, G.D., E.A.M. Duffy and S.S. Miles (1985) A conveyor belt task for assessing visuo-motor coordination in the marmoset (*Callithrix jacchus*): effects of diazepam, chlorpromazine, pentabarbitol and d-amphetamine. *Psychopharmcology* **86**, 125–31.

Doyle, G.A. and R.D. Martin, eds (1979) *The Study of Prosimian Behaviour.* Academic Press: London.

Drickamer, L.C. (1973) Semi-natural and enclosed groups of *Macaca mulatta*: a behavioral comparison. *Amer. J. Phys. Anthropol.* **39**, 249–54.

Drickamer, L.C. (1974) A ten-year summary of reproductive data for free-ranging *Macaca mulatta. Folia Primatol.* **21**, 61–80.

Drickamer, L.C. and S.H. Vessey (1973) Group-changing behaviour of free-ranging rhesus monkeys. *Primates* **14**, 359–68.

Drucker, G.R. (1984) The feeding ecology of the Barbary macaque and cedar forest conservation in the Moroccan Moyen Atlas. In: *The Barbary Macaque – A Case Study in Conservation* (J.E. Fa, ed.). Plenum Press, New York, 135–64.

Dubey, A.K. and T.M. Plant (1985) A suppression of gonadotrophin secretion by cortisol in castrated male rhesus monkeys (*Macaca mulatta*) mediated by the interruption of hypothalamic gonadotrophin-releasing hormone release. *Biol. Reprod.* **33**, 423–31.

Dubey, A.K., J.L. Cameron, R.A. Steiner and T.M. Plant (1986) Inhibition of gonadotrophin secretion in castrated male rhesus monkeys (*Macaca mulatta*) induced by dietary restriction: analogy with the prepubarted hiatus of gonadotrophin release. *Endocrinology* **118**, 518–25.

Dunbar, R.I.M. (1980) Determinants and evolutionary consequences of dominance among female gelada baboons. *Behav. Ecol. Sociobiol.* **7**, 253–65.

Dunbar, R.I.M. (1984a) *Reproductive Decisions: An Economic Analysis of Gelada Baboon Social Strategies.* Princeton University Press, Princeton, NJ.

Dunbar, R.I.M. (1984b) Theropithecines and hominids: contrasting solutions to the same ecological problem. *J. Hum. Evol.* **12**, 647–58.

Dunbar, R.I.M. (1986) The social ecology of gelada baboons. In: *Ecological Aspects of Social Evolution* (D. Rubenstein and R. Wrangham, eds). Princeton University Press, Princeton, NJ, 332–51.

Dunbar, R.I.M. (1988) *Primate Social Systems.* Croom Helm, Beckenham, Kent.

Dunbar, R.I.M. (1989) Reproductive strategies of female gelada baboons. In: *The Sociobiology of Sexual and Reproductive Strategies* (A. Rasa, C. Vogel and E. Voland, eds). Chapman and Hall, London, 74–92.

Dunbar, R.I.M. and E.P. Dunbar (1977) Dominance and reproductive success among female gelada baboons. *Nature, Lond.* **266**, 351–2.

Dutrillaux, B. (1979) Chromosomal evolution in primates: tentative phylogeny from *Microcebus murinus* (prosimian) to man. *Hum. Genet.* **48**, 251–314.

Eastman, S.A.K., D.W. Makawati, W.P. Collins and J.K. Hodges (1984) Pattern of excretion of urinary steroid metabolites during the ovarian cycles and pregnancy in the marmoset monkey. *J. Endocrinol.* **102**, 19-26.

Eberhart, J.A. and E.B. Keverne (1979) Influence of the dominance hierarchy and luteinizing hormone, testosterone and prolactin in male talapoin monkeys. *J. Endocrinol.* **83**, 42–3.

Edwards, S.D. (1982) Social potential expressed in captive group-living orang utans. In: *The Orang Utan: Its Biology and Conservation* (L.E.M. Boer, ed.). Junk, The Hague, 249–55.

Ehlinger, T.J. (1986) Learning, sampling and the role of individual variability in the foraging behaviour of bluegill sunfish. PhD dissertation, Michigan State University.

Ehlinger, T.J. (1989) Learning and individual variation in bluegill foraging: habitat specific techniques. *Anim. Behav.* **38**, 643–58.

Ehrlich, A. and A. Musicant (1977) Social and individual behaviors in captive slow lorises. *Behaviour* **60**, 195–220.

Eisenberg, J.F. (1981) *The Mammalian Radiations: An Analysis of Trends in Evolution, Adaptation and Behavior.* University of Chicago Press, Chicago.

Eley, D. and J.G. Else (1984) Primate pest problems in Kenya hotels and lodges. *Int. J. Primatol.* **5**, 334.

Else, J.G. and D. Eley (1985) Please don't feed the monkeys. *SAWAR* **8**(4), 31–2.

Emmons, L.H., A. Gautier-Hion and G. Dubost (1983) Community structure of the frugivorous/folivorous forest mammals of Gabon. *J. Zool. Lond.* **199**, 209–22.

Epple, G. (1967) Vergleichende Untersuchungen über Sexual und Sozialverhalten der Krallenaffen (Hapalidae). *Folia Primatol.* **7**, 37–65.

Epple, G. (1975) The behavior of marmoset monkeys (Callithricidae). In: *Primate Behavior: Developments in Field and Laboratory Research,* Vol. 4 (L.A. Rosenblum ed.). Academic Press, New York, 195–239.

Epple, G. and Y. Katz (1984) Social influences on estrogen excretion and ovarian cyclicity in saddle-back tamarins (*Saguinus fuscicollis*). *Amer. J. Primatol.* **6**, 215–28.

Erwin, J. (1986) Environments for captive propagation of primates: interaction of social and physical factors. In: *Primates: The Road to Self-Sustaining Populations* (W.K. Benirschke ed.). Springer-Verlag, New York, 297–305.

Erwin, J. and R. Deni (1979) Strangers in a strange land: abnormal behaviors or abnormal environments. In: *Captivity and Behaviour: Primates in Breeding Colonies, Laboratories and Zoos* (J. Erwin, T.L. Maple and G. Mitchell, eds). Van Nostrand Reinhold, New York, 1–28.

Essock-Vitale, S. and R.M. Seyfarth (1987) Intelligence and social cognition. In: *Primate Societies* (B.B. Smuts, D.L. Cheney, R.M. Seyfarth, R.W. Wrangham and T.T. Struhsaker, eds). University of Chicago Press, Chicago, 452–61.

Eudey, A.A. (1980) Pleistocene glacial phenomena and the evolution of Asian macaques. In: *The Macaque: Studies in Ecology, Behavior and Evolution* (D.G. Lindburg, ed.). Van Nostrand Reinhold, New York, 52–83.

Eudey, A.A. (1987) *Action Plan for Asian Primate Conservation: 1987–91.* IUCN/SSC Primate Specialist Group, New York.

Evans, S. (1981) Monogamy in the common marmoset (*Callithrix jacchus jacchus*). PhD thesis, University of Wales, Aberystwyth.

Evans, S. and J.K. Hodges (1984) Reproductive status of adult daughters in family groups of common marmosets (*Callithrix jacchus jacchus*). *Folia Primatol.* **42**, 127–33.

Fa, J.E. (1981) The apes on the Rock. *Oryx* **16**, 73–6.

Fa, J.E. (1982) A survey of population and habitat of the Barbary macaque (*Macaca sylvanus* L.) in North Morocco. *Biol. Conserv.* **24**, 45–66.

Fa, J.E. (1984) Structure and dynamics of the Barbary macaque population in Gibraltar. In: *The Barbary Macaque – A Case Study in Conservation* (J.E. Fa, ed.). Plenum Press, New York, 263–306.

Fa, J.E. (1986) Use of time and resources in provisioned troops of monkeys: social behaviour, time and energy in the Barbary macaque (*Macaca sylvanus* L.) at Gibraltar. *Contr. Primatol.* **23**. Karger, Basel.

Fa, J.E. (1988) Supplemental food as an extranormal stimulus in Barbary macaques (*Macaca sylvanus*) at Gibraltar – its impact on activity budgets. In: *Ecology and Behavior of Food-Enhanced Primate Groups* (J.E. Fa and C.H. Southwick, eds). Alan R. Liss, New York, 53–78.

Fa, J.E. (1989) Influence of people on the behavior of display primates. In: *Psychological Wellbeing of Captive Primates* (E.F. Segal, ed.). Noyes Publications, Park Ridge, NJ, 52–64.

Fa, J.E. and C. Southwick, eds (1988) *The Ecology and Behaviour of Food-Enhanced Primate Groups*. Alan R. Liss, New York.

Fagen, R. (1981) *Animal Play Behaviour*. Oxford University Press, London.

Fagen, R. (1982) Evolutionary issues in the development of behavioural flexibility. In: *Perspectives in Ethology* (P.P.G. Bateson and P.H. Klopfer, eds). Plenum Press, New York, 365–83.

Fagen, R. (1987) Phenotypic plasticity and social environment. *Evol. Ecol.* **1**, 263–71.

Fairbanks, L.A. and M.T. McGuire (1988) Long-term effects of early mothering behaviour on the responsiveness to the environment in vervet monkeys. *Dev. Psychobiol.* **21**, 711–24.

Fischer, A.C. (1982) *Economic Analysis and the Extinction of Species*. Department of Energy and Resources, University of California, Berkeley, California.

Fleagle, J.G. (1977) Locomotor behaviour and muscular anatomy of sympatric Malaysian leaf-monkeys (*Presbytis obscura* and *Presbytis melalophos*). *Amer. J. Phys. Anthropol.* **46**, 297–308.

Fleagle, J.G. (1978) Locomotion, posture and habitat utilization in two sympatric Malaysian leaf-monkeys (*Presbytis obscura* and *Presbytis melalophos*). In: *The Ecology of Arboreal Folivores* (G.G. Montgomery, ed.). Smithsonian Institution Press, Washington, DC, 243–52.

Fleagle, J.G. (1986) Early anthropoid evolution in Africa and South America. In: *Primate Evolution* (J.G. Else and P.C. Lee, eds). Cambridge University Press, Cambridge, 133–42.

Fleagle, J.G. and R.A. Mittermeier (1980) Locomotor behaviour, body size and comparative ecology of seven Surinam monkeys. *Amer. J. Phys. Anthropol.* **52**, 301–14.

Fleagle, J.G., J.T. Stern, W.L. Jungers, R.L. Susman, A.K. Vangor and J.P. Wells (1981) Climbing: a biomechanical link with brachiation and with bipedalism. *Symp. Zool. Soc. Lond.* **48**, 359–75.

Foley, R. (1987) *Another Unique Species*. Longman, Harlow, Essex.

Foley, R.A. and P.C. Lee (1989) Finite social space, evolutionary pathways and reconstructing hominid behaviour. *Science* **243**, 901–6.

Fooden, J. (1963). A revision of the woolly monkeys (genus *Lagothrix*). *J. Mammal.* **44**, 213–47.

Fooden, J. (1964) Stomach contents and gastro-intestinal proportions in wild shot Guianan monkeys. *Amer. J. Phys. Anthropol.* **22**, 227–32.

Foose, T.J., U.S. Seal and N.R. Flesness (1987) Captive propagation as a component of conservation strategies for endangered primates. In: *Primate*

Conservation in the Tropical Rain Forest (C.W. Marsh and R.A. Mittermeier, eds). Alan R. Liss, New York, 263–99.

Ford, S.M. (1986) Systematics of New World monkeys. In: *Comparative Primate Biology*, Vol. 1: *Systematics, Evolution, and Anatomy* (D.R. Swindler and J. Erwin, eds). Alan R. Liss, New York, 73–135.

Forthman-Quick, D.L. (1984) An integrative approach to environmental engineering in zoos. *Zoo Biol.* **3**, 65–77.

Forthman-Quick, D.L. (1986) Activity budgets and the consumption of human food in two troops of baboons, *Papio anubis*, at Gilgil, Kenya. In: *Primate Ecology and Conservation* (J.G. Else and P.C. Lee, eds). Cambridge University Press, Cambridge, 221–8.

Fossey, D. (1981) The imperilled mountain gorilla. *Natn. Geogr.* **159**, 500–23.

Fossey, D. (1983) *Gorillas in the Mist*. Houghton Miflin, Boston.

Foster, S.A. and C.H. Janson (1985) The relationship between seed size and establishment conditions in tropical woody plants. *Ecology* **66**, 773–80.

Fowler, H. (1965) *Curiosity and Exploratory Behaviour*. Macmillan, New York.

Fox, M.W. (1972) Socioecological implications of individual differences in wolf litters: a developmental and evolutionary perspective. *Behaviour* **41**, 298–313.

Fragaszy, D.M. (1978) Contrasts in feeding behaviour in squirrel and titi monkeys. In: *Recent Advances in Primatology: Behaviour*, Vol. 1 (D.J. Chivers and J. Herbert, eds). Academic Press, London, 363–7.

Fragaszy, D.M. (1979) Squirrel and titi monkeys in a novel environment. In: *Captivity and Behaviour* (J. Irwin, T. Maple and G. Mitchell, eds). Van Nostrand, New York, 172–216.

Fragaszy, D.M. (1980) Comparative studies of squirrel monkeys (*Saimiri*) and titi monkeys (*Callicebus*) in travel tasks. *Z. Tierpsychol.* **54**, 1–36.

Fragaszy, D.M. (1981) Comparative performance in discrimination learning tasks in two New World primates (*Saimiri sciureus* and *Callicebus moloch*). *Anim. Learn. Behav.* **9**, 127–34.

Fragaszy, D.M. (1985) Cognition in squirrel monkeys: a contemporary perspective. In: *Handbook of Squirrel Monkey Research* (C. Coe and L. Rosenblum, eds). Plenum Press, New York, 55–98.

Fragaszy, D.M. (1986) Time budgets and foraging behavior in wedge-capped capuchins (*Cebus olivaceous*): age and sex differences. In: *Current Perspectives in Primate Social Dynamics* (D. Taub and F. King, eds). Van Nostrand Reinhold, New York, 159–74.

Fragaszy, D.M. and L.E. Adams-Curtis (1987) Manipulative behaviors in capuchin monkeys (*Cebus apella*): patterns of development and social context. Paper presented at the International Ethological Conference XX, Madison, WI.

Fragaszy, D.M. and H.O. Box (1986) Primate behaviour and cognition in nature. Introduction. In: *Primate Ontogeny, Cognition and Social Behaviour* (J.G. Else and P.C. Lee, eds). Cambridge University Press, Cambridge, 73–8.

Fragaszy, D.M. and W.A. Mason (1978) Response to novelty in *Saimiri* and *Callicebus*: influence of social context. *Primates* **19**, 311–31.

Fragaszy, D.M. and W.A. Mason (1983) Comparisons of feeding behavior in captive squirrel and titi monkeys. *J. Comp. Psychol.* **97**, 310–26.

Fragaszy, D. and E. Visalberghi (1989) Social influences on the acquisition of tool-using behaviors in tufted capuchin monkeys (*Cebus apella*). *J. Comp. Psychol.* **103**, 159–70.

Fragaszy, D.M., H.O. Box. and E. Visalberghi (1989) The game of social chess, not trivial pursuit. *Primate Eye* **37**, 10–11. Paper read at the Primate Society of Great Britain, November 30, 1988.

Frankel, O.H. and M.E. Soulé (1981) *Conservation and Evolution*. Cambridge University Press, Cambridge.

Fraser, D.G. and J.F. Gilliam (1987) Feeding under predation hazard: response of the guppy and Hart's rivulus from sites with contrasting predation hazard. *Behav. Ecol. Sociobiol.* **21**, 203–9.

Freeman, B.J. and O.S. Ray (1972) Strain, sex and environment effects on appetitively and aversively motivated learning tasks. *Dev. Psychobiol.* **5**, 101–10.

Freese, C.H., P.G. Heltne, R.N. Castro and G. Whitesides (1982) Patterns and determinants of monkey densities in Peru and Bolivia, with notes on distributions. *Int. J. Primatol.* **3**, 53–90.

French, J.A. (1983) Lactation and fertility: an examination of nursing and interbirth intervals in tamarins (*Saguinus oedipus*). *Folia Primatol.* **40**, 276–82.

French, J.A. (1987) Reproductive suppression in marmosets and tamarins: absence of social effects in the lion tamarin. *Amer. J. Primatol.* **12**, 342.

French, J.A. and J. Cleveland (1984) Scent marking in tamarins (*Saguinus oedipus*): sex differences and ontogeny. *Anim. Behav.* **32**, 615–23.

French, J.A. and B.J. Inglett (1989) Female–female aggression and male indifference in response to unfamiliar intruders in lion tamarins. *Anim. Behav.* **37**, 487–97.

French, J.A. and C.T. Snowdon (1981) Sexual dimorphism in responses to unfamiliar intruders in the tamarin *Saguinus oedipus*. *Anim. Behav.* **29**, 822–9.

French, J.A. and J.A. Stribley (1985) Patterns of urinary oestrogen excretion in female golden lion tamarins (*Leontopithecus rosalia*). *J. Reprod. Fert.* **75**, 537–46.

French, J.A. and J.A. Stribley (1987) Synchronisation of ovarian cycles within and between social groups in the golden lion tamarin (*Leontopithecus rosalia*). *Amer. J. Primatol.* **12**, 469–78.

French, J.A. and J.A. Stribley (in press) The reproductive endocrinology and behaviour of the golden lion tamarin. In: *A Case Study in Conservation: the Golden Lion Tamarin* (D.G. Kleiman, ed.). Smithsonian Institution Press, Washington, DC.

French, J.A., D.H. Abbott and C.S. Snowdon (1984) The effects of social environment on oestrogen excretion, scentmarking and sociosexual behaviour in tamarins (*Saguinus oedipus*). *Amer. J. Primatol.* **6**, 156–67.

French, J.A., B.J. Inglett and T.M. Dethlefs (1989) The reproductive status of nonbreeding group members in captive golden lion tamarin social groups. *Amer. J. Primatol.* **18**, 73–86.

Frisch, R.E. and J.W. MacArthur (1974) Menstrual cycles: fatness as a determinant of minimum weight for height necessary for their maintenance or onset. *Science* **185**, 949–51.

Fritz, J. and L.T. Nash (1983) Rehabilitation of chimpanzees: captive population crises. *Lab. Primate Newsl.* **22**(1), 4–8.

Fuchs, E. (1987) Physiological changes in tree shrews (*Tupaia belangeri*) monitored by non-invasive techniques. *Int. J. Primatol.* **8**, 462.

Furuichi, T. (1983) Interindividual distance and influence of dominance on feeding in a natural Japanese macaque troop. *Primates* **24**, 445–55.

Galdikas, B. (1975) Orang utans, Indonesia's people of the forest. *Natn. Geogr.* **148**, 444–73.
Galdikas, B. (1980) Indonesia's orang utans: living with the great orange apes. *Natn. Geogr.* **157**, 830–53.
Galef, B.G. (1988) Imitation in animals: history, definition and interpretation of data from the psychological laboratory. In: *Social Learning: Psychological and Biological Perspectives* (T. Zentall and B.G. Galef, eds). Lawrence Erlbaum Associates, Hillsdale, NJ, 3–28.
Gandini, G. and P.J. Baldwin (1978) An encounter between chimpanzees and a leopard in Senegal. *Carnivore* **1**, 107–9.
Garber, P.A., L. Moya and C. Malaga (1984) A preliminary field study of the moustached tamarin monkey (*Saguinus mystax*) in north-eastern Peru: questions concerned with the evolution of a communal breeding system. *Folia Primatol.* **42**, 17–32.
Gardin, J.F., C.P. Jerome, M.J. Jayo and D.S Weaver (1989) Maternal factors affecting reproduction in a breeding colony of cynomolgus macaques (*Macaca fascicularis*). *Lab. Anim. Sci.* **39**, 205–12.
Gaulin, S.J.C. and M. Konner (1977) On the natural diet of primates, including humans. In: *Nutrition and the Brain* (R.J. Wurtman and J.J. Wurtman, eds). Raven Press, New York, 1–86.
Gautier-Hion, A. (1980) Seasonal variations of diet related to species and sex in a community of cercopithecus monkeys. *J. Anim. Ecol.* **49**, 237–69.
Geist, V. (1978) *Life Strategies, Human Evolution, Environmental Design: Toward a Biological Theory of Health*. Springer-Verlag, Berlin.
Ghiglieri, M.P. (1984) *The Chimpanzees of Kibale Forest*. Columbia University Press, New York.
Gibbons, J.L. (1968) The adrenal cortex and psychological distress. In: *Endocrinology and Human Behavior* (R.P. Michael, ed.). Oxford University Press, London, 220–36.
Gindoff, P.R. and M. Ferin (1987) Endogenous opioid peptides modulate the effect of corticotrophin-releasing factor on gonadotropin release in the primate. *Endocrinology* **121**, 837–42.
Glickman, S.E. and R.W. Sroges (1966) Curiosity in zoo animals. *Behaviour* **26**, 151–88.
Goldfoot, D.A. and D.A. Neff (1985) On measuring behavioural sex differences in social contexts. *Hand. Behav. Neuriobiol.* **7**, 767–83.
Goldizen, A.W. (1986) Tamarins and marmosets: communal care of offspring. In: *Primate Societies* (B.B. Smuts, D.L. Cheney, R.M. Seyfarth, R.O. Wrangham and T.T. Struhsaker, eds). University of Chicago Press, Chicago, 34–43.
Goldizen, A.W. (1987) Facultative polyandry and the role of infant-carrying in wild saddle-back tamarins (*Saguinus fuscicollis*). *Behav. Ecol. Sociobiol.* **20**, 99–109.
Goldizen, A.W. and J. Terborgh (1986) Cooperative polyandry and helping behaviour in saddle-backed tamarins (*Saguinus fuscicollis*). In: *Primate Ecology and Conservation* (J.G. Else and P.C. Lee, eds). Cambridge University Press, Cambridge, 191–8.
Goldman, L., G.D. Coover and S. Levine (1973) Bidirectional effects of reinforcement shifts on pituitary-adrenal activity. *Physiol. Behav.* **10**, 209–14.
Golub, M.S., E.N. Sassenrath and G.P. Goo (1979) Plasma cortisol levels and dominance in peer groups of rhesus weanlings. *Horm. Behav.* **12**, 50–9.

Golub, M.S., J.H. Anderson, G.P. Goo and E.N. Sassenrath (1981) Plasma cortisol response to different methods of weaning in rhesus monkey (*Macaca mulatta*) infants. *Lab. Anim. Sci.* **31**, 401–2.

Gonzalez, C.A., N.R. Gunnar and S. Levine (1981a) Behavioral and hormonal responses to social disruption and infant stimuli in female rhesus monkeys. *Psychoneuroendocrinology* **6**, 53–64.

Gonzalez, C.A., H.R. Gunnar, B.L. Goudlin and S. Levine (1981b) Behavioural and hormonal responses to social disruption and infant stimuli in female rhesus monkeys. *Psychoneuroendocrinology* **6**, 65–75.

Goodall, J. (1963) Feeding behaviour of wild chimpanzees: a preliminary report. *Symp. Zool. Soc. Lond.* **10**, 39–48.

Goodall, J. van Lawick (1968) The behaviour of free-living chimpanzees in the Gombe Stream Reserve. *Anim. Behav. Monogr.* **1**, 161–311.

Goodall, J. (1979) *The Wandering Gorillas*. Collins, London.

Goodall, J. (1983) Population dynamics during a fifteen year period in one community of free-living chimpanzees in the Gombe National Park, Tanzania. *Z. Tierpsychol.* **61**, 1–60.

Goodall, J. (1985) Part 2. In: H. Kummer and J. Goodall, Conditions of innovative behaviour in primates. *Phil. Trans. R. Soc. Lond.* B **308**, 205–13.

Goodall, J. (1986) *The Chimpanzees of Gombe: Patterns of Behaviour*. Harvard/Belknap, Cambridge, MA.

Goosen, C. (1989) Influence of age of weaning on the behaviour and well-being of rhesus monkeys. In: *Laboratory Animal Welfare Research: Primates*. UFAW Symposium. Universities Federation for Animal Welfare, Potter's Bar, Hertfordshire, 17–22.

Graham, C.A. and W.C. McGrew (1980) Menstrual synchrony in female undergraduates living on a coeducational campus. *Psychoneuroendocrinolgy* **5**, 245–52.

Gray, G.D., A.M. Bergfors, R. Levin and S. Levine (1978) Comparison of the effects of restricted morning or evening water intake on adrenocortical activity in female rats. *Neuroendocrinology* **25**, 236–46.

Gray, J.A. (1982) *The Neuropsychology of Anxiety: An Enquiry into the Functions of the Septo-hippocampal System*. Oxford, Oxford University Press.

Gross, M. and R. Charnov (1980) Alternative male life histories in bluegill sunfish. *Proc. Natn. Acad. Sci.* **77**, 6937–40.

Grossman, A. (1988) Opioids and stress in man. *J. Endocrinology* **119**, 377–81.

Grossman, A., P.J.A. Moult, H. McIntyre, J. Evans, T. Silverstone, L.H. Reese and G.M. Besser (1982) Opiate mediation of amennorhea in hyperprolactinaemia and in weight-loss related amennorhea. *Clin. Endocrinol.* **17**, 379–88.

Gruter, M. and R.D. Masters (1986) Ostracism as a social and biological phenomenon: an introduction. *Ethol. Sociobiol.* **7**, 149–58.

Grzimek, B. (1970) *Among Animals of Africa*. Collins, London.

Gunnar, M.R., C.A. Gonzalez and S. Levine (1980) The role of peers in modifying behavioral distress and pituitary adrenal response to a novel environment in year-old rhesus monkeys. *Physiol. Behav.* **25**, 795–8.

Hainsworth, F.R. and L.L. Wolf (1980) Feeding – an ecological approach. *Adv. Study Behav.* **9**, 53–96.

Hall, K.R.L. (1963) Observational learning in monkeys and apes. *Brit. J. Psychol.* **54**, 201–26.

Hamilton, W.D. and R. May (1977) Dispersal in stable habitats. *Nature, Lond.* **269**, 578–81.

Hamilton, W.J. III (1986) Namib Desert chacma baboon (*Papio ursinus*) use of food and water resources during a food shortage. *Madoqu* **14**, 397–407.

Hamm, T., J. Kaplan, T. Clarkson and B. Bullock (1983) Effects of gender and social status on coronary artery atherosclerosis in cynomolgus monkeys. *Atherosclerosis* **48**, 221–33.

Hampton, J.K., S.H. Hampton and B.T. Landwehr (1966) Observations on a successful breeding colony of the marmoset, *Oedipomidas oedipus*. *Folia Primatol.* **4**, 265–87.

Hannah, A.C. (1986) Observations on a group of captive chimpanzees released into a natural environment. *Primate Eye* **29**, 16–20.

Hannah, A.C. and W.C. McGrew (1987) Chimpanzees using stones to crack open oil palm nuts in Liberia. *Primates* **28**, 31–46.

Harcourt, A.H. (1979) Social relationships between adult male and adult female mountain gorillas in the wild. *Anim. Behav.* **27**, 325–42.

Harcourt, A.H. (1986) Gorilla conservation: anatomy of a campaign. In: *Primates: The Road to Self-Sustaining Populations* (W.K. Benirschke, ed.). Springer-Verlag, New York, 31–46.

Harcourt, A.J. (1987a) Dominance and fertility among female primates. *J. Zool. Lond.* **213**, 471–87.

Harcourt, A.J. (1987b) Options for unwanted or confiscated primates. *Primate Conservation* **8**, 111–13.

Harcourt, A.H. (1987c) Behaviour of wild gorillas and their management in captivity. *Int. Zoo Yearb.* **26**, 248–55.

Harcourt, A.H. (1988) Alliances in contests and social intelligence. In: *Machiavellian Intelligence: Social Expertise and the Evolution of Intellect in Monkeys, Apes and Humans* (R. Byrne and A. Whiten, eds). Clarendon Press, Oxford, 132–52.

Harcourt, A.H. and K.J. Stewart (1981) Gorilla male relationships: can differences during immaturity lead to contrasting reproductive tactics in adulthood? *Anim. Behav.* **29**, 206–10.

Harlow, C.R., S. Gems, J.K. Hodges and J.P. Hearn (1983) The relationship between plasma progesterone and the timing of ovulation and early embryonic development in the marmoset monkey (*Callithrix jacchus*). *J. Zool. Lond.* **201**, 273–82.

Harlow, C.R., S.G. Hillier and J.K. Hodges (1986) Androgen modulation of follicle stimulating hormone-induced granulosa cell steroidogenesis in the primate ovary. *Endocrinology* **110**, 1403–5.

Harlow, H.F., N.C. Blazek, and G.E. McClearn (1956) Manipulatory motivation in the infant rhesus monkey. *J. Comp. Physiol. Psychol.* **49**, 449–53.

Harper, D., ed. (1983) *Proceedings of the Symposium on the Conservation of Primates and their Habitats*. Vaughan Paper, No. 31. University of Leicester, Leicester.

Harris, D. (1988) *Welfare and Housing of Laboratory Primates*. Universities Federation for Animal Welfare, Potter's Bar, Hertfordshire.

Harrisson, B. (1963) Education to wild-living of young Orang utans at Bako National Park, Sarawak. *Sarawak Mus. J.* **11**(21), 220–58.

Harvey, S.M. (1987) Female sexual behaviour: fluctuations during the menstrual cycle. *J. Psychosom. Res.* **31**, 101–10.

Harvey, P.H., R.D. Martin and T.H. Clutton-Brock (1987) Life histories in comparative perspective. In: *Primate Societies* (B.B. Smuts, D.L. Cheney, R.M. Seyfarth, R.W. Wrangham and T.T. Struhsaker, eds). University of Chicago Press, Chicago, 181–96.

Hauser, M.D. and L.A. Fairbanks (1988) Mother–offspring conflict in vervet monkeys: variations in response to ecological conditions. *Anim. Behav.* **36**, 802–13.

Hauser, M.D., D.L. Cheney and R.M. Seyfarth (1986) Group extinction and fusion in free-ranging vervet monkeys. *Amer. J. Primatol.* **11**, 63–77.

Hauser, W.D., V. Reinhardt, D. Cowley, S. Eisele and R. Vertein (1987) Socialising individually caged female rhesus monkeys for the purpose of environmental enrichment. *Lab. Anim. Sci.* **37**, 509 (Abstract).

Hausfater, G. (1975) Dominance and reproduction in baboons (*Papio cynocephalus*). *Contr. Primatol.* **7**, 1–150. Karger, Basel.

Hausfater, G., J. Altmann and S.A. Altmann (1982) Long-term consistency of dominance relations among female baboons (*Papio cynocephalus*). *Science* **217**, 752–5.

Hawkes, P. (1970) Group formation in four species of macaques studied in captivity. PhD dissertation, Department of Anthropology, University of California, Davis.

Hearn, J.P. (1983) The common marmoset (*Callithrix jacchus*). In: *Reproduction in New World Primates* (J.P. Hearn, ed.). MTP, Lancaster, 181–215.

Hearn, J.P., D.H. Abbott, P.C. Chambers, J.K. Hodges and S.F. Lunn (1978) Use of the common marmoset, *Callithrix jacchus* in reproductive research. In: *Primates in Medicine*, Vol. 10 (E.I. Goldsmith and J. Moor-Jankowski, eds). Karger, Basel, 40–9.

Heath, M. (1989) The training of cynomolgus monkeys and how the human/animal relationship improves with environmental and mental enrichment. *Anim. Technol.* **40**, 11–22.

Hediger, H. (1950) *Wild Animals in Captivity*. Butterworth, London.

Hediger, H. (1955) *Studies of the Psychology and Behaviour of Captive Animals in Zoos and Circuses*. Butterworth, London.

Heltne, P.G., J.F. Wojcik and A.G. Pook (1981) Goeldi's monkey, genus *Callimico*. In: *Ecology and Behaviour of Neotropical Primates* (A.F. Coimbra-Filho and R.A. Mittermeier, eds). Academia Brasiliera de Ciências, Rio de Janeiro, 169–209.

Hennessy, J.W. and S. Levine (1979) Stress, arousal, and the pituitary-adrenal system: a psychoendocrine hypothesis. In: *Progress in Psychobiology and Physiological Psychology*, Vol. 6 (J.M. Sprague and A.N. Epstein, eds). Academic Press, New York, 133–78.

Hennessy, M.B. (1986) Multiple, brief maternal separations in the squirrel monkey: changes in hormonal and behavioural responsiveness. *Physiol. Behav.* **36**, 245–50.

Hennessy, M.B., S.P. Mendoza and J.N. Kaplan (1982) Behavior and plasma cortisol following brief peer separation in juvenile squirrel monkeys. *Amer. J. Primatol.* **3**, 143–51.

Herbert, J. (1984) Behaviour and the limbic system with particular reference to sexual and aggressive interactions. In: *Psychopharmacology of the Limbic System* (R. Trimble and E. Zarifian, eds). Oxford, Oxford University Press, 51–67.

Herndon, J.G., M. Ruiz de Elvira and J.J. Turner (1987) Influence of female behaviour and endocrine status on sexual initiation in rhesus monkey groups. *Primate Rep.* **16**, 21–6.

Hershkovitz, P. (1977) *Living New World Primates (Platyrrhini) with an Introduction to Primates*, Vol. 1. University of Chicago Press, Chicago.

Hicks, S. (1976) Nutritional Farm Scheme. *Jersey Wildl. Preserv. Trust Ann. Rep.* **13**, 112–15.

Hiiemae, K. (1984) Functional aspects of primate jaw morphology. In: *Food Acquisition and Processing in Primates* (D.J. Chivers, B.A. Wood and A. Bilsborough, eds). Plenum Press, New York, 257–81.

Hill, C.W., W.E. Greer and O. Felsenfeld (1967) Psychological stress, early response to foreign protein, and blood cortisol in vervets. *Psychosom. Med.* **29**, 273–83.

Hill, W.C.O. (1958) Pharynx, oesophagus, stomach, small and large intestine. In: *Primatologia: Handbook of Primatology* , Vol. 3 (1) (H. Hofer, A.H. Schultz and D. Starck, eds). Karger, Basel, 139–207.

Hillier, S.G., C.R. Harlow, H.J. Shaw, E.J. Wickings, A.F. Dixson and J.K. Hodges (1987) Granulosa cell differentiation in primate ovaries: the marmoset monkey (*Callithrix jacchus*) as a laboratory model. In: *The Primate Ovary* (R.L. Stouffer, ed.). Plenum Press, New York, 61–73.

Hinde, R.A. (1976) Interactions, relationships and social structure. *Man (Lond.)* **11**, 1–17.

Hinde, R.A. (1983a) *Primate Social Relationships: An Integrated Approach*. Blackwell Scientific, Oxford.

Hinde, R.A. (1983b) A conceptual framework. In: *Primate Social Relationships* (R.A. Hinde, ed.). Blackwell Scientific, Oxford, 1–7.

Hinde, R.A. (1987) *Individuals, Relationships and Culture*. Cambridge University Press, Cambridge.

Hinde, R.A. and P. Bateson (1984) Discontinuities in behavioural development and the neglect of process. *Int. J. Behav. Dev.* **7**, 129–43.

Hinde, R.A., Y. Spencer-Booth and M. Bruce (1966) Effects of 6-day maternal deprivation on rhesus monkey infants. *Nature, Lond.* **210**, 1021–3.

Hiraiwa-Hasegawa, M., R.W. Byrne, H. Takasaki and J.M.E. Byrne (1986) Aggression towards large carnivores by wild chimpanzees of Mahale Mountains National Park, Tanzania. *Folia Primatol.* **47**, 8–13.

Hirshhorn, K., F. Bach, R. Kolodney, I. Firschein and N. Hashem (1963) Immune response and mitosis of human peripheral blood lymphocytes in vitro. *Science* **142**, 1185–6.

Hladik, C.M. (1973) Alimentation et activité d'un groupe de chimpanzés réintroduits en forêt Gabonaise. *Terre et la Vie* **27**, 343–413.

Hladik, C.M. (1974) La vie d'un groupe de chimpanzés dans la forêt du Gabon. *Science et Nature* **121**, 5–14.

Hladik, C.M. and D.J. Chivers (1978) Concluding discussion: ecological factors and specific behavioural patterns determining primate diet. In: *Recent Advances in Primatology*, Vol. 1 *Behaviour* (D.J. Chivers and J. Herbert, eds). Academic Press, London, 433–44.

Hobbs, K.R., M.D. Welshman, J.B. Nazareno and R.G. Resuello (1987) Conditioning and breeding facilities for the cynomolgus monkey (*Macaca fascicularis*) in the Philippines: a progress report on the SICONBREC project. *Lab. Anim.* **21**, 131–7.

Hodges, J.K., B.A. Gulick, N.M. Czelka and B.L. Lasley (1981) Comparison of urinary oestrogen excretion in South American primates. *J. Reprod. Fert.* **61**, 83–90.

Home Office (1989) Code of practice for the housing and care of animals used

in scientific procedures. HMSO, London.
Horr, D.A. (1975) The Bornean orang utan: population structure and dynamics in relation to ecology and reproductive strategy. *Primate Behav.* **4**, 307–23.
House, J.S., K.R. Landis and D. Umberson (1988) Social relationships and health. *Science* **241**, 540–4.
Hrdy, S.B. (1976) Care and exploitation of nonhuman primate infants by conspecifics other than the mother. *Adv. Study Behav.* **6**, 101–58.
Hrdy, S.B. (1977) *The Langurs of Abu*. Harvard University Press, Cambridge.
Hrdy, S.B. and P.L. Whitten (1987) Patterning of sexual activity. In: *Primate Societies* (B.B. Smuts, D.L. Cheney, R.M. Seyfarth, R.W. Wrangham and T.T. Struhsaker, eds). University of Chicago Press, Chicago, 370–84.
Hubrecht, R.C. (1984) Field observations on group size and composition of the common marmoset (*Callithrix jacchus jacchus*), at Tapacura, Brazil. *Primates* **25**, 13–21.
Humphrey, N. (1976) The social function of intellect. In: *Growing Points in Ethology* (P.P.G. Bateson and R.A. Hinde, eds). Cambridge University Press, Cambridge, 303–17.
Huntingford, F.A. (1976) The relationship between anti-predator behaviour and aggression among conspecifics in the three-spined stickleback (*Gasterosteus aculeatus*). *Anim. Behav.* **24**, 245–60.
Iltis, H.H. (1983) Tropical forests: what will be their fate? *Environment* **25**(10), 55–60.
Inglett, B.J., J.A. French, L.G. Simmons and K. Vires (1989) Dynamics of intrafamily aggression and social reintegration in lion tamarins. *Zoo Biol.* **8**, 67–78.
Ingram, J.C. (1975) Parent–infant interactions in the common marmoset (*Callithrix jacchus*) and the development of young. PhD thesis, University of Bristol.
International Union for the Conservation of Nature and Natural Resources (1980) *World Conservation Strategy. IUCN/UNEP/WWF*, Gland, Switzerland.
Ishida, H., T. Kimura and M. Okada (1975) Patterns of bipedal walking in anthropoid primates. In: *Proceedings of Symposia of the 5th Congress of the International Primatological Society* (S. Kondo, M. Kawai, A. Ehara and S. Kawamura, eds). Japan Science Press, Tokyo, 287–301.
Itani, J. and A. Nishimura (1973) The study of infrahuman culture in Japan. In: *Precultural Primate Behaviour* (E.W. Menzel, ed.). Karger, Basel.
Iwamoto T. (1988) Food and energetics of provisioned wild Japanese macaques (*Macaca fuscata*). In: *Ecology and Behavior of Food-Enhanced Primate Groups* (J.E. Fa and C.H. Southwick, eds). Alan R. Liss, New York, 79–94.
Izawa, K. (1976) Group size and composition of monkeys in the upper Amazon basin. *Primates* **17**, 367–99.
Izawa, K. (1979) Foods and feeding behavior of wild black-capped capuchins (*Cebus apella*). *Primates* **20**, 57–76.
Jablonski, N.G. (1986) The hand of *Theropithecus brumpti*. In: *Primate Evolution* (J.G. Else and P.C. Lee, eds). Cambridge University Press, Cambridge, 173–82.
Jaeckel, J. (1989) The benefits of training rhesus monkeys living under laboratory conditions. In: *Laboratory Animal Welfare Research: Primates*. UFAW Symposium. Universities Federation for Animal Welfare, Potter's Bar, Hertfordshire, 23–5.
Janson, C. (1985) Aggressive competition and individual food consumption in

wild brown capuchin monkeys (*Cebus apella*). *Behav. Ecol. Sociobiol.* **18**, 125–38.
Janzen, D.H. (1970) Herbivores and the number of tree species in tropical forests. *Amer. Nat.* **104**, 501–28.
Janzen, D.H. (1978) Complications in interpreting the chemical defence of trees against tropical arboreal plant-eating vertebrates. In: *The Ecology of Arboreal Folivores* (G.G. Montgomery, ed.). Smithsonian Institution Press, Washington, DC, 73–84.
Jerkins, T. (1972) Free range breeding of squirrel monkeys on Santa Sofia Island, Colombia. In: *Breeding Primates* (W. Beveridge, ed.). Karger, Basel, 144.
Jerome, C.P. and L. Szostak (1987) Environmental enrichment for adult female baboons (*Papio anubis*) (Abstract). *Lab. Anim. Sci.* **37**, 508.
Johns, A.D. (1986a) Effects of habitat disturbance on rainforest wildlife in Brazilian Amazonia. Unpublished report to World Wildlife Fund US, Washington, DC.
Johns, A.D. (1986b) Notes on the ecology and current status of the buffy saki, *Pithecia albicans*. *Primate Conserv.* **7**, 26–9.
Johns, A.D. (1986c) Effects of selective logging on the behavioural ecology of West Malaysian primates. *Ecology* **67**, 684–94.
Johns, A.D. (in press). Effects of 'selective' timber extraction on rainforest structure and composition and some consequences for frugivores and folivores. *Biotropica*.
Johns, A.D. and J.P. Skorupa (in press) Responses of rainforest primates to habitat disturbance: a review. *Int. J. Primatol.*
Johnstone Scott, R. (1984) Integration and management of a group of lowland gorillas. *Dodo, J. Jersey Wildl. Preserv. Trust* **21**, 67–79.
Jolly, A. (1964) Prosimians' manipulation of simple object problems. *Anim. Behav.* **12**, 560–70.
Jolly, A. (1966) Lemur social behaviour and primate intelligence. *Science* **153**, 501–6.
Jolly, C.J. (1972) The classification and natural history of *Theropithecus* (*Simopithecus*) (Andrews, 1916) baboons of the African Plio-Pleistocene. *Bull. Br. Mus. Nat. Hist. Soc.* **22**, 1–123.
Jones, C. and J. Sabater-Pi (1971) Comparative ecology of *Gorilla gorilla* (Savage and Wyman) and *Pan troglodytes* (Blumenbach) in Rio Muni, West Africa. *Biblioth. Primatol.* **13**, 1–96.
Jordan, C.F. (1986) Local effects of tropical deforestation. In: *Conservation Biology: The Science of Scarcity and Diversity* (M. Soulé, ed.). Sinauer Associates, Sunderland, MA, 410–26.
Joubert, A. and J. Vauclair (1986) Reaction to novel objects in a troop of Guinea baboons: approach and manipulation. *Behaviour* **96**, 92–104.
Juraska, J.M. (1984) Sex differences in dendritic response to differential experience in the rat visual cortex. *Brain Res.* **295**, 27–34.
Juraska, J.M. (1986) Sex differences in developmental plasticity of behaviour and the brain. In: *Developmental Neuropsychology* (W.T. Greenough and J.M. Juraska, eds). Academic Press, New York, 409–22.
Juraska, J.M., J. Fitch, C. Henderson and N. Rivers (1985) Sex differences in the dendritic branching of dentate granule cells following differential experience. *Brain Res.* **333**, 73–80.
Kagan, J. (1980) Perspectives on continuity. In: *Constancy and Change in Human*

Development (O.G. Brim, Jr and J. Kagan, eds). Harvard University Press, Cambridge, 26–74.
Kagan, J. (1984) *On the Nature of the Child*. Basic Books, New York.
Kagan, J., J.S. Reznick and N. Snidman (1988) Biological bases of childhood shyness. *Science* **240**, 167–71.
Kagan, J., J.S. Reznick and J. Gibbons (1989) Inhibited and uninhibited types of children. *Child. Dev.* **60**, 838–45.
Kalra, S.P., L.G. Allen and P.S. Kalra (1989) Opioids in the steroid adrenergic circuit regulating LH secretion: dynamics and diversities. In: *Brain Opioid Systems in Reproduction* (R.G. Dyer and R.J. Bicknell, eds). Oxford University Press, Oxford, 95–111.
Kaplan, J.R. (1986) Psychological stress and behavior in nonhuman primates. In: *Comparative Primate Biology*, Vol.2A: *Behavior, Conservation, and Ecology* (G. Mitchell and J. Erwin, eds). Alan R. Liss, New York, 455–92.
Kaplan, J.R., S.B. Manuck, T.B. Clarkson, F.M. Lusso and D.M Taub (1982) Social status, environment and atherosclerosis in cynomolgus monkeys. *Arteriosclerosis* **2**, 359–68.
Kaplan, J.R., S.B. Manuck, T.B. Clarkson, F.M. Lusso, D.M. Taub and C.W. Miller (1983) Social factors and coronary artery atherosclerosis in normocholesterolemic monkeys. *Science* **220**, 733–5.
Kaplan, J.R., M.R. Adams, D.R. Koritnik, J.C. Rose and S.B. Manuck (1986) Adrenal responsiveness and social status in intact and ovariectomised *Macaca fascicularis*. *Amer. J. Primatol.* **11**, 181–93.
Katz, R.J. (1983) Stress, conflict and depression. In: *The Origins of Depression: Current Concepts and Approaches* (J. Angst, ed.). Springer-Verlag, Berlin, 121–32.
Katzir, G. (1982) Relationships between social structure and response to novelty in captive Jackdaws, *Corus monedula* L., I. Response to novel space. *Behaviour* **81**, 231–63.
Katzir, G. (1983) Relationships between social structure and response to novelty in captive Jackdaws, *Corus monedula* L., II. Response to novel palatable food. *Behaviour* **87**, 183–208.
Kaufman, I.C. and L.A. Rosenblum (1967) The reaction to separation in infant monkeys: anaclitic depression and conservation-withdrawal. *Psychosom. Med.* **29**, 648–75.
Kavanagh, M. (1980) Invasion of the forest by an African savannah monkey: behavioural adaptations. *Behaviour* **73**, 238–60.
Kawai, M. (1958) On the system of social ranks in a natural troop of Japanese monkeys. II. Ranking order as observed among the monkeys on or near the test box. In: *Japanese Monkeys: A Collection of Translations* (trans. S. Takada: selected by K. Imanishi and S.A. Altmann, eds.). Published by the editor (S.A. Altmann), Atlanta, Georgia, 1965.
Kawai, M. (1960) A field experiment on the process of group formation in Japanese monkeys and the releasing of the groups at Ohirayama. *Primates* **2**, 181–253.
Kawamura, S. (1965) Sub-culture among Japanese macaques. In: *Monkeys and Apes: Sociological Studies* (S. Kawamura and J. Itani, eds). Chuokoronsha, Tokyo, 237–89.
Kay, R.F. (1975) The functional adaptation of primate molar teeth. *Amer. J. Phys. Anthropol.* **43**, 195–216.
Kay, R.F. and H.H. Covert (1984) Anatomy and behaviour of extinct primates.

In: *Food Acquisition and Processing in Primates* (D.J. Chivers, B.A. Wood and A. Bilsborough, eds). Plenum Press, New York, 467–508.

Kay, R.F. and W.L. Hylander (1978) The dental structure of mammalian folivores with special reference to Primates and Phalangeroidea (Marsupalia). In: *The Ecology of Arboreal Folivores* (G.G. Montgomery, ed.). Smithsonian Institution Press, Washington, DC, 173–91.

Kelleher, R.T., W.H. Morse and J.A. Herd (1972) Effects of propranolol, phentolamine and methyl atropine on cardiovascular function in the squirrel monkey during behavioral experiments. *J. Pharmacol. Exper. Ther.* **182**, 204–17.

Kendrick, K.M. and A.F. Dixson (1984) Ovariectomy does not abolish proceptive behaviour cyclicity in the common marmoset (*Callithrix jacchus*). *J. Endocrinol.* **101**, 155–62.

Kenyon, E.R. (1938) *Gibraltar under Moor, Spaniard and Briton*. Methuen, London.

Keverne, E.B. (1979) Sexual and aggressive behaviour in social groups of talapoin monkeys. In: *Sex, Hormones and Behaviour*. Ciba Foundation Symposium (New Series), Vol. 62. Excerpta Medica, Amsterdam, 271–97.

Keverne, E.B. (1985) Hormones and the sexual behaviour of monkeys. In: *Neurobiology* (R. Gilles and J. Balthazart, eds). Springer-Verlag, Berlin 37–47.

Keverne, E.B. (1987) Processing of environmental stimuli and primate reproduction. *J. Zool. Lond.* **213**, 395–408.

Keverne, E.B., R.E. Meller and J.A. Eberhardt (1982) Dominance and subordination: concepts or physiological states? In: *Advanced Views in Primate Biology* (O. Chiarelli, ed.). Springer-Verlag, New York, 81–94.

Keverne, E.B., J.A. Eberhart, U. Yodyingyuad and D.H. Abbott (1984) Social influences on sex differences in the behavior and endocrine state of talapoin monkeys. *Prog. Brain Res.* **61**, 331–47.

King, F. and P.C. Lee (1987) A brief survey of human attitudes to a pest species of primate – *Cercopithecus aethiops*. *Primate Conserv.* **8**, 82–4.

King, G. (1975) Feeding and nutrition of the Callitrichidae. *Jersey Wildl. Preserv. Trust Ann. Rep.* **12**, 81–90.

King, G.J. (1976) Investigation into marmoset wasting syndrome. *Jersey Wildl. Preserv. Trust Ann. Rep.* **13**, 97–107.

King, G.J. and J.W.P. Rivers (1976) The affluent anthropoid. *Jersey Wildl. Preserv. Trust Ann. Rep.* **13**, 86–95.

Kingdon, J. (1980) The role of visual signals and face patterns in African forest monkeys (guenons) of the genus *Cercopithecus*. *Trans. Zool. Soc. Lond.* **35**, 425–75.

Kinzey, W.G. (1981). The titi monkeys, genus *Callicebus*. In: *Ecology and Behaviour of Neotropical Primates* (A.F. Coimbra-Filho and R.A. Mittermeier, eds). Academia Brasiliera de Ciências, Rio de Janeiro, 241–76.

Kinzey, W.G. and A.H. Gentry (1979) Habitat utilization in two species of *Callicebus*. In: *Primate Ecology: Problem-oriented Field Studies* (R.W. Sussman ed.). Wiley, New York, 89–100.

Kittenger, G.W. and N.B. Beamer (1968) Quantitative gas chromatography of squirrel monkey (*Saimiri sciureus*) corticosteroids. *Steroids* **12**, 275–89.

Kleiber, M. (1961) *The Fire of Life*. Wiley, New York.

Kleiman, D.G. (1977) Monogamy in mammals. *Quart. Rev. Biol.* **52**, 39–69.

Kleiman, D.G. (1979) Parent–offspring conflict and sibling competition in a monogamous primate. *Amer. Nat.* **194**, 753–60.

Kleiman, D.G. and D.S. Mack (1980) Effects of age, sex, and reproductive status on scent marking frequencies in the golden lion tamarin, *Leontopithecus rosalia*. *Folia Primatol.* **33**, 1–14.

Kleiman, D.G., B.B. Beck, J.M. Dietz, L.A. Dietz, J.A. Ballou and A.F. Coimbra-Filho (1986) Conservation programme for the golden lion tamarin: captive research and management, ecological strategies and reintroduction. In: *Primates: The Road to Self-Sustaining Populations* (W.K. Benirschke, ed.). Springer-Verlag, New York, 959–79.

Kleiman, D.G., R.T. Hoage and K.M. Green (1988) The lion tamarins, genus *Leontopithecus*. In: *Ecology and Behavior of Neotropical Primates*, Vol. 2 (R.A. Mittermeier, A.B. Rylands, A. Coimbra-Filho and G.A.B. Fonseca, eds). World Wildlife Fund, Washington, DC, 299–347.

Klopfer, P.H. (1969) *Habitats and Territories*. New York, Basic Books.

Kluver, H. (1937) Re-examination of implement-using behavior in a cebus monkey after an interval of three years. *Acta Psychol.* **2**, 347–95.

Koford, C.B. (1963) Rank of mothers and sons in bands of rhesus monkeys. *Science* **141**, 356–7.

Kohler, W. (1927) *The Mentality of Apes*. Vintage Books, New York.

Konstant, W.R. and R.A. Mittermeier (1982) Introduction, reintroduction and translocation of Neotropical primates: past experience and future possibilities. *Int. Zool. Yearb.* **22**, 69–77.

Kortlandt, A. (1962) Chimpanzees in the wild. *Scient. Amer.* **206**(5), 128–38.

Koyama, N., K. Norikoshi and T. Mano (1975) Population dynamics of Japanese monkeys at Arashiyama. In: *Proceedings of the 5th International Congress of Primatology, Nagoya, 1974*. Karger, New York, 411–17.

Kraemer, G.W. and W.T. McKinney (1979) Interactions of pharmacological agents which alter biogenic amine metabolism and depression. *J. Affective Dis.* **1**, 33–54.

Kram, J., H. Bourne, H. Maibach and K. Melmon (1975) Cutaneous immediate hypersensitivity in man: effects of systemically administered drugs. *J. Aller. Clin. Immunol.* **56**, 387–92.

Krebs, C.J. (1978) *Ecology: The Experimental Analysis of Distribution and Abundance*. Harper and Row, New York.

Kummer, H. (1971) *Primate Societies*. Aldine, Chicago.

Kummer, H. (1982) Social knowledge in free-ranging primates. In: *Animal Mind – Human Mind* (D.R. Griffin, ed.). Springer-Verlag, Berlin, 113–30.

Kummer, H. and F. Kurt (1965) A comparison of social behaviour in captive and wild hamadryas baboons. In: *The Baboon in Medical Research* (H. Vagfborg, ed.). University of Texas Press, Austin, 1–46.

Kummer, H. and J. Goodall (1985) Conditions of innovative behavior in primates. *Phil. Trans. R. Soc. Lond.* B. **308**, 203–14.

Langer, P. (1987) Evolutionary patterns of Perissodactyla and Artiodactyla (Mammalia) with different types of digestion. *Z. Zool. Syst. Evol.-forsch.* **25**, 212–36.

Lea, S.E.G. (1979) Foraging and reinforcement schedules in the pigeon: optimal and non-optimal aspects of choice. *Anim. Behav.* **27**, 875–86.

Lee, P.C. (1981) Ecological influence on social development of vervet monkeys (*Cercopithecus aethiops*). *Primate Eye* **17**, 4–5.

Lee, P.C. (1983) Context-specific unpredictability in dominance interactions. In: *Primate Social Relationships* (R.A. Hinde, ed.). Blackwell Scientific, Oxford, 35–44.

Lee, P.C. (1986) Environmental influences on development: play, weaning and social structure. In: *Primate Ontogeny, Cognition and Social Behaviour* (J.G. Else and P.C. Lee, eds). Cambridge University Press, Cambridge, 227–38.

Lee, P.C. (1987a) Nutrition, fertility and maternal investment in primates. *J. Zool. Lond.* **213**, 409–22.

Lee, P.C. (1987b) Sibship: cooperation and competition among immature vervet monkeys. *Primates* **28**, 47–59.

Lee, P.C. (1988) Ecological constraints and opportunities: interactions, relationships and social organization of primates. In: *The Ecology and Behaviour of Food-Enhanced Primate Groups* (J.E. Fa and C. Southwick, eds). Alan R. Liss, New York, 297–312.

Lee, P.C. and R.A. Foley (in press) Ecological energetics and the extinction of the giant gelada. In: *Theropithecus: A Case Study in Primate Evolutionary Biology* (N. Jablonki, ed.). Cambridge University Press, Cambridge.

Lee, P.C., J. Thornback and E.L. Bennett (1988) *Threatened Primates of Africa. The IUCN Red Data Book*. International Union for the Conservation of Nature and Natural Resources, Gland, Switzerland.

Lee, P.C., E.J. Brennan, J.G. Else and J. Altmann (1986) Ecology and behaviour of vervet monkeys in a tourist lodge habitat. In: *Primate Ecology and Conservation* (J.G. Else and P.C. Lee, eds). Cambridge University Press, Cambridge, 229–36.

Leighton, M. and D.R. Leighton (1983) Vertebrate responses to fruiting seasonality within a Bornean rain forest. In: *Tropical Rain Forest: Ecology and Management* (S.L. Sutton, T.C. Whitmore and A.C. Chadwick, eds). Blackwell Scientific, Oxford, 181–96.

Le Poivre, H. and B. Pallaud (1985) Social facilitation in a troop of Guinea baboons (*Papio papio*) living in an enclosure. *Behav. Process.* **11**, 405–18.

Leshner, A.I. and D.K. Candland (1972) Endocrine effects of grouping and dominance rank in squirrel monkeys. *Physiol. Behav.* **8**, 441–5.

Lessertisseur, J. and F.K. Jouffroy (1975) Comparative osteometry of the foot of man and facultatively bipedal primates. In: *Primate Functional Morphology and Evolution* (R.H. Tuttle, ed.). Mouton, The Hague, 327–40.

Leutenegger, W. (1973) Maternal–fetal weight relationships in primates. *Folia Primatol.* **20**, 280–94.

Levine, S. (1962) Plasma-free corticosteroid response to electric shock in rats stimulated in infancy. *Science* **135**, 795–6.

Levine, S. (1965) Maturation of the neuroendocrine response to stress. *Medica Int. Congr. Ser.* **99**, E.103.

Levine, S. and R.F. Mullins (1968) Hormones on infancy. In: *Early Experience and Behavior* (G. Newton and S. Levine, eds). Charles C. Thomas, Springfield, IL, 168–97.

Levine, S., L. Goldman and G.D. Coover (1972) Expectancy and the pituitary adrenal system. In: *Physiology, Emotion and Psychosomatic Illness*, CIBA Foundation Symposium, Vol. 8. Elsevier, Amsterdam, 281–96.

Levins, R. (1968) *Evolution in Changing Environments*. Princeton University Press, Princeton, NJ.

Lewis, J.K. and W.T. McKinney (1976) The effect of electrically induced convulsions on the behavior of normal and abnormal rhesus monkeys. *Dis. Nerv. Sys.* **37**, 687–93.

Lewis, J.K., L.D. Young, W.T. McKinney and G.W. Kraemer (1976) Mother–infant separation as a model of human depression: a reconsideration. *Arch.*

Gen. Psychiat. **33**, 699–705.

Lindburg, D.G., ed (1980) *The Macaques: Studies in Ecology, Behavior and Evolution.* Van Nostrand Reinhold, New York.

Lindsley, D.B. (1960) Attention, consciousness, sleep and wakefulness. In: *Handbook of Physiology: Neurophysiology,* Vol. 3 (J. Field, ed.). American Physiological Society, Washington, DC.

Line, S.W., A.S. Clarke, Ellman and H. Markowitz (1987) Behavioral and physiologic responses of rhesus macaques to an environmental enrichment device (Abstract). *Lab. Anim. Sci.* **37**, 509.

Lorenz, R. and W.A. Mason (1971) Establishment of a colony of titi monkeys. *Int. Zoo Yearb.* **11**, 168–75.

Lucas, J. (1983) The role of foraging time constraints and variable prey encounter in optimal diet choice. *Amer. Nat.* **2**, 191–209.

Lucas, P.W. and D.A. Luke (1984) Chewing it over: basic principles of food breakdown. In: *Food Acquisition and Processing in Primates* (D.J. Chivers, B.A. Wood and A. Bilsborough, eds). Plenum Press, New York, 283–301.

Lynch, J.J. (1979) *The Broken Heart.* Basic Books, New York.

McClintock, M.K. (1971) Menstrual synchrony and suppression. *Nature, Lond.* **229**, 244–5.

McClintock, M.K. (1983) Pheromonal regulation of the ovarian cycle: enhancement, suppression and synchrony. In: *Pheromones and Reproduction in Mammals* (J.G. Vandenbergh, ed.). Academic Press, New York, 113–49.

MacDonald, D., ed. (1984) *The Encyclopaedia of Mammals.* George Allen & Unwin.

MacDonald, D.W. and G.M. Carr (1989) Food security and the rewards of tolerance. In: *Comparative Socioecology: The Behavioural Ecology of Humans and Other Mammals* (V. Standen and R.A. Foley, eds). Blackwell Scientific. Oxford, 75–99.

McEwen, P. (1986) *Environmental Enrichment: An Artificial Termite Mound for Orang Utans.* Universities Federation for Animal Welfare, Potter's Bar, Hertfordshire.

McGrew, W.C. (1977) Socialization and object manipulation of wild chimpanzees. In: *Primate Biosocial Development* (S. Chevalier Skolnikoff and F. Poirier, eds). Garland, New York, 261–88.

McGrew, W.C. (1983) Chimpanzees can be rehabilitated. *Lab. Primate Newsl.* **22**(2), 2–3.

McGrew, W.C. (1989) Why is ape tool use so confusing? In: *Comparative Socioecology: The Behavioural Ecology of Humans and Other Mammals* (V. Standen and R. Foley, eds). Blackwell Scientific, Oxford, 457–72.

McGrew, W.C. and E.C. McLuckie (1986) Philopatry and dispersion in the cotton-top tamarin, *Saguinus* (*o.*) *oedipus*: an attempted laboratory simulation. *Int. J. Primatol.* **7**, 399–420.

McGrew, W.C. and C.E.G. Tutin (1978) Evidence for a social custom in wild chimpanzees? *Man* **13**, 234–51.

McGrew, W.C., C.E.G. Tutin and P.J. Baldwin (1979) Chimpanzees, tools and termites: cross-cultural comparisons of Senegal, Tanzania and Rio Muni. *Man* **14**, 185–215.

McGrew, W.C., J.A. Brennan and J. Russell (1986) An artificial 'gum-tree' for marmosets (*Callithrix j. jacchus*). *Zoo Biol.* **5**, 45–50.

McKenzie, S.M., A.S. Chamove and A.T.C. Felstner (1986) Floor-coverings and hanging screens alter arboreal monkey behavior. *Zoo Biol.* **5**, 339–48.

McKey, D. (1978) Soils, vegetation and seed-eating by colobus monkeys. In:

The Ecology of Arboreal Folivores (G.G. Montgomery, ed.). Smithsonian Institution Press, Washington, DC, 423–33.

MacKinnon, J.R. (1974) The behaviour and ecology of wild orangutans (*Pongo pygmaeus*). *Anim. Behav.* **22**, 3–74.

MacKinnon, J. (1977) Pet orang utans: should they return to the forest? *New Scientist* **74**, 697–9.

MacKinnon, J.R. and K.S. MacKinnon (1980a) The behaviour of wild spectral tarsiers. *Int. J. Primatol.* **1**, 361–79.

MacKinnon, J.R. and K.S. MacKinnon (1980b) Niche differentiation in a primate community. In: *Malayan Forest Primates*, (D.J. Chivers, ed.). Plenum Press, New York, 167–90.

McLanahan E.B. and K.M. Green (1978) The vocal repertoire and an analysis of the contexts of vocalizations in *Leontopithecus rosalia*. In: *The Biology and Conservation of the Callitrichidae* (D. Kleiman, ed.). Smithsonian Institution Press, Washington, DC, 251–70.

MacLarnon, A.M., D.J. Chivers and R.D. Martin (1986) Gastro-intestinal allometry in primates and other mammals including new species. In: *Primate Ecology and Conservation* (J.G. Else and P.C. Lee, eds). Cambridge University Press, Cambridge, 75–85.

Mager, W.B. and T. Griede (1986) Using outside areas for tropical primates in the Northern Hemisphere: Callitrichidae: *Saimiri* and *Gorilla*. In: *Primates: The Road to Self-Sustaining Populations* (W.K. Benirschke, ed.). Springer-Verlag, New York, 471–7.

Maier, W. (1984) Tooth morphology and dietary specialization. In: *Food Acquisition and Processing in Primates* (D.J. Chivers, B.A. Wood and A. Bilsborough, eds). Plenum Press, New York, 303–30.

Malik, I. (1986) Increased home range for a self-sustaining free-ranging rhesus population at Tughlaqabad, India. In: *Primates: The Road to Self-Sustaining Populations* (W.K. Benirschke, ed.). Springer-Verlag, New York, 190–5.

Malik, I. and C.H. Southwick (1988) Feeding behavior and activity patterns of rhesus monkeys at Tughlaqabad, India. In: *Ecology and Behavior of Food-Enhanced Primate Groups* (J.E. Fa and C.H. Southwick, eds). Alan R. Liss, New York, 95–112.

Malik, I., P.K. Seth and C.H. Southwick (1985) Group fission in free-ranging rhesus monkey at Tughlaqabad, Northern India. *Int. J. Primatol.* **6**, 411–21.

Mallinson, J.J.C. (1964) Notes on the nutrition, social behaviour and reproduction of *Hapalidae* in captivity. *Int. Zoo Yearb.* **5**, 137–40.

Mallinson, J.J.C. (1971) The breeding and maintenance of marmosets at Jersey Zoo. *Int. Zoo. Yearb.* **11**, 79–83.

Mallinson, J.J.C. (1975a) Breeding marmosets in captivity. In: *Breeding Endangered Species in Captivity* (R.D. Martin ed.). Academic Press, London, 203–12.

Mallinson, J.J.C. (1975b) The design of two marmoset complexes at the Jersey Zoological Park. *Jersey Wildl. Preserv. Trust Ann. Rep.* **12**, 21–6.

Mallinson, J.J.C. (1977) Acclimatisation of tropical animals in the northern hemisphere, with special reference to the family Callitrichidae. *Dodo, J. Jersey Wildl. Preserv. Trust* **14**, 34–40.

Mallinson, J.J.C. (1980) The concept behind and design of the new gorilla environment at the Jersey Wildlife Preservation Trust. *Dodo, J. Jersey Wildl. Preserv. Trust* **17**, 79–85.

Mallinson, J.J.C. (1986a) The importance of an interdisciplinary approach:

getting the conservation act together. In: *Primates: The Road to Self-Sustaining Populations* (W.K. Benirschke, ed.). Springer-Verlag, New York, 995–1003.
Mallinson, J.J.C. (1986b) The Wildlife Preservation Trust's support for the conservation of the genus *Leontopithecus*. *Dodo, J. Jersey Wildl. Preserv. Trust* **23**, 6–18.
Mallinson, J.J.C. (1987) Lion tamarins fly in and out. *On the Edge* No. 54, 7.
Mallinson, J.J.C., P. Coffey and J. Usher Smith (1973) Maintenance, breeding and hand-rearing of lowland gorilla *Gorilla g. gorilla* at the Jersey Zoological Park. *Jersey Wildl. Preserv. Trust Ann. Rep.* **10**, 5–28.
Manzolillo, D.L. (1986) Factors affecting intertroop transfer by adult male *Papio anubis*. In: *Primate Ontogeny, Cognition and Social Behaviour* (J.G. Else and P.C. Lee, eds). Cambridge University Press, Cambridge, 371–80.
Maple, T.L. (1979) Great apes in captivity: the good, the bad and the ugly. In: *Captivity and Behaviour* (J. Erwin, T. Maple and G. Mitchell, eds). Van Nostrand Reinhold, New York, 239–72.
Maples, W.R. (1969) Adaptive behavior of baboons. *Amer. J. Phys. Anthropol.* **31**, 107–9.
Maples, W.R., M.K. Maples, W.F. Greenhood and M.L. Walek (1976) Adaptations of crop-raiding baboons in Kenya. *Amer. J. Phys. Anthropol.* **45**, 309–16.
Marchlewska-Koj, A. (1984) Pheromones and mammalian reproduction. In: *Oxford Reviews of Reproductive Biology*, Vol. 6 (J. Clarke, ed.). Oxford University Press, Oxford, 266–302.
Markowitz, H. (1979) Environmental enrichment and behavioural engineering for captive primates. In: *Captivity and Behaviour* (J. Erwin, T. Maple and G. Mitchell, eds). Van Nostrand Reinhold, New York, 231–8.
Markowitz, H. (1982) *Behavioural Enrichment in the Zoo*. Van Nostrand Reinhold, New York.
Markowitz, H. and J.S. Spinelli (1986) Environmental engineering for primates. In: *Primates: The Road to Self-Sustaining Populations* (W.K. Benirschke, ed.). Springer-Verlag, New York, 489–98.
Marler, P. (1976) On animal aggression: the roles of strangeness and familiarity. *Amer. Psychol.* **31**, 239–46.
Marriott, B.M. (1988) Time budgets of rhesus monkeys (*Macaca mulatta*) in a forest habitat in Nepal and on Cayo Santiago. In: *Ecology and Behavior of Food-Enhanced Primate Groups* (J.E. Fa and C.H. Southwick, eds). Alan R. Liss, New York, 125–52.
Marsh, C.W. (1985) A resurvey of Tana River primates. Cyclostyled report to the Institute of Primate Research and Department of Wildlife Conservation and Management, Kenya, 31pp.
Marsh, J.T. and A.F. Rasmussen (1960). Response of adrenal, thymus, spleen and leucocytes to shuttle box and confinement stress. *Proc. Soc. Exp. Biol. Med.* **104**, 180–3.
Marsh, C.W., A.D. Johns and J.M. Ayres (1987) Effects of habitat disturbance on rain forest primates. In: *Primate Conservation in Tropical Rain Forest* (C.W. Marsh and R.A. Mittermeier, eds). Alan R. Liss, New York, 83–107.
Martensz, N.D., S.V. Vellucci, J. Herbert and E.B. Keverne (1986) β-endorphin in the cerebrospinal fluid of male talapoin monkeys related to dominance status in social groups. *Neuroscience* **18**, 651–8.
Martin, R.D. (1972) A preliminary field study of the lesser mouse lemur (*Microcebus murinus*, Miller, 1777). *Z. Tierpsychol. Suppl.* **9**, 42–90.
Martin, R.D. (1973) A review of the behaviour and ecology of the lesser mouse

lemur (*Microcebus murinus*). In: *Comparative Ecology and Behaviour of Primates* (R.P. Michael and J.H. Crook, eds). Academic Press, London, 1–168.
Martin, R.D. (1975) Strategies of reproduction. *Nat. Hist.* **84**(9), 48–57.
Martin, R.D. (1984) Scaling effects and adaptive strategies in mammalian lactation. *Symp. Zool. Soc. Lond.* **51**, 87–117.
Martin, R.D. (1984b) Dwarf and mouse lemurs. In: *The Encyclopedia of Mammals,* Vol. 1 (D.W. MacDonald, ed.). George Allen & Unwin, London, 326–7.
Martin, R.D. (1986) Primates: a definition. In: *Major Topics in Primate and Human Evolution* (B. Wood, L. Martin and P. Andrews, eds). Cambridge University Press, Cambridge, 1–31.
Martin, R.D., D.J. Chivers, A.M. MacLarnon and C.M. Hladik (1985) Gastrointestinal allometry in primates and other mammals. In: *Size and Scaling in Primate Biology* (W.L. Jungers ed.). Plenum Press, New York, 61–89.
Masau, J.M. and S.C. Strum (1984) Response of wild baboon troops to incursion of agriculture at Gilgil, Kenya. *Int. J. Primatol.* **5**, 364.
Mason, J.W. (1968) A review of psychoendocrine research on the sympathetic-adrenal medullary system. *Psychosom. Med.* **30**, 631–53.
Mason, J.W. (1975) A historical view of the stress field. *J. Hum. Stress* **1**, 6–12.
Mason, J.W., W.J.H. Nauta, J.V. Brady, J.A. Robinson and J.S. Thack (1960) Limbic system influences on the pituitary-adrenal cortical system. *Psychosom. Med.* **22**, 322–36.
Mason, J.W., W.J.H. Nauta, J.V. Brady and E.D. Taylor (1961) The role of the limbic system structures in the regulation of ACTH secretion. *Acta Neuroveg.* (Vienna), **23**, 4–10.
Mason, J.W., J.V. Brady and G.A. Tolliver (1968) Plasma and urinary 17–hydroxycorticosteroid responses to 72–hr. avoidance sessions in the monkey. *Psychosom. Med.* **30**, 608–30.
Mason, W.A. (1966) Social organization of the South American monkey *Callicebus moloch*: a preliminary report. *Tulane Stud. Zool.* **13**, 23–8.
Mason, W.A. (1968) Use of space by *Callicebus* groups. In: *Primates: Studies in Adaptation and Variability* (P.C. Jay, ed.). Holt Rinehart and Winston, New York, 200–16.
Mason, W.A. (1971) Field and laboratory studies of social organization in *Saimiri* and *Callicebus*. In: *Primate Behavior: Developments in Field and Laboratory Research,* Vol. 2 (L.A. Rosenblum, ed.). Academic Press, New York, 107–37.
Mason, W.A. (1974) Comparative studies of social behavior in *Callicebus* and *Saimiri*: behaviour of male–female pairs. *Folia Primatol.* **22**, 1–8.
Mason, W.A. (1975) Comparative studies of social behavior in *Callicebus* and *Saimiri*: strength and specificity of attraction between male–female cagemates. *Folia Primatol.* **23**, 11–123.
Mason, W.A. (1976) Primate social behavior: pattern and process. In: *Evolution of Brain and Behavior in Vertebrates* (R.B. Masterton, M.E. Betterman, C.B.G. Campbell and N. Hotton, eds). Lawrence Erlbaum Associates, Hillsdale, NJ, 425–55.
Mason, W.A. (1978) Ontogeny of social systems. In: *Recent Advances in Primatology* Vol. 1: *Behaviour* (D.J. Chivers and J. Herbert, eds). Academic Press, London, 5–14.
Mason, W.A. (1984) Animal learning: experience, life modes and cognitive style. *Verh. Dtsch. Zool. Ges.* **77**, 45–56.
Mason, W.A. and G. Epple (1969) Social organization in experimental groups of *Saimiri* and *Callicebus*. *Proceedings of the Second International Congress of*

Primatology, Atlanta, GA, Vol. 1. Karger, Basel, 59–65.

Mason, W.A. and R. Lorenz (1988) Social influences on feeding behavior in *Callicebus* and *Saimiri*. *Amer. J. Primatol.* **14**, 433.

May, R.M. (1979) The structure and dynamics of ecological communities. In: *Population Dynamics* (R.M. Anderson, B.R. Turner and L.R. Taylor, eds). Blackwell, Oxford, 385–407.

Maynard Smith, J. (1982) *Evolution and the Theory of Games*. Cambridge University Press, New York, 385–407.

Mehlman, P.T. (1984) Aspects of the ecology and conservation of the Barbary macaque in the fir forest habitat of the Moroccan Rif Mountains. In: *The Barbary Macaque – A Case Study in Conservation* (J.E. Fa, ed.). Plenum Press, New York, 165–99.

Meikle, D.B., B.L. Tilford and S.H. Vessey (1984) Dominance rank, secondary sex ratio and reproduction of offspring in polygymous primates. *Amer. Nat.* **124**, 173–88.

Meller, R.E., E.B. Keverne and J. Herbert (1980) Behavioural and endocrine effects of naltrexone in male talapoin monkeys. *Pharmacol. Biochem. Behav.* **13**, 663–72.

Melmon, K.L., Y. Weinstein, H.R. Bourne, G. Shearer, T. Poon, L. Kraany and S. Segal (1976) Isolation of cells with specific receptors for amines. In: *Opportunities and Problems in Cell Membrane Receptors for Viruses, Antigens and Antibodies* (R.F. Beers and R.G. Basset, eds). Academic Press, New York, 117–34.

Melnick, D.J. and M.C. Pearl (1987) Cercopithecines in multi-male groups: genetic diversity and population structure. In: *Primate Societies* (B.B. Smuts, D.L. Cheney, R.M. Seyfarth, R.W. Wrangham and T.T. Struhsaker, eds). University of Chicago Press, Chicago, 121–34.

Mendoza, S.P. (1984) The psychobiology of social relationships. In: *Social Cohesion: Essays Toward a Sociophysiological Perspective* (P.R. Barchas and S.P. Mendoza, eds). Greenwood Press, Westport, CT, 3–29.

Mendoza, S.P. (1987) Formation of male–male relationships in squirrel monkeys reduces heart rate. *Amer. J. Primatol.* **12**, 360.

Mendoza, S.P. and W.A. Mason (1984) Rambunctious *Saimiri* and reluctant *Callicebus*: systemic contrasts in stress physiology. *Amer. J. Primatol.* **6**, 415.

Mendoza, S.P. and W.A. Mason (1986a) Contrasting responses to intruders and to involuntary separation by monogamous and polygynous New World monkeys. *Physiol. Behav.* **38**, 795–801.

Mendoza, S.P. and W.A. Mason (1986b) Parental division of labour and differentiation of attachments in a monogamous primate (*Callicebus moloch*). *Anim. Behav.* **34**, 1336–47.

Mendoza, S.P. and W.A. Mason (1989a) Behavioral and endocrine consequencs of heterosexual pair formation in squirrel monkeys. *Physiol. Behav.* **46**, 597–603.

Mendoza, S.P. and W.A. Mason (1989b) Primate relationships: social dispositions and physiological responses: In: *Perspectives in Primate Biology*, Vol. 2: *Neurobiology* (P.K. Seth and S. Seth, eds). Today and Tomorrow's Printers and Publishers, New Delhi, 129–43.

Mendoza, S.P. and W.A. Mason (submitted). Breeding readiness in squirrel monkeys: II. Female-primed females are triggered by males. *Horm. Behav.*

Mendoza, S.P. and G.P. Moberg (1985) Species differences in adrenocortical activity of New World Primates: responses to dexamethasone suppression.

Amer. J. Primatol. **8**, 215–24.
Mendoza, S.P., E.L. Lowe and S. Levine (1978) Social organization and social behavior in two subspecies of squirrel monkeys (*Saimiri sciureus*). *Folia Primatol.* **30**, 126–44.
Mendoza, S.P., C.L. Coe, E.L. Lowe and S. Levine (1979) The physiological response to group formation in adult male squirrel monkeys. *Psychoneuroendocrinology* **3**, 221–9.
Mendoza, S.P., C.L. Coe, W.P. Smotherman, J.N. Kaplan and S. Levine (1980) Functional consequences of attachment: a comparison of two species. In: *Maternal Influences and Early Behavior* (R.W. Bell and W.P. Smotherman, eds). Spectrum Publications, Jamaica, 235–52.
Menzel, C.R. (1986) Structural aspects of arboreality in titi monkeys (*Callicebus moloch*). *Amer. J. Phys. Anthropol.* **70**, 167–76.
Menzel, E.W. (1971) Group behaviour in young chimpanzees: responsiveness to cumulative novel changes in a large outdoor enclosure. *J. Comp. Physiol. Psychol.* **74**, 46–51.
Menzel, E.W. Jr (1974) A group of young chimpanzees in a one-acre field. In: *Behaviour of Nonhuman Primates* (A.M. Schrier and F. Stollnitz, eds). Academic Press, New York, 83–153.
Menzel, E.W. Jr and C. Juno (1982) Marmosets (*S. fuscicollis*): are learning sets learned? *Science* **217**, 750–2.
Menzel, E.W. Jr and C.R. Menzel (1979) Cognitive, developmental and social aspects of responsiveness to novel objects in a family group of marmosets (*Saguinus fuscicollis*). *Behaviour* **70**, 251–79.
Miller, L.C., K.A. Bard, C.J. Juno and R.D. Nadler (1986) Behavioural responsiveness of young chimpanzees (*Pan troglodytes*) to a novel environment. *Folia Primatol.* **47**, 128–42.
Miller, N.E. (1980) A perspective on the effects of stress and coping on disease and health. In: *Coping and Health* (S. Levine and H. Ursin, eds). Plenum Press, New York, 323–53.
Milton, K. (1984) The role of food processing factors in primate food choice. In: *Adaptations for Foraging in Non-human Primates* (P.S. Rodman and J.G.H. Cant, eds). Columbia University Press, New York, 249–79.
Mineka, S. and S.J. Suomi (1978) Social separation in monkeys. *Psychol. Bull.* **85**, 1376–400.
Mineka, S., S.J. Suomi and R. DeLizio (1981) Multiple separations in adolescent monkeys: an opponent-process interpretation. *J. Exp. Psychol. General* **110**, 56–85.
Misslin, R. and M. Cigrang (1986) Does neophobia necessarily imply fear or anxiety? *Behav. Process.* **12**, 45–50.
Mittermeier, R.A. (1986a) A global overview of primate conservation. In: *Primate Ecology and Conservation* (J.G. Else and P.C. Lee, eds). Cambridge University Press, Cambridge, 325–40.
Mittermeier, R.A. (1986b) Strategies for the conservation of highly endangered primates. In: *Primates: The Road to Self-Sustaining Populations* (W.K. Benirschke, ed.). Springer-Verlag, New York, 1013–22.
Mittermeier, R.A. and D.L. Cheney (1987) Conservation of primates and their habitats. In: *Primate Societies* (B.B. Smuts, D.L. Cheney, R.M. Seyfarth, R.W. Wrangham and T.T. Struhsaker, eds). University of Chicago Press, Chicago, 477–90.
Mittermeier, R.A. and M.G.M. van Roosmalen (1981) Preliminary observations

on habitat utilization and diet in eight Surinam monkeys. *Folia Primatol.* **36**, 1–39.
Mittermeier, R.A., J.F. Oates, A.A. Eudey and J. Thornback (1986) Primate conservation. In: *Comparative Primate Biology*, Vol. 2A: *Behavior, Conservation and Ecology* (G. Mitchell and J. Erwin, eds). Alan R. Liss, New York, 3–72.
Moberg, G.P. (1985) Biological response to stress: key to assessment of animal well-being? In: *Animal Stress* (G.P. Moberg, ed.). American Physiological Society, Washington, DC, 27–49.
Mori, A. (1979) Analysis of population changes by measurement of body weight in the Koshima troop of Japanese monkeys. *Primates* **20**, 371–97.
Mori, U. and R.I.M. Dunbar (1985) Changes in the reproductive constitution of female gelada baboons following the takeover of one-male units. *Z. Tierpsychol.* **67**, 215–24.
Morse, D. (1980) *Behavioral Mechanisms in Ecology*. Harvard University Press, Cambridge, MA.
Moss, C. (1988) *Elephant Memories*. Elm Tree Books, London.
Moynihan, M. (1970) Some behavior patterns of platyrrhine monkeys: II. *Saguinus geoffroyi* and some other tamarins. *Smithson. Contr. Zool.* **28**, 1–77.
Myers, N. (1980) The present state and future prospects of tropical moist forests. *Env. Conserv.* **7**, 101–14.
Myers, N. (1983) *A Wealth of Wild Species: Storehouse for Human Welfare*. Westview, Boulder, CO.
Myers, N. (1984) *The Primary Source: Tropical Forests and Our Future*. W.W. Norton, New York.
Myers, N. (1986) Tropical deforestation and a mega-extinction spasm. In: *Conservation Biology: The Science of Scarcity and Diversity* (M. Soulé, ed.). Sinauer Associates, Sunderland, MA, 394–409.
Myers, N. (1988) Mass extinction – profound problem, splendid opportunity. *Oryx* **22**, 205–10.
Napier, J.R. (1971) *The Roots of Mankind*. Allen & Unwin, London.
Napier, J.R. and P.H. Napier (1967) *A Handbook of Living Primates*. Academic Press, London.
Napier, J.R. and A.C. Walker (1967) Vertical clinging and leaping – a newly recognised category of locomotor behaviour of primates. *Folia Primatol.* **6**, 204–19.
National Research Council (1981) *Techniques for the Study of Primate Population Ecology*. National Academy Press, Washington, DC.
Neyman, P.R. (1978) Aspects of the ecology and social organization of free-ranging cotton-top tamarins (*Saguinus oedipus*) and the conservation status of the species. In: *The Biology and Conservation of the Callitrichidae* (D. Kleiman, ed.). Smithsonian Institution Press, Washington, DC, 39–71.
Nicolson, N.A. (1977) A comparison of early behavioral development in wild and captive chimpanzees. In: *Primate Biosocial Development* (S. Chevalier-Skolnikoff and F.E. Poirier, eds). Garland, New York, 529–63.
Nishida, T. (1979) The social structure of chimpanzees of the Mahale Mountains. In: *The Great Apes* (D.A. Hamburg and E.R. McCown, eds). Benjamin/Cummings, Palo Alta, CA, 73–122.
Nishida, T. (1987) Local traditions and cultural transmissions. In: *Primate Societies* (B.B. Smuts, D.L. Cheney, R.M. Seyfarth, R.W. Wrangham and T.T. Struhsaker, eds). University of Chicago Press, Chicago, 462–74.
Nishida, T. and M. Hiraiwa-Hasegawa (1987) Chimpanzees and bonobos:

cooperative relationships among males. In: *Primate Societies* (B.B. Smuts, D.L. Cheney, R.M. Seyfarth, R.W. Wrangham and T.T. Struhsaker, eds). University of Chicago Press: Chicago, 165–80.

Noe, R. (1986) Lasting alliances among adult male savannah baboons. In: *Primate Ontogeny, Cognition and Social Behaviour* (J.G. Else and P.C. Lee, eds). Cambridge University Press, Cambridge, 380–92.

Novak, M.A. and S.J. Suomi (1989) Psychological well-being of primates in captivity. *ILAR News*, **31**(3), 5–14.

Novotny, M., B. Jemiolo, S. Harvey, D. Weisler and A. Marchlewska-Koj (1986) Adrenal-mediated endogenous metabolites inhibit puberty in female mice. *Science* **231**, 722–5.

Nur, N. (1987) Alternative reproductive tactics in birds: individual variation in clutch size. In: *Perspectives in Ethology* (P.P.G. Bateson and P. Klopfer, eds). Plenum Press, New York, 49–77.

Oates, J.F. (1985) *Action Plan for African Primate Conservation: 1986–90.* IUCN/SSC Primate Specialist Group, New York.

Oates, J.F., P.G. Waterman and G.M. Choo (1980) Food selection by the South Indian leaf monkey, *Presbytis johnii*, in relation to leaf chemistry. *Oecologia (Berl.)* **45**, 45–56.

Oates, J.F., J.S. Gartlan and T.T. Struhsaker (1987) A framework for African primate conservation. In: *Primate Conservation in the Tropical Rainforest* (C.W. Marsh and R.A. Mittermeier, eds). Alan R. Liss, New York, 321–7.

Olster, D.H. and M. Ferin (1987) Corticotrophin-releasing hormone inhibits gonadotropin secretion in the ovariectomised rhesus monkey. *J. Clin. Endocrin. Metab.* **65**, 262–7.

Ortony, A., G.L. Clore and A. Collins (1988) *The Cognitive Structure of Emotions*. Cambridge University Press, New York.

Osborne, S.R. (1977) The free food (contra-freeloading) phenomenon: a review and analysis. *Anim. Learn. Behav.* **5**, 221–35.

Overmier, J.B., J. Patterson and R.M. Wielkiewicz (1980) Environmental contingencies as sources of stress in animals. In: *Coping and Health* (S. Levine and H. Ursin, eds). Plenum Press, New York, 1–38.

Oxnard, C. (1975) Primate locomotor classifications for evaluating fossils: their inutility and an alternative. In: *Symposia of the Fifth Congress of the International Primatological Society* (S. Kondo, M. Kawai, A. Ehara and S. Kawamura, eds). Japan Science Press, Tokyo, 269–86.

Oxnard, C.E. (1979) The morphological–behavioral interface in extant primates: some implications for systematics and evolution. In: *Environment, Behavior and Morphology: Dynamic Interactions in Primates* (M.E. Morbeck, H. Preuschoft and N. Gomberg, eds). Fischer, New York, 183–208.

Oxnard, C.E. (1987) Comparative anatomy of primates: old and new. In: *Comparative Primate Biology*, Vol. 1: *Systematics, Evolution and Anatomy* (D.R. Swindler and J. Erwin, eds). Alan R. Liss, New York, 719–63.

Packer, C. (1977) Reciprocal altruism in *Papio anubis*. *Nature*, **265**, 441–3.

Packer, C. (1979) Inter-troop transfer and inbreeding avoidance in *Papio anubis*. *Anim. Behav.* **27**, 1–36.

Patkai, P. (1971) Catecholamine excretion in pleasant and unpleasant situations. *Acta Psychol.* **35**, 352–63.

Paul, A. and D. Thommen (1984) Timing of birth, female reproduction success and infant sex ratio in semi-free-ranging Barbary macaques (*Macaca sylvanus*). *Folia Primatol.* **42**, 2–16.

Payne J. (1987) Surveying orang utan populations by counting nests from a helicopter: a pilot survey in Sabah. *Primate Conserv.* **8**, 92–103.

Periera, M.E. and J. Altmann (1985) Development of social behaviour in free-living non-human primates. In: *Nonhuman Primate Models for Human Growth and Development* (E.S. Watts, ed.). Alan R. Liss, New York, 217–309.

Perret, M. (1986) Social influences on oestrous cycle length and plasma progesterone concentrations in the female lesser mouse lemur (*Microcebus murinus*). *J. Reprod. Fertil.* **77**, 303–11.

Peters, R.H. (1983) *The Ecological Implications of Body Size.* Cambridge University Press, Cambridge.

Petter-Rousseaux, A. (1980) Seasonal activity rhythms, reproduction and body weight variations in five sympatric nocturnal prosimians in simulated light and climatic conditions. In: *Nocturnal Malagasy Primates: Ecology, Physiology and Behaviour* (P. Charles-Dominique, H.M. Cooper, A. Hladik, C.M. Hladik, C. Pages, G. Pariente, A. Petter-Rousseaux, J.J. Petter and A. Schilling, eds). Academic Press, New York, 137–52.

Pfeiffer, A.J. and L.J. Koebner (1978) The resocialization of single-caged chimpanzees and establishment of an island colony. *J. Med. Primatol.* **7**, 70–81.

Phillips, M.J. and W.A. Mason (1976) Comparative studies of social behavior in *Callicebus* and *Saimiri*: social looking in male–female pairs. *Bull. Psychonomic Soc.* **7**, 55–6.

Plimpton, E.H. and L.A. Rosenblum (1983) Detachment-avoidance responses to mother following a separation: a comparative perspective. In: *Advances in the Study of Primate Social Development* (H. Harlow, L. Rosenblum and S. Suomi, eds). Academic Press, New York, 146–59.

Pollard, I., J.R. Bassett and K.D. Cairncross (1976) Plasma glucocorticoid elevation and ultrastructural changes in the adenohypophysis of the male rat following prolonged exposure to stress. *Neuroendocrinology* **21**, 312–30.

Poole, T.B. (1987) Social behaviour of a group of orang utans (*Pongo pygmaeus*) on an artificial island in Singapore Zoological Gardens. *Zoo Biol.* **6**, 315–30.

Poole, T.B. (1988) Normal and abnormal behaviour in captive primates. *Primate Rep.* **22**, 3–12.

Popp, J.L. (1983) Ecological determinism in the life histories of baboons. *Primates* **24**, 198–210.

Post, D. (1978) Feeding and ranging behaviour of yellow baboons *Papio cynocephalus*. PhD thesis, Yale University, New Haven, CT.

Post, D., G. Hausfater and S.A. McCuskey (1980) Feeding behavior of yellow baboons (*Papio cynocephalus*): relationship to age, gender and dominance rank. *Folia Primatol.* **34**, 170–95.

Post, W. and J. Baulu (1978) Time budgets of *Macaca mulatta*. *Primates* **19**, 125–39.

Preti, G., W.B. Cutler, C.R. Garcia, G.R. Huggins and H.J. Lawley (1986) Human axillary secretions influence women's menstrual cycles: the role of donor extract of females. *Horm. Behav.* **20**, 474–82.

Pribram, K.H. (1967) The neurology and biology of emotion: a structural approach. *Amer. Psychol.* **22**, 830–8.

Quadagno, D.M., H.E. Shubeita, J. Deck and D. Fraucoer (1981) Influence of male social contacts, exercise and all-female living conditions on the female menstrual cycle. *Psychoneuroendocrinology* **6**, 239–44.

Raemaekers, J.J. and D.J. Chivers (1980) Socio-ecology of Malayan forest primates. In: *Malayan Forest Primates* (D.J. Chivers, ed.). Plenum Press, New York, 279–316.

Rasmussen, D.R. and K.L. Rasmussen (1979) Social ecology of adult males in a confined troop of Japanese macaques (*Macaca fuscata*). *Anim. Behav.* **27**, 434–45.

Rathbun, C.D. (1979) Description and analysis of the arch display in the golden lion tamarin, *Leontopithecus rosalia rosalia*. *Folia Primatol.* **32**, 125–48.

Rawlins, R.G. and M.J. Kessler (1986) *The Cayo Santiago Macaques: History, Behaviour and Biology*. State University of New York Press, New York.

Redican, W.K. and G. Mitchell (1973) The social behaviour of adult male–infant pairs of rhesus macaques in a laboratory environment. *Amer. J. Phys. Anthropol.* **38**, 523–6.

Redshaw, M.E. (1978) Cognitive development in human and gorilla infants. *J. Hum. Evol.* **7**, 133–41.

Reinhardt, V., D. Hauser, S. Eisele, D. Cowley and R. Vertein (1988) Behavioural responses to unrelated rhesus monkey females paired for the purpose of environmental enrichment. *Amer. J. Primatol.* **14**, 135–40.

Reite, M. (1987) Infant abuse and neglect: lessons from the primate laboratory. *Child Abuse and Neglect* **11**, 347–55.

Reite, M. and R. Short (1980) A biobehavioral developmental profile (B, D, P) for the pigtailed monkey. *Dev. Psychobiol.* **13**, 243–85.

Reite, M., R. Short, I.C. Kaufman, A.J. Stynes and J.D. Pauley (1978) Heart rate and body temperature in separated monkey infants. *Biol. Psychiat.* **13**, 91–105.

Reite, M., R. Short, C. Seiler and J.D. Pauley (1981) Attachment, loss and depression. *J. Child Psychol. Psychiat.* **22**, 141–69.

Richard, A.F. (1985) *Primates in Nature*. W.H. Freeman, New York.

Richardson, R., M.A. Siegel and B.A. Campbell (1988) Unfamiliar environments impair information processing as measured by behavioural and cardiac orienting responses to auditory stimuli in preweanling and adult rats. *Dev. Psychobiol.* **21**, 491–503.

Ridley, R. and H.F. Baker (1982) Stereotypy in monkeys and humans. *Psychol. Med.* **12**, 61–72.

Ridley, R.M., H.F. Baker and P.R. Scraggs (1979) The time course of the behavioural effects of amphetamine and their reversal by haloperidol in a primate species. *Biol. Psychiat.* **14**, 753–65.

Ridley, R., A.J. Haystead and H.F. Baker (1981) An analysis of visual object reversal learning in the marmoset after amphetamine and haloperidol. *Pharmac. Biochem. Behav.* **14**, 345–51.

Rijksen, H.D. (1978) *A Field Study on Sumatran Orang utans* (*Pongo pygmaeus abelii*), Lesson 1827. H. Veenman, Wageningen.

Rijksen, H.D. and A.G. Rijksen-Graatsma (1975) Orang utan rescue work in North Sumatra. *Oryx* **13**, 63–73.

Ripley, S. (1979) Environmental grain, niche diversification and positional behaviour in Neogene primates: an evolutionary hypothesis. In: *Environment, Behavior and Morphology: Dynamic Interactions in Primates* (M.E. Morbeck, H. Preuschoft and N. Gomberg, eds). Fischer, New York, 37–74.

Ripley, S. (1984) Environmental grain, niche diversification and feeding behaviour in primates. In: *Food Acquisition and Processing in Primates* (D.J. Chivers, B.A. Wood and A. Bilsborough, eds). Plenum Press, New York, 33–72.

Ritchie, B. and D.M. Fragaszy (1988) Treatment of her infant's wound with tools by a capuchin monkey (*Cebus apella*). *Amer. J. Primatol.* **16**, 345–8.
Robbins, T.C. (1983) *Wildlife Feeding and Nutrition*. Academic Press, New York.
Robinson, J.G. (1986) Seasonal variation in use of time and space by the wedge-capped capuchin monkey *Cebus olivaceus*: implications for foraging theory. *Smithson. Contr. Zool.* **431**, 1–60.
Romer, A.S. (1966) *Vertebrate Paleontology*. University of Chicago Press, Chicago.
Rose, M.D. (1974) Postural adaptations in New and Old World monkeys. In: *Primate Locomotion* (F.A. Jenkins, ed.). Academic Press, New York, 201–22.
Rosenberger, A.L. (1981) Systematics: the higher taxa. In: *Ecology and Behaviour of Neotropical Primates* (A.F. Coimbra-Filho and R.A. Mittermeier, eds). Academia Brasiliera de Ciências, Rio de Janeiro, 9–27.
Rosenberger, A.L. (1983) Aspects of the systematics and evolution of the marmosets. In: *A Primatologia no Brasil* (M. Thiago de Mello, ed.). Universidade Federale de Brasilia Press, Brasilia, 159–80.
Rosenblum, L.A. (1971) Infant attachment in monkeys. In: *Origins of Human Social Relations* (H.R. Schaffer, ed.). Academic Press, London, 85–109.
Rothbart, M.K. and D. Derryberry (1981) Theoretical issues in temperament. In: *Developmental Disabilities: Theory, Assessment and Intervention* (M. Lewis and L.T. Taft, eds). Spectrum Publications, New York, 383–400.
Rothe, H. (1975) Some aspects of sexuality and reproduction in groups of captive marmosets (*Callithrix jacchus*). *Z. Tierpsychol.* **37**, 255–73.
Rowell, T.E. (1978) How female reproductive cycles affect interaction patterns in groups of patas monkeys. In: *Recent Advances in Primatology*, Vol. 1 (D.J. Chivers and J. Herbert, eds). Academic Press, London, 409–20.
Rowell, T.E. and A.F. Dixson (1975) Changes in social organisation during the breeding season of wild talapoin monkeys. *J. Reprod. Fert.* **43**, 419–34.
Rowell, T.E. and K. Hartwell (1978) The interaction of behaviour and reproductive cycles in patas monkeys. *Behav. Biol.* **24**, 141–67.
Royal Society/Universities Federation for Animal Welfare (1986) *Guidelines for the Care of Laboratory Animals and their use for Scientific Procedures*. Part I. *Housing and Care*. Royal Society/UFAW, Potters Bar, Hertfordshire.
Ruiz de Elvira, M.C. and D.H. Abbott (1986) A backpack system for long-term osmotic minipump infusions into unrestrained marmoset monkeys. *Lab. Anim.* **20**, 329–34.
Ruiz de Elvira, M.C., J.G. Herndon and M.E. Wilson (1982) Influence of oestrogen-treated females on sexual behaviour and male testosterone levels of a social group of rhesus monkeys during the non-breeding season. *Biol. Reprod.* **26**, 825–34.
Rupniak, N.M.J. and S.D. Iverson (1989) Psychological welfare, behaviour and housing conditions for laboratory primates: a discussion. *Laboratory Animal Welfare Research: Primates*. UFAW Symposium. Universities Federation for Animal Welfare, Potters Bar, Hertfordshire, 31–8.
Russell, M.J., G.M. Switz and K. Thompson (1980) Olfactory influences on the human menstrual cycle. *Pharmacol. Biochem. Behav.* **13**, 737–8.
Russell, P.A. (1983) Psychological studies of exploration in animals: a reappraisal. In: *Exploration in Animals and Humans* (J. Archer and L. Birke, eds). Van Nostrand Reinhold, Wokingham, Berkshire, 22–54.
Rylands, A.B. (1982) The behaviour and ecology of three species of marmosets

and tamarins (Callitrichidae, Primates) in Brazil. Unpublished PhD thesis, University of Cambridge.
Sabater Pi, J. (1966) Gorilla attacks against humans in Rio Muni, West Africa. *J. Mammal.* **47**, 123–4.
Sachar, E.J., L. Hellman, D. Fukushima and T. Gallaher (1967) Corticosteroid responses to psychotherapy of depression. *Arch. Gen. Psychiat.* **16**, 461–70.
Sackett, G.P. (1974) Sex differences in rhesus monkeys following varied rearing experiences. In: *Sex Differences in Behavior* (R.C. Friedman, R.M. Richart and R.L. Vande Wiele, eds). Wiley, New York, 99–122.
Sade, D.S. (1967) Determinants of dominance in a group of free-ranging rhesus monkeys. In: *Social Communication among Primates* (S.A. Altmann, ed.). University of Chicago Press, Chicago, 99–114.
Sade, D., K. Cushing, P. Cushing, J. Dunaif, A. Figuerola, J. Kaplan, C. Lauer, D. Rhodes and J. Schneider (1977) Population dynamics related to social structure on Cayo Santiago. *Yearb. Phys. Anthropol.* **20**, 253–62.
Sadleir, R.M.F.S. (1969a) *The Ecology of Reproduction in Wild and Domestic Mammals.* Methuen, London.
Sadleir, R.M.F.S. (1969b) The role of nutrition in the reproduction of wild mammals. *J. Reprod. Fertil.* (Suppl.) **6**, 39–48.
Sailer, L.D., S.J.C. Gaulin, J.S. Boster and J.A. Kurland (1985) Measuring the relationship between dietary quality and body size in primates. *Primates* **26**, 14–27.
Samuels, A., J.B. Silk and J. Altmann (1987) Continuity and change in dominance relations among female baboons. *Anim. Behav.* **35**, 785–93.
Sapolsky, R.M. (1986) Endocrine and behavioural correlates of drought in wild olive baboons (*Papio anubis*). *Amer. J. Primatol.* **11**, 217–27.
Savage, A., T.E. Ziegler and C.T. Snowdon (1988) Sociosexual development, pair bond formation and mechanisms of fertility suppression in female cotton-top tamarins (*Saguinus oedipus oedipus*). *Amer. J. Primatol.* **14**, 345–59.
Scallet, A.S., S.J. Suomi and R.E. Bowman (1981) Sex differences in adrenocortical response to controlled agonistic encounters in rhesus monkeys. *Physiol. Behav.* **1**, 385–90.
Scanlon, C.E., N.R. Chalmers and M.A. Monteiro da Cruz (1988) Changes in size, composition and reproductive condition of wild marmoset groups (*Callithrix jacchus jacchus*) in north-east Brazil. *Primates* **29**, 295–305.
Scanlon, J.M., S.J. Suomi, J.D. Higley, A.S. Scallet and G.W. Kraemer (1982) Stress and heredity in adrenocortical response in rhesus monkeys (*Macaca mulatta*). *Soc. Neurosci. Abstr.* **8**, 461–2.
Schaller, G.B. (1963) *The Mountain Gorilla: Ecology and Behavior.* University of Chicago Press, Chicago.
Schilling, A. (1979) Olfactory communication in prosimians. In: *The Study of Prosimian Behaviour* (G.A. Doyle and R.D. Martin, eds). Academic Press, New York, 461–542.
Schilling, A., M. Parrett and J. Predine (1984) Sexual inhibition in a prosimian primate: a pheromone-like effect. *J. Endocrinology* **102**, 143–51.
Schmale, A.H. (1958). The relation of separation and depression to disease. *Psychosom. Med.* **20**, 259–63.
Schmutzler, W. and G.P. Freundt (1975). The effect of glucocorticoids and catecholamines on cyclic AMP and allergic histamine response in guinea pig lung. *Int. Arch. Allergy Appl. Immunol.* **49**, 209–12.

Schneirla, T.C. (1949) Levels in the psychological capacities of animals. *Philosophy for the Future*. Macmillan, New York.
Sclater, P.L. (1900) Mr. Sclater on *Macaca inuus*. *Trans. Zool. Soc. Lond.* November 20, 773–4.
Scraggs, P.R. and R.M. Ridley (1979) Behavioural effects of amphetamine in a small primate: relative potencies of the D- and L-isomers. *Psychopharmacology* **59**, 343–5.
Seal, U.S. (1986) Goals of captive propagation for the conservation of endangered species. *Int. Zoo Yearb.* **24/25**, 174–9.
Seay, B., E. Hanson and H.F. Harlow (1962) Mother–infant separation in monkeys. *J. Child Psychol. Psychiat.* **3**, 123–32.
Selye, H. (1937) Studies on adaptation. *Endocrinology* **21**, 169–75.
Seth, P.K. and S. Seth (1986) Ecology and behaviour of rhesus monkeys in India. In: *Primate Ecology and Conservation* (J.G. Else and P.C. Lee, eds). Cambridge University Press, Cambridge, 89–104.
Seyfarth, R.M., D.L. Cheney and R.A. Hinde (1978) Some principles relating social interactions and social structure among primates. In: *Recent Advances in Primatology*, Vol. 1 (D.J. Chivers and J. Herbert, eds). Academic Press, London, 39–51.
Shepherdson, D. (1989) Environmental enrichment in zoos: 2. *Ratel* **16**, 68–73.
Shettleworth, S.J. (1978) Reinforcement and organization of behaviour in golden hamsters: punishment of three action patterns. *Learn. Motiv.* **9**, 99–123.
Shively C., S. Clarke, N. King, S. Schapiro and G. Mitchell (1982) Patterns of sexual behaviour in male macaques. *Amer. J. Primatol.* **2**, 373–84.
Shopland, J. (1987) Food quality, spatial deployment and intensity of feeding interference in yellow baboons (*Papio cynocephalus*). *Behav. Ecol. Sociobiol.* **21**, 149–56.
Sigg H. (1980) Differentiation of female positions in hamadryas one-male units. *Z. Tierpsychol.* **53**, 265–302.
Silk, J.B. (1986) Eating for two: behavioural and environmental correlates of gestation length among free-ranging baboons (*Papio cynocephalus*). *Int. J. Primatol.* **7**, 583–602.
Silk, J.B. (1987) Social behaviour in evolutionary perspective. In: *Primate Societies* (B.B. Smuts, D.L. Cheney, R.M. Seyfarth, R.W. Wrangham and T.T. Struhsaker, eds). University of Chicago Press, Chicago, 318–29.
Silk, J.B. and R. Boyd (1983) Cooperation, competition and mate choice in matrilineal macaque groups. In: *Social Behaviour of Female Vertebrates* (S.K. Wasser, ed.). Academic Press, New York, 315–47.
Simpson, M.J.A. (1985) Effects of early experience on the behaviour of yearling rhesus monkeys (*Macaca mulatta*) in the presence of a strange object: classification and correlation approaches. *Primates* **26**, 57–72.
Simpson, M.J.A. and A.E. Simpson (1982) Birth, sex ratio and social rank in rhesus monkey mothers. *Nature, Lond.* **300**, 440–1.
Singh, S.D. (1969) Urban monkeys. *Scient. Amer.* **221**, 108–15.
Skandhan, K.P., A.K. Pandya, S. Skandhan and V.B. Metha (1979) Synchronization of menstruation among inmates and kindreds. *Pan Minerva Medica.* **21**, 131–4.
Slobodkin, L.B. (1961) *Growth and Regulation of Animal Populations*. Holt, New York.

Small, M.F. (1983) Females without infants: mating strategies in two species of captive macaques. *Folia Primatol.* **40**, 125–33.
Smith, R.J. (1984) Comparative functional morphology of maximum mandibular opening (gape) in primates. In: *Food Acquisition and Processing in Primates* (D.J. Chivers, B.A. Wood and A. Bilsborough, eds). Plenum Press, New York, 231–55.
Smith-Gill, S.J. (1983) Developmental plasticity. Developmental conversion versus phenotypic modulation. *Amer. Zool.* **23**, 47–55.
Smotherman, W.P., L.E. Hunt, L.M. McGinnis and S. Levine (1979) Mother–infant separation in group living rhesus macaques: a hormonal analysis. *Dev. Psychobiol.* **12**, 211–17.
Smuts, B. (1985) *Sex and Friendship in Baboons*. Aldine, Hawthorne, NY.
Smuts, B.B., D.L. Cheney, R.M. Seyfarth, R.W. Wrangham and T.T. Struhsaker, eds (1987) *Primate Societies*. University of Chicago Press, Chicago.
Snowdon, C.T. and P. Soini (1988) The tamarins, genus *Saguinus*. In: *Ecology and Behavior of Neotropical Primates*, Vol. 2 (R.A. Mittermeier, A.B. Rylands, A. Coimbra-Filho and G.A.B. Fonseca, eds). World Wildlife Fund, Washington, DC, 223–98.
Snowdon, C.T. and S.J. Suomi (1982) Paternal behaviour in primates. In: *Child Nurturance* (H.E. Fitzgerald, J.A. Mullins and P. Gage, eds). Plenum Press, New York, 63–108.
Snowdon, C.T., A. Savage and P.B. McConnell (1985). A breeding colony of cotton-top tamarins (*Saguinus oedipus*). *Lab. Anim. Sci.* **35**, 477–80.
Soave, O. (1982) The rehabilitation of chimpanzees and other apes. *Lab. Primate Newsl.* **21**(4), 3–8.
Soini, P. (1982) Ecology and population dynamics of the pygmy marmoset, *Cebuella pygmaea*. *Folia Primatol.* **39**, 1–21.
Soini, P. (1986) A synecological study of a primate community in the Pacaya-Samiria National Reserve, Peru. *Primate Conserv.* **7**, 63–71.
Soini, P. (1988) The pygmy marmoset, genus *Cebuella*. In: *Ecology and Behaviour of Neotropical Primates*, Vol. 2 (R.A. Mittermeier, A.B. Rylands, A. Coimbra-Filho and G.A.B. Fonseca, eds). World Wildlife Fund, Washington, DC, 79–129.
Sokolov, E.N. (1963) *Perception and the Conditioned Reflex*. Pergamon Press, Oxford.
Sommer, A. (1976) Attempt at an assessment of the world's tropical forests. *Unasylva* **28**, 5–24.
Soulé, M.E., ed. (1986) *Conservation Biology: The Science of Scarcity and Diversity*. Sinauer Associates, Sunderland, MA.
Southwick, C.H. (1967) An experimental study of intra-group agonistic behaviour in rhesus monkeys (*Macaca mulatta*). *Anim. Behav.* **28**, 182–209.
Southwick, C.H., M.F. Siddiqi, M.Y. Farooqui and B.C. Pal (1976) Effects of artificial feeding on aggressive behaviour of rhesus monkeys in India. *Anim. Behav.* **24**, 11–15.
Southwick, C.H., T. Richie, H. Taylor, H.J. Teas and M.F. Siddiqi (1980) Rhesus monkey populations in India and Nepal: patterns of growth, decline and natural regulation. In: *Biosocial Mechanisms of Population Regulation* (M.N. Cohen, R.S. Malpass and H.G. Klein, eds). Yale University Press, New Haven, CT, 151–70.
Southwick, C.H., M.F. Siddiqi and J.R. Oppenheimer (1983) Twenty-year

changes in rhesus monkey populations in agricultural areas of Northern India. *Ecology* **64**, 434–9.

Southwick, C.H., M.F. Siddiqi and R. Johnson (1984) Subgroup relocation of rhesus monkeys in India as a conservation measure (Abstract). *Amer. J. Primatol.* **6**, 423.

Spinelli, J.S. and H. Markowitz (1987) Clinical recognition and anticipation of situations likely to induce suffering in animals. *J. Amer. Vet. Med. Ass.* **191**, 1216–18.

Stark, R., I.J. Roper, A.M. MacLarnon and D.J. Chivers (1987) Gastro-intestinal anatomy of the European badger *Meles meles* L: a comparative study. *Z. Säugetierk.* **52**, 88–96.

Stearns, S.C. (1977) The evolution of life history traits: a critique of the theory and a review of the data. *Ann. Rev. Ecol. Syst.* **8**, 145–71.

Stellar, E. (1960) The marmoset as a laboratory animal: maintenance, general observations of behavior and simple learning. *J. Comp. Physiol. Psychol.* **53**, 1–10.

Stern, J.T. (1971) Functional myology of the hip and thigh muscles of cebid monkeys, and its implications for the evolution of erect posture. *Biblioth. Primatol.* **14**, 1–318. Karger, Basel.

Stern, J.T. and C.E. Oxnard (1973) Primate locomotion: some links with evolution and morphology. *Primatologia* **4**(ii), 1–94. Karger, Basel.

Stern, J.T., J.P. Wells, A.K. Vangor and J.G. Fleagle (1976) Electromyography of some muscles of the upper limb in *Ateles* and *Lagothrix*. *Yearb. phys. Anthropol.* **20**, 498–507.

Stern, J.T., J.P. Wells, W.L. Jungers and A.K. Vangor (1980) An electromyographic study of serratus anterior in Atelines and *Alouatta*: implications for hominoid evolution. *Amer. J. Phys. Anthropol.* **52**, 323–34.

Stevenson, M.F. (1983) The captive environment: its effect on exploratory and related behavioural responses in wild animals. In: *Exploration in Animals and Humans* (J. Archer and L. Birke, eds). Van Nostrand Reinhold, Wokingham, Berkshire, 176–97.

Stevenson, M.F. and A.B. Rylands (1988) The marmoset monkeys, genus *Callithrix*. In: *Ecology and Behavior of Neotropical Primates*, Vol. 2 (R.A. Mittermeier, A.B. Rylands, A. Coimbra-Filho and G.A.B. Fonseca, eds). World Wildlife Fund, Washington, DC, 131–222.

Stevenson, M.F., D.J. Chivers and J.C. Ingram, eds (1986) *Current Issues in Primate Conservation*. The Primate Society of Great Britain, Bristol.

Stevenson-Hinde, J., R. Stillwell-Barnes and M. Zunz (1980) Individual differences in young rhesus monkeys: consistency and change. *Primates* **21**, 498–509.

Stribley, J.A., J.A. French and B.J. Inglett (1989) Mating patterns in the golden lion tamarin (*Leontopithecus rosalia*): continuous receptivity and concealed ovulation. *Folia Primatol.* **43**, 133–50.

Struhsaker, T.T. (1973) A recensus of vervet monkeys in the Masai-Amboseli Game Reserve, Kenya. *Ecology* **54**, 930–2.

Struhsaker, T.T. (1976) A further decline in numbers of Amboseli vervet monkeys. *Biotropica* **8**, 211–14.

Struhsaker, T.T. (1981) Forest and primate conservation in East Africa. *Afr. J. Ecol.* **19**, 99–114.

Strum, S. (1982) Agonistic dominance in male baboons: an alternative view. *Int. J. Primatol.* **3**, 175–202.

Strum, S.C. (1986a) A role for long-term primate field research in source countries. In: *Primate Ecology and Conservation* (J.G. Else and P.C. Lee, eds). Cambridge University Press, Cambridge, 215–20.

Strum, S.C. (1986b) Activist conservation: the human factor. In: *Primate Ecology and Conservation* (J.G. Else and P.C. Lee, eds). Cambridge University Press, Cambridge, 367–82.

Strum, S.C. and C.H. Southwick (1986) Translocation of primates. In: *Primates: The Road to Self-Sustaining Populations* (W.K. Benirschke, ed.). Springer-Verlag, New York, 949–57.

Strum, S.C. and D. Western (1982) Variations in fecundity with age and environment in olive baboons (*Papio anubis*). *Amer. J. Primatol.* **3**, 61–76.

Sugiyama, Y. and H. Ohsawa (1982) Population dynamics of Japanese monkeys with special reference to the effect of artificial feeding. *Folia Primatol.* **39**, 238–63.

Summers, P.M., C.J. Wennick and J.K. Hodges (1985) Cloprostenol-induced luteolysis in the marmoset monkey (*Callithrix jacchus*). *J. Reprod. Fert.* **73**, 133–8.

Suomi, S.J. (1981) Genetic, maternal, and environmental influences on social development in rhesus monkeys. In: *Primate Behavior and Sociobiology* (B. Chiarelli and R. Corruccini, eds). Springer-Verlag, Berlin, 81–7.

Suomi, S.J. (1982) Abnormal behavior and primate models of psychopathology. In: *Primate Behavior* (J.L. Fobes and J.E. King, eds). Academic Press, New York, 172–215.

Suomi, S.J. (1983) Social development in rhesus monkeys. In: *The Behaviour of Human Infants* (A. Oliverio and M. Zappella, eds). Plenum Press, New York, 71–92.

Suomi, S. (1985) Biological response styles: experiential effects. In: *Biologic Response Styles: Clinical Implications* (H. Klas and L.J. Siever, eds). American Psychiatric Press, Washington, DC, 2–17.

Suomi, S. (1987) Genetic and maternal contributions to individual differences in rhesus monkey biobehavioural development. In: *Perinatal Development: A Psycho-biological Perspective* (N.A. Krasnegor, E.M. Blass, M.A. Hofer and W.P. Smotherman, eds). Academic Press, New York, 397–419.

Suomi, S.J., H.F. Harlow and C.J. Domek (1970) Effect of repetitive infant–infant separation of young monkeys. *J. Abnorm. Psychol.* **76**, 161–72.

Suomi, S.J., S.F. Seaman, J.K. Lewis, R. DeLizio and W.T. McKinney (1978) Antidepressant effects of imipramine treatment on separation-induced social disorders in rhesus monkeys. *Arch. Gen. Psychiat.* **35**, 321–5.

Suomi, S.J., G.W. Kraemer, C.M. Baysinger and R.D. DeLizio (1981) Inherited and experiential factors associated with individual differences in anxious behaviour displayed by rhesus monkeys. In: *Anxiety: New Research and Changing Concepts* (D.F. Klein and J. Rabkin, eds). Raven Press, New York, 179–200.

Sussman, R.W. and P.A. Garber (1987) A new interpretation of the social organisation and mating system of the Callitrichidae. *Int. J. Primatol.* **8**, 73–92.

Sussman, R.W. and W.G. Kinzey (1984) The ecological role of the Callitrichidae: a review. *Amer. J. Phys. Anthropol.* **64**, 419–49.

Sutton, S.L., T.C. Whitmore and A.C. Chadwick, eds (1983) *Tropical Rain Forest: Ecology and Management*. Blackwell, Oxford.

Tabor, B.A. (1986) The development and structure of *Saimiri* social relationships. Unpublished PhD dissertation, University of California, Davis.

Teas, J., T.L. Richie, H.G. Taylor, M.F. Siddiqi and C.H. Southwick (1981) Natural regulation of rhesus monkey populations in Kathmandu, Nepal. *Folia Primatol.* **35**, 117–23.

Temerlin, M.K. (1975) *Lucy: Growing-up Human*. Science and Behaviour Books, California.

Terasawa, E., T.E. Nass, R.R. Yeoman, M.D. Loose and N.J. Schultz (1983) Hypothalamic control of puberty in the female rhesus macaque. In: *Neuroendocrine Aspects of Reproduction* (R.L. Norman, ed.). Academic Press, New York, 149–82.

Terborgh, J. (1983) *Five New World Primates: A Study in Comparative Ecology*. Princeton University Press, Princeton, NJ.

Terborgh, J. (1986a) Conserving New World primates: present problems and future solutions. In: *Primate Ecology and Conservation* (J.G. Else and P.C. Lee, eds). Cambridge University Press, Cambridge, 355–66.

Terborgh, J. (1986b) Keystone plant resources in the tropical forest. In: *Conservation Biology: The Science of Scarcity and Diversity* (M. Soulé, ed.). Sinauer Associates, Sunderland, MA, 330–44.

Terborgh, J. and A.W. Goldizen (1985) On the mating system of the cooperatively breeding saddle-backed tamarin (*Saguinus fuscicollis*). *Behav. Ecol. Sociobiol.* **16**, 293–9.

Thomas, A. and S. Chess (1977) *Temperament and Development*. Brunner Mazel, New York.

Thornhill, R. and J. Alcock (1983) *The Evolution of Insect Mating Systems*. Harvard University Press, Cambridge, MA.

Toates, F. (1987) The relevance of models of motivation and learning to animal welfare. In: *Biology of Stress in Farm Animals: An Integrative Approach* (P.R. Wieipkema and P.W.M. Van Adrichem, eds). Martinus Nijhoff: The Hague, 153–86.

Torigoe, T. (1985) Comparison of object manipulation among 74 species of non-human primates. *Primates* **26**, 182–94.

Torrence, R. (1983) Time budgeting and hunter–gatherer technology. In: *Hunter–Gatherer Economy in Prehistory* (G. Bailey, ed.). Cambridge University Press, Cambridge, 11–22.

Trivers, R.L. (1971) The evolution of reciprocal altruism. *Quart. Rev. Biol.* **46**, 35–57.

Tutin, C.E.G. and M. Fernandez (1987) Sympatric gorillas and chimpanzees in Gabon. *Anthroquest* **37**, 3–6.

Tutin, C.E.G., W.C. McGrew and P.J. Baldwin (1981) Responses of wild chimpanzees to potential predators. In: *Primate Behavior and Sociobiology* (A.B. Chiarelli and R.S. Corrucini, eds). Springer, Berlin, 136–41.

Tutin, C.E.G., P.J. Baldwin and W.C. McGrew (1983) Social organisation of savanna-dwelling chimpanzees *Pan troglodytes verus* at Mt. Assirik. *Senegal Primates* **24**, 154–73.

Tuttle, R.H. (1969) Quantitative and functional studies on the hands of the Anthropoidea. I. Hominoidea. *J. Morph.* **128**, 309–63.

Ursin, H., E. Baade and S. Levine (1978) *Psychobiology of Stress: A Study of Coping Men*. Academic Press, New York.

Vaitl, E. (1978) Nature and implications of the complexly organized social system in nonhuman primates. In: *Recent Advances in Primatology*, Vol. 1: *Behaviour* (D.J. Chivers and J. Herbert, eds). Academic Press, London, 17–30.

Vaitl, E., W.A. Mason, D.M. Taub and C.O. Anderson (1978) Contrasting effects of living in heterosexual pairs and mixed groups on the structure of social attraction in squirrel monkeys (*Saimiri*). *Anim. Behav.* **23**, 358–67.

van Noordwijk, M.A. and C.P. van Schaik (1985) Male migration and rank acquisition in wild long-tailed macaques (*Macaca fascicularis*). *Anim. Behav.* **33**, 849–61.

van Noordwijk, M.A. and C.P. van Schaik (1988) Male careers in Sumatran long-tailed macaques (*Macaca fascicularis*). *Behaviour* **107**, 24–43.

van Roosmalen, M.G.G. (1981) Habitat preferences, diet, feeding strategy and social organization of the black spider monkey (*Ateles paniscus paniscus*, Linnaeus 1758) in Surinam. Unpublished PhD thesis, Agricultural University of Wageningen, Leersum, The Netherlands.

Vessey, S.H. and D.B. Meikle (1987) Factors affecting social behaviour and reproductive success of male rhesus monkeys. *Int. J. Primatol.* **8**, 281–92.

Victor M. (1975) *A Devastacão Florestal*. Sociedade Brasileira de Silvicultura: São Paũlo.

Visalberghi, E. (1986) The acquisition of tool-use behavior in two capuchin monkey groups (*Cebus apella*). *Primate Rep.* **14**, 226–7.

Visalberghi, E. (1988) Responsiveness to objects in two social groups of tufted capuchin monkeys (*Cebus apella*). *Amer. J. Primatol.* **15**, 349–60.

Visalberghi, E. and F. Antinucci (1986) Tool use in the exploitation of food resources in *Cebus apella*. In: *Primate Ecology and Conservation* (J.G. Else and P.C. Lee, eds). Cambridge University Press, Cambridge, 57–62.

Visalberghi E. and W.A. Mason (1983) Determinants of problem-solving success in *Saimiri* and *Callicebus*. *Primates* **24**, 385–96.

Vogt, J.L. and S. Levine (1980) Response of mother and infant squirrel monkeys to separation and disturbance. *Physiol. Behav.* **24**, 829–32.

Vogt, J.L., C.L. Coe and S. Levine (1981) Behavioral and adrenocortical responsiveness of squirrel monkeys to a live snake: is flight necessarily stressful? *Behav. Neur. Biol.* **32**, 39–405.

Vrba, E. (1985) Ecological and adaptive changes associated with early hominid evolution. In: *Ancestors: The Hard Evidence* (E. Delson, ed.). Alan R. Liss, New York, 63–71.

Walker, E.P. (1964) *Mammals of the World* (2 vols). Johns Hopkins Press, Baltimore, MD.

Walker, M.L., T.P. Gordon and M.E. Wilson (1983) Menstrual cycle characteristics of seasonally breeding rhesus monkeys. *Biol. Reprod.* **29**, 841–8.

Wallis, J. (1985) Synchrony of oestrus swelling in group-living chimpanzees (*Pan troglodytes*). *Int. J. Primatol.* **6**, 335–50.

Walters, J.R. (1980) Interactions and the development of dominance relationships in female baboons. *Folia Primatol.* **34**, 61–89.

Walters, J.R. (1987a) Kin recognition in non-human primates. In: *Kin Recognition in Animals* (D.J.C. Fletcher and C.D. Michener, eds). Wiley, Chichester, West Sussex, 359–93.

Walters J.R. (1987b) Transition to adulthood. In: *Primate Societies* (B.B. Smuts, D.L. Cheney, R.M. Seyfarth, R.W. Wrangham and T.T. Struhsaker, eds). University of Chicago Press, Chicago, 358–69.

Warburton, D.M. (1975) *Brain, Behaviour and Drugs*. Wiley, London.

Warburton, D.M. (1979a) Physiological aspects of information processing and stress: In: *Human Stress and Cognition* (V.H. Hamilton and D.M. Warburton, eds). Wiley, London, 33–66.

Warburton, D.M. (1979b) Stress and the processing information. In: *Human Stress and Cognition* (V.H. Hamilton and D.M. Warburton, eds). Wiley, London, 469–75.
Wasser, S.K. and D.P. Barash (1983) Reproductive suppression among female mammals: implications for biomedicine and sexual selection theory. *Quart. Rev. Biol.* **58**, 513–38.
Wasser, S.K. and A.K. Starling (1988) Proximate and ultimate causes of reproductive suppression among female yellow baboons at Mikumi National Park, Tanzania. *Amer. J. Primatol.* **16**, 97–121.
Waterman, P.G. (1984) Food acquisition and processing as a function of plant chemistry. In: *Food Acquisition and Processing in Primates* (D.J. Chivers, B.A. Wood and A. Bilsborough, eds). Plenum Press, New York, 117–211.
Weatherhead, P.J. (1986) How unusual are unusual events? *Amer. Nat.* **128**, 150–4.
Weber, A.W. (1987) Socioecologic factors in the conservation of afromontane forest reserves. In: *Primate Conservation in the Tropical Rain Forest* (C.W. Marsh and R.A. Mittermeier, eds). Alan R. Liss, New York, 205–29.
Webley, G.E., D.H. Abbott, L.M. George, J.P. Hearn and H. Mehl (1989) The circadian pattern of plasma melatonin concentrations in the marmoset monkey (*Callithrix jacchus*). *Amer. J. Primatol.* **17**, 73–9.
Webster, A.F. (1984) *Calf Husbandry, Health and Welfare*. Collins, London.
Wehrenberg, W.B. and J. Dyrenfurth (1983) Photoperiod and menstrual cycles in female macaque monkeys. *J. Reprod. Fert.* **68**, 119–212.
Weinberg, J. and S. Levine (1980) Psychobiology of coping in animals: the effects of predictability. In: *Coping and Health* (S. Levine and H. Ursin, eds). Plenum Press, New York, 39–59.
Wemelsfelder, F. (1984) Animal boredom: is a scientific study of the subjective experiences of animals possible? *Advances in Animal Welfare Science 1984/85*. Humane Society of the United States, Washington, DC.
West Eberhard, M.J. (1979) Sexual selection, social competition and evolution. *Proc. Amer. Phil. Soc.* **123**, 222–34.
Westergaard, G.C. (1987) Long-tailed macaques manufacture and use of tools. *Amer. J. Primatol.* **12**, 376.
Westergaard G.G. and D.M. Fragaszy (1985) Effects of manipulatable objects in the activity of captive capuchin monkeys (*Cebus apella*). *Zoo Biology* **4**, 317–27.
Westergaard, G.C. and D.M. Fragaszy (1987a) The manufacture and use of tools by capuchin monkeys (*Cebus apella*). *J. Comp. Psychol.* **101**, 159–68.
Westergaard, G.C. and D.M. Fragaszy (1987b) Self-treatment of wounds by a capuchin monkey (*Cebus apella*). *J. Hum. Evol.* **2**, 557–62.
Western, D. (1986) Introduction: primate conservation in the broader realm. In: *Primate Ecology and Conservation* (J.G. Else and P.C. Lee, eds). Cambridge University Press, Cambridge, 343–53.
Western, D. and M. Pearl, eds (1989) *Conservation Biology for the 21st Century*. Oxford University Press, Oxford.
Whitten, P.L. (1983) Diet and dominance among female vervet monkeys (*Cercopithecus aethiops*). *Amer. J. Primatol.* **5**, 139–59.
Whitten, P.L. (1984) Competition among female vervet monkeys. In: *Female Primates: Studies by Women Primatologists* (M.F. Small, ed.). Alan R. Liss, New York, 137–40.
Williams, J.B. (1987) *Behavior of Captive Marmosets and Tamarins (Callitrichidae): A Bibliography, 1975–1987*. Primate Information Center, Washington, DC.

Williamson, E.A. (1988) Behavioural ecology of western lowland gorillas in Gabon. PhD thesis, University of Stirling, Scotland.
Wilson, E.O. (1975) *Sociobiology: The New Synthesis*. Belknap Press, Cambridge, MA.
Wilson, E.O. (1988) *Biodiversity*. National Academy Press, Washington, DC.
Wilson, M.E. (1981) Social dominance and female reproductive behavior in rhesus monkeys (*Macaca mulatta*). *Anim. Behav.* **29**, 472–82.
Wilson, M.E., T.P. Gordon and I.S. Bernstein (1978) Timing of births and reproductive success in rhesus monkey social groups. *J. Med. Primatol.* **7**, 202–12.
Wilson, M.E., M.L. Walker and T.P. Gordon (1983) Consequences of first pregnancy in rhesus monkeys. *Amer. J. Phys. Anthropol.* **61**, 103–10.
Wilson, M.M. (1985) Hippocampal inhibition of the pituitary adrenocortical response to stress. In: *Stress: Psychology and Physiological Interactions* (S.R. Burchfield ed.). Hemisphere, Washington, DC, 163–84.
Wilson, S.F (1982) Environmental influences on the activity of captive apes. *Zoo Biol.* **1**, 201–9.
Wilson, V. (1980) Operation monkey release. *IPPL Newsl.* **7**(2), 12–13.
Witt, R., C. Schmidt and J. Schmitt (1981) Social rank in a multimale group of Barbary macaques (*Macaca sylvanus* Linnaeus, 1758). *Folia Primatol.* **36**, 201–11.
Wolf, L.L., F.R. Hainsworth and F.B. Gill (1975) Foraging efficiencies and time budgets of nectar feeding birds. *Ecologist* **56**, 117–28.
Wolfe, L.D. (1979) Sexual maturation among members of a transplanted troop of Japanese macaques (*Macaca fuscata*). *Primates* **20**, 411–18.
Wolfe, L.D. (1986) Reproductive biology of rhesus and Japanese macaques. *Primates* **27**, 95–101.
Wolfheim, J.H. (1976) The perils of primates. *Nat. Hist.* **85**(8), 90–9.
Wolfheim, J.H. (1983) *Primates of the World: Distribution, Abundance and Conservation*. Harwood Academic, Chur, Switzerland.
Wolters, H.-J. (1978) Some aspects of role taking behaviour in captive family groups of the cotton-top tamarin (*Saguinus oedipus oedipus*). In: *The Biology and Behaviour of Marmosets* (H. Rothe, H.-J. Wolters and J. P. Hearn, eds). Eigenverlag-H. Rothe, Göttingen, 259–78.
Wood, B., L. Martin and P. Andrews, eds (1986) *Major Topics in Primate and Human Evolution*. Cambridge University Press, Cambridge.
Wood, B.S., W.A. Mason and M.D. Kenney (1979) Contrasts in visual responsiveness and emotional arousal between rhesus monkeys raised with living and inanimate substitute mothers. *J. Comp. Physiol. Psychol.* **93**, 368–77.
Woodcock, A.J. (1982) The first weeks of cohabitation of newly-formed heterosexual pairs of common marmosets (*Callithrix jacchus*). *Folia Primatol.* **37**, 228–54.
Woodruff, G. and D. Premak (1979) Intentional communication in the chimpanzee: the development of deception. *Cognition* **7**, 333–62.
Wrangham, R.W. (1974) Artificial feeding of chimpanzees and baboons in their natural habitat. *Anim. Behav.* **22**, 83–93.
Wrangham, R.W. (1980) An ecological model of female-bonded primate groups. *Behaviour* **75**, 262–300.
Wrangham, R.W. (1981) Drinking competition among vervet monkeys. *Anim. Behav.* **29**, 904–10.
Wrangham, R.W. (1983) Social relationships in comparative perspective. In:

Primate Social Relationships: An Integrated Approach (R.A. Hinde, ed.). Blackwell Scientific, Oxford, 255–62.

Wrangham, R.W. (1986) Ecology and social relations in two species of chimpanzee. In: *Ecological Aspects of Social Evolution* (D. Rubenstein and R.W. Wrangham, eds). Princeton University Press, Princeton, NJ, 352–78.

Wrangham, R.W. (1987) Evolution of social structure. In: *Primate Societies* (B.B. Smuts, D.L. Cheney, R.M. Seyfarth, R.W. Wrangham and T.T. Struhsaker, eds). University of Chicago Press, Chicago, 282–96.

Wrenshall, E. and S.G. Gilbert (1986) Environmental enrichment of cynomolgus monkeys used for behavioural toxicology studies. *Amer. J. Primatol.* **10**, 442.

Wright, P.C. (1985) Costs and benefits of nocturnality for *Aotus* (the night monkey). Unpublished PhD thesis, City University of New York.

Wright, P.C. (1986) Ecological correlates of monogamy in *Aotus* and *Callicebus*. In: *Primate Ecology and Conservation* (J.G. Else and P.C. Lee, eds). Cambridge University Press Cambridge, 159–67.

Wünschmann, A. (1963) Quantitative untersuchungen zum Neugier verhalten on Wirbelitien. *Z. Tierpsychol.* **20**, 80–109.

Wynne-Edwards, V.C. (1962) *Animal Dispersion in Relation to Social Behaviour*. Oliver and Boyd, London.

Yates, F.E. and J. Urquhart (1962) Control of plasma concentrations of adrenocortical hormones. *Physiol. Rev.* **42**, 359–433.

Yen, S.S.C. (1986) Chronic anovulation due to CNS-hypothalamic-pituitary dysfunction. In: *Reproductive Endocrinology, Physiology, Pathophysiology and Clinical Management*, 2nd edn (S.S.C. Yen and R.B. Jaffe, eds). W.B. Saunders, Philadelphia, 500–45.

Ziegler, T.E., W.E. Bridson, C.T. Snowdon and S. Eman (1987a) Urinary gonadotropin and oestrogen excretion during the postpartum oestrus, conception and pregnancy in the cotton-top tamarin *(Saguinus oedipus oedipus)*. *Amer. J. Primatol.* **12**, 127–40.

Ziegler, T.E., A. Savage, G. Scheffler and C.T. Snowdown (1987b) The endocrinology of puberty and reproductive functioning in female cotton-top tamarins (*Saguinus oedipus*) under varying social conditions. *Biol. Reprod.* **37**, 618–27.

Ziegler, T.E., A. Savage, G. Scheffler and C.T. Snowdon (1987c) The endocrinology of puberty and reproductive functioning in female cotton-top tamarins (*Saguinus oedipus*) under varying social conditions. *Biol. Reprod. 37*: 410–27.

Zumpe, D. and R.P. Michael (1986) Dominance index: a simple measure of relative dominance status in primates. *Amer. J. Primatol.* **10**, 291–300.

Index